# Security and Resilience in Intelligent Data-Centric Systems and Communication Networks

# Security and Resilience in Intelligent Data-Centric Systems and Communication Networks


Edited by

**Massimo Ficco**

Università degli Studi della Campania Luigi Vanvitelli, Italy

**Francesco Palmieri**

University of Salerno, Italy

Series Editor

**Fatos Xhafa**

Universitat Politècnica de Catalunya, Spain

Academic Press is an imprint of Elsevier
125 London Wall, London EC2Y 5AS, United Kingdom
525 B Street, Suite 1800, San Diego, CA 92101-4495, United States
50 Hampshire Street, 5th Floor, Cambridge, MA 02139, United States
The Boulevard, Langford Lane, Kidlington, Oxford OX5 1GB, United Kingdom

Copyright © 2018 Elsevier Inc. All rights reserved.

No part of this publication may be reproduced or transmitted in any form or by any means, electronic or mechanical, including photocopying, recording, or any information storage and retrieval system, without permission in writing from the publisher. Details on how to seek permission, further information about the Publisher's permissions policies and our arrangements with organizations such as the Copyright Clearance Center and the Copyright Licensing Agency, can be found at our website: www.elsevier.com/permissions.

This book and the individual contributions contained in it are protected under copyright by the Publisher (other than as may be noted herein).

**Notices**
Knowledge and best practice in this field are constantly changing. As new research and experience broaden our understanding, changes in research methods, professional practices, or medical treatment may become necessary.

Practitioners and researchers must always rely on their own experience and knowledge in evaluating and using any information, methods, compounds, or experiments described herein. In using such information or methods they should be mindful of their own safety and the safety of others, including parties for whom they have a professional responsibility.

To the fullest extent of the law, neither the Publisher nor the authors, contributors, or editors, assume any liability for any injury and/or damage to persons or property as a matter of products liability, negligence or otherwise, or from any use or operation of any methods, products, instructions, or ideas contained in the material herein.

**Library of Congress Cataloging-in-Publication Data**
A catalog record for this book is available from the Library of Congress

**British Library Cataloguing-in-Publication Data**
A catalogue record for this book is available from the British Library

ISBN: 978-0-12-811373-8

For information on all Academic Press publications visit our website at https://www.elsevier.com/books-and-journals

*Publisher:* Mara Conner
*Acquisition Editor:* Sonnini R. Yura
*Editorial Project Manager:* Ana Claudia A. Garcia
*Production Project Manager:* Sruthi Satheesh
*Cover Designer:* Vicky Pearson

Typeset by SPi Global, India

# Contents

# Contributors

**Ahmed Alzubi**
Girne American University, Kyrenia, Cyprus

**Alba Amato**
University of Campania "Luigi Vanvitelli", Aversa, Italy

**Alessandra De Benedictis**
University of Naples Federico II, Naples, Italy

**Antonio Scarfò**
Next-Era Prime Spa, Naples, Italy

**Arif Sari**
Girne American University, Kyrenia, Cyprus

**Chang Choi**
Chosun University, Gwangju, South Korea

**Christian Esposito**
University of Salerno, Salerno; National Inter-university Consortium for Information Technology (CINI), Naples, Italy

**David Sembroiz**
Technical University of Catalonia (UPC) - BarcelonaTech, Barcelona, Spain

**Davide Careglio**
Technical University of Catalonia (UPC) - BarcelonaTech, Barcelona, Spain

**Domenico Cotroneo**
National Inter-university Consortium for Information Technology (CINI), Naples, Italy

**Ersin Caglar**
Girne American University, Kyrenia, Cyprus

**Francesco Palmieri**
University of Salerno, Salerno, Italy

**Gaetano Papale**
University of Naples "Parthenope", Naples, Italy

**Gianfranco Cerullo**
University of Naples "Parthenope", Naples, Italy

**Gianni D'Angelo**
University of Sannio, Benevento, Italy

**Giovanni Mazzeo**
University of Naples "Parthenope", Naples, Italy

**Jian Shen**
Nanjing University of Information Science & Technology, Nanjing, China

**Joan Arnedo-Moreno**
Open University of Catalonia, Internet Interdisciplinary Institute (IN3), Barcelona, Spain

**Luigi Sgaglione**
University of Naples “Parthenope”, Naples, Italy

**Luis Gómez-Miralles**
Open University of Catalonia, Internet Interdisciplinary Institute (IN3), Barcelona, Spain

**Marcello Cinque**
National Inter-university Consortium for Information Technology (CINI), Naples, Italy

**Mario Fiorentino**
National Inter-university Consortium for Information Technology (CINI), Naples, Italy

**Massimiliano Rak**
University of Campania “Luigi Vanvitelli”, Aversa, Italy

**Massimo Ficco**
University of Campania “Luigi Vanvitelli”, Aversa, Italy

**Mauro Iacono**
Università degli Studi della Campania “Luigi Vanvitelli”, Caserta, Italy

**Michał Choraś**
UTP University of Science and Technology in Bydgoszcz, Bydgoszcz, Poland

**Petar Kochovski**
University of Ljubljana, Ljubljana, Slovenia

**Rafał Kozik**
UTP University of Science and Technology in Bydgoszcz, Bydgoszcz, Poland

**Salvatore Rampone**
University of Sannio, Benevento, Italy

**Salvatore Venticinque**
University of Campania “Luigi Vanvitelli”, Aversa, Italy

**Sergio Ricciardi**
Technical University of Catalonia (UPC) - BarcelonaTech, Barcelona, Spain

**Stefano Marrone**
Università degli Studi della Campania “Luigi Vanvitelli”, Caserta, Italy

**Stefano Russo**
National Inter-university Consortium for Information Technology (CINI), Naples, Italy

**Umberto Villano**
University of Sannio, Benevento, Italy

**Valentina Casola**
University of Naples Federico II, Naples, Italy

**Vlado Stankovski**
University of Ljubljana, Ljubljana, Slovenia

**Bruno Ragucci**
University of Naples “Parthenope”, Naples, Italy

# INTELLIGENT DATA-CENTRIC CRITICAL SYSTEMS: SECURITY AND RESILIENCE KEY CHALLENGES

**Massimo Ficco*, Francesco Palmieri†**

*University of Campania "Luigi Vanvitelli", Aversa, Italy* University of Salerno, Salerno, Italy†*

## 1 MOTIVATION

Intelligent data-centric systems play a key role in several fundamental human activities. In particular, in recent years, complex and distributed intelligent data-centric systems have often been used in highly critical contexts, including airports; seaports; nuclear and chemical plants; water, oil, and energy supply systems; telecommunication systems; banking and financial services; and government and emergency services.

In general, such systems have been adopted for supporting advanced monitoring and control facilities. The consequences of their malfunctions and outages can be catastrophic in terms of public safety, economical losses, and consumer dissatisfaction (Ficco et al., 2016). Therefore the security and dependability of these systems are of paramount importance in both the industrial and public domain. Accordingly, they have to be highly resilient against attacks and malicious activities in order to reduce the risk of severe failures, hijacking, and sensitive data compromises. The practicality of such systems poses new challenges for computer engineers, who must develop more robust solutions to ensure a high level of protection while containing both cost and development time (ElMaraghy et al., 2012; Ficco et al., 2017).

On the other hand the security and resilience of intelligent data-centric critical systems is hard to achieve due to the structural and dynamic complexity of such systems (Ferrario, 2015). First, they are designed as the composition of several off-the-shelf and, often, open-source components, primarily for their ability to reduce development costs. In general, such components are originally designed for different application domains, which potentially may introduce unpredictable security vulnerabilities when used in other contexts. Second, their size has significantly grown on a large scale, and their operational environment, originally planned to be "closed," becomes more and more "open" to allow interoperability as well as remote control and mobile access (Ficco et al., 2014). This may potentially introduce new security threats exploitable by malicious intruders, which may compromise the system operations and access to confidential data. In particular, such systems use and manage a huge amount of heterogeneous, complex, and sensitive data (sensor data, mobile data, monitoring data, forensics data), which can be the target of many potential threats. Third, dynamic complexity that manifests through the emergence of unexpected system behaviors in response to changes in the environmental and operational conditions of its components, as well as the unpredictable fluctuations both in workloads and in availability of computational and communication resources that support the execution of these workloads, can introduce additional complexity in their monitoring and

Security and Resilience in Intelligent Data-Centric Systems and Communication Networks. https://doi.org/10.1016/B978-0-12-811373-8.09990-0

Copyright © 2018 Elsevier Inc. All rights reserved.

control activities. Moreover, such systems are not isolated but highly interconnected and mutually interdependent, which may increase the potential for cascading deliberated failures and amplify the impact of both large and small scale initial intrusions into events that may produce large consequences on a regional or national scale (Zio and Ferrario, 2013). Integration among systems may introduce unexpected vulnerability that usually manifests later, at the systems' operation time, thus affecting their overall security and dependability. In detail, ensuring the trustworthiness of an individual component system does not guarantee that the whole system will be secure. Additionally, as they cannot be detected earlier, they require on-site maintenance operations resulting in increased maintenance costs and overspending in terms of personnel resources. Therefore identifying, understanding, and analyzing complex interactions and interdependency between them, represent a challenge associated to the evaluation of the real vulnerability of each system in consequence of a malicious event (Pederson et al., 2006).

Nowadays, the trend is to reorganize such systems by applying a System of Systems (SoS) concept, where the sparse islands are progressively interconnected by proper middleware solutions through wide-area networks (Maier, 1998). SoSes are large-scale concurrent and distributed systems, which are composed of complex systems made by a huge number of physically and functionally heterogeneous elements, interacting in a network structure, and organized in a hierarchy of subsystems that contributes to the overall system function (Guckenheimer and Ottino, 2008).

Furthermore, the new needs and the extreme dynamism of modern organizations and companies (for example, the necessity of system administrators and insiders to be connected to their systems, applications, and files anywhere and at anytime), as well as the emerging information technologies and software paradigms (including virtualization, cloud computing, Internet of Things (IoT), big data analytics technologies, mobile devices, and applications, etc.), and wired and wireless communication networks (which enable big data collection and transmission cheap and widespread), make private and sensitive data immediately available and more mobile; and therefore, more vulnerable to possible attacks. Moreover, since the SoS context is often dynamic, system data deemed "secure" in the specific context in which the system was designed to work may exhibit new vulnerabilities when used in a slightly changed or totally new context. In other words, many organizations do not have the perception of the security risks nor do they know where copies of their sensitive data are, in which context they will be used, and who has access to them.

The distributed nature and huge complexity of such SoSes, even spread among different hosting infrastructures, often geographically distributed on many sites, as well as the continuing evolution and extension of the concept of large-scale critical system supporting the extreme dynamism and adaptivity of modern organizations, implies an evolution of the traditional concept of monitoring, protection, and control, as well as the integration of independently developed software components, especially combined with real-time performance demands, becomes one of the key challenges.

Traditional security enforcement tools, including malware gateway appliances, antivirus, firewalls, and intrusion detection systems, could not be enough to detect and mitigate new emerging and more sophisticated attacks that circumvent security controls and steal sensitive data for the purpose of perpetrating illicit profits and industrial espionage. Standard security technologies can only be effective for detecting and preventing the spread of obvious malware and the occurrence of external attacks, but there are great limits associated to detecting and responding to stealth distributed attacks and internal malicious behaviors (insider threats). Indeed, it is not unusual for intruders to circumvent these tools and techniques by performing attacks that are more difficult to reveal, that is, based on low-rate attack patterns designed to look like normal traffic performed by authorized users (Ficco and Rak, 2015). In the same manner, a malicious individual, for example an employee, contractor, business partner, etc., with legitimate access to sensitive data can conduct a slow internal attack by siphoning off a few sensitive data each day. The above attacks may adversely affect the traditional assets (computing, storage, and connectivity resources) characterizing application service providers on the Internet, resulting in a more or less perceivable Denial of Service (DoS) against the involved organizations. Even worse, they may focus on new less evident targets and attack goals, such as increasing the energy consumption of individual data centers or complex federated cloud organizations (Palmieri et al., 2011, 2015; Ficco and Palmieri, 2015). The security administrators frequently remain unaware of the above activity until the malicious task has been almost completely accomplished so that severe damages have already occurred (Oltsik, 2015).

## 2 CONTENTS

In new data-centric architectures and technologies, enabling security and resiliency become essential for managing sensitive data and supporting situational awareness and risk management in complex and critical systems. The development of new intelligent data-centric critical systems requires the use of appropriate techniques, architectures, and tools, that are in line with both the state-of-the-art solutions in this area and with the development of novel, more secure and resilient solutions. Therefore this book will present current advances in the field, collecting both theoretical and practical results and exploring the potential of the above architectures and techniques in future developments. In particular, it aims at disseminating recent research efforts in the security and resilience of intelligent data-centric critical systems in order to support research evolution in this area.

This book also aims to introduce novel security and resilience aspects and challenges characterizing the new IoT-empowered intelligent data-centric critical systems and communication networks, including techniques and tools that can be useful for preventing and avoiding accidental problems as well as malicious behaviors. Its overall objectives essentially focus on presenting and explaining state-of-the-art technological solutions for the main issues hindering the development of monitoring and self-reacting solutions for supporting security and resilience in these sophisticated and heterogeneous scenarios. In particular, successful approaches and methodologies to detect potential tampering risks and attacks against such systems and communication networks are presented, in order to familiarize readers with the right concepts and technologies that can support the implementation of more robust, reliable, and reactive mission critical systems. In particular, the book includes several chapters that have been carefully selected according to their subject and accepted on their merit and contribution basis. It is organized according to the following structure:

- "Dependability of Container-Based Data-Centric Systems" chapter: Presents the main aspects of dependability in data-centric systems and analyzes to what extent various Cloud computing technologies can be used to achieve this property.
- "Risk Assessment and Monitoring in Intelligent Data-Centric Systems" chapter: Focuses on the risk assessment methods developed for traditional large-scale computer based systems and on the necessity to extend such approaches to Data Centric Systems.

- "The Cyber Security Challenges in the IoT Era" chapter: Provides both an overview of current trends about cyber-security concerns in the new IoT arena and an idea of what we could expect in the coming years.
- "IoT and Sensor Networks Security" chapter: Presents a comprehensive view of current security, safety, and privacy issues due to the adoption of IoT technologies.
- "Smart Access Control Models in Sensor Network" chapter: Faces security challenges in collecting sensitive data though sensor networks, and deals with it by means of a semantic approach focused on ontologies, as well as on matching different ontological descriptions of access control models in order to make them interoperable.
- "Smart Sensor and Big Data Security and Resilience" chapter: Provides a survey about the intersection of IoT and Big-Data technologies by focusing on security issues and resilience, which must be taken in mind when addressing system design and data management.
- "Load Balancing Algorithms and Protocols to Enhance Quality of Service and Performance in Data of WSN" chapter: Elaborates the variety of up-to-date load-balancing algorithms and protocols proposed in the literature to enhance quality of service (QoS) and performance in wireless sensor networks.
- "Machine Learning Techniques for Threat Modeling and Detection" chapter: Investigates practical solutions for the evolutionary-based optimization techniques, collective intelligence, and techniques that mimic social behavior of species in order to prove that the bio-inspired techniques can be really implemented to support cyber security.
- "Cognitive Distributed Application Area Networks" chapter: Presents a distributed cognition-based architecture for pervasive computing in a framework that has to estimate the trustworthiness of the entities interacting in a network.
- "A Novel Cloud-Based IoT Architecture for Smart Building Automation" chapter: Presents an overview on cloud-based IoT architectures for smart building automation.
- "Monitoring Data Security in the Cloud: A Security SLA-Based Approach" chapter: Outlines the problems related to monitoring security in the cloud and illustrates a SLA-based security monitoring approach, as well as provides concrete examples related to offering services protected against Denial of Service (DoS) attacks, provided with continuous scanning and management of existing software vulnerabilities.
- "Hardening iOS Devices Against Remote Forensic Investigation" chapter: Presents a proof of concept tool to prevent malicious abuse against smart sensors, by disabling unnecessary services in hardening devices, and analyzes its antiforensic consequences, privacy implications, and possible countermeasures (anti-antiforensics).
- "Path Loss Algorithms for Data Resilience in Wireless Body Area Networks for Healthcare Framework" chapter: Provides a systematic review of existing literature related with proposed path loss algorithms for data centric critical system protection and data resilience in Wireless Body Area Networks (WBAN) for healthcare frameworks.
- "Designing Resilient and Secure Large-Scale Crisis Information Systems" chapter: Determines the requirements to be satisfied in order to provide reliability and security when designing federated large-scale distributed systems. As a concrete case, the chapter focuses on crisis information systems, as they are typical examples of large-scale federated systems.

We are sure that the works presented in this book are significant contributions to the field of intelligent data-centric critical systems by introducing new approaches, experiences, and studies conducted by academic researchers, industry professionals, and students interested in this subject.

## ACKNOWLEDGMENTS AND THANKS

We would like to express our sincere appreciation of the valuable contributions made by all the authors and our deep gratitude to all the highly qualified anonymous reviewers who have carefully analyzed the assigned papers and significantly contributed to improve their quality.

## REFERENCES

ElMaraghy, W., ElMaraghya, H., Tomiyamac, T., Monostorid, L., 2012. Complexity in engineering design and manufacturing. CIRP Ann. Manuf. Technol. 61 (2), 793–814.

Ferrario, E., 2015. System-of-systems modeling and simulation for the risk analysis of industrial installations and critical infrastructures. Available at: https://tel.archives-ouvertes.fr/tel-01127194.

Ficco, M., Palmieri, F., 2015. Introducing fraudulent energy consumption in cloud infrastructures: a new generation of denial-of-service attacks. IEEE Syst. J. 1–11.

Ficco, M., Rak, M., 2015. Stealthy denial of service strategy in cloud computing. IEEE Trans. Cloud Comput. 3 (1), 80–94.

Ficco, M., Avolio, G., Battaglia, L., Manetti, V., 2014. Hybrid simulation of distributed large-scale critical infrastructures. In: Proceedings of the International Conference on Intelligent Networking and Collaborative Systems (INCoS 2014), pp. 616–621.

Ficco, M., Avolio, G., Palmieri, F., Castiglion, A., 2016. An HLA-based framework for simulation of large-scale critical systems. Concurrency Computat. Pract. Exper. 28 (2), 400–419.

Ficco, M., Di Martino, B., Pietrantuono, R., Russo, S., 2017. Optimized task allocation on private cloud for hybrid simulation of large-scale critical systems. Futur. Gener. Comput. Syst. Available at: http://www.sciencedirect.com/science/article/pii/S0167739X16300061.

Guckenheimer, J., Ottino, J., 2008. Foundations for complex systems research in the physical sciences and engineering. Report from NSF Workshop.

Maier, M., 1998. Architecting principles for systems-of-systems. Syst. Eng. 1 (4), 267–284.

Oltsik, J., 2015. Data-centric security: a new information security perimeter, pp. 1–4. Available at: https://www.informatica.com/content/dam/informatica-com/global/amer/us/collateral/analyst-report/en_esg-data-centric-security_analyst-report_2876.pdf.

Palmieri, F., Ricciardi, S., Fiore, U., 2011. Evaluating network-based dos attacks under the energy consumption perspective: new security issues in the coming green ICT area, pp. 374–379.

Palmieri, F., Ricciardi, S., Fiore, U., Ficco, M., Castiglione, A., 2015. Energy-oriented denial of service attacks: an emerging menace for large cloud infrastructures. J. Supercomput. 71 (5), 1620–1641.

Pederson, P.D.D., Hartley, S., Permann, M., 2006. Critical infrastructure interdependency modeling: a survey of us and international research. Idaho National Laboratory, Idaho Falls, pp. 1–126, INL/EXT-06-11464.

Zio, E., Ferrario, E., 2013. A framework for the system-of-systems analysis of the risk for a safety-critical plant exposed to external events. Reliab. Eng. Syst. Saf. 114, 114–125.

CHAPTER

# DEPENDABILITY OF CONTAINER-BASED DATA-CENTRIC SYSTEMS

# 1

**Petar Kochovski***, **Vlado Stankovski***
*University of Ljubljana, Ljubljana, Slovenia**

## 1 INTRODUCTION

The overall computing capacity at disposal to humanity has been exponentially increasing in the past decades. With the current speed of progress, supercomputers are projected to reach 1 Exa Floating Point Operations per Second (EFLOPS) in 2018. The Top500.org website presents an interesting overview of various achievements in the area. However, there are some new developments, which would result in even greater demands for computing resources in the very near future.

While exascale computing is certainly the new buzzword in the High Performance Computing (HPC) domain, the trend of Big Data is expected to dramatically influence the design of the World's leading computing infrastructures, supercomputers, and cloud federations. There are various areas where data is increasingly important to applications. These include research, environment, electronics, telecommunication, automotive, weather and climate, biodiversity, geophysics, aerospace, finance, chemistry, logistics, and energy. Applications in these areas have to be designed to process exabytes of data in many different ways. All this poses some new requirements on the design of future computing infrastructures.

The recent high development and implementation of the Internet of Things (IoT) has led to the installation of billions of devices that can sense, actuate, and even compute large amounts of data. The data coming from these devices pose a threat to the existing, traditional data management approaches. The IoT produces a continuous stream of data that constantly needs to be stored in some data center storage. However, this technology is not intended for storing the data, but processing it as well, and giving the necessary response to the devices. Big Data is not a new idea, because it was firstly mentioned in an academic paper in 1999 (Bryson et al., 1999). Though much time has passed since then, the active use of the Big Data has just recently begun. As the amount of data produced by the IoT kept rising, organizations were forced to adapt technologies to be able to map with IoT data. Therefore it could be stated that the rise of the IoT forced the development and implementation of the Big Data technologies.

Big Data has no clear definition, despite the fact that it is first thought of in relation to size. Big Data is based on four main characteristics, also known as the 4Vs: volume, variety, velocity, and veracity. The **volume** is related to the size of data, which is growing exponentially in time. Thus none knows the exact amount of data that is being generated, but everyone is aware that it is an enormous quantity of

**Security and Resilience in Intelligent Data-Centric Systems and Communication Networks. https://doi.org/10.1016/B978-0-12-811373-8.00001-X**

Copyright © 2018 Elsevier Inc. All rights reserved.

information. Although most of the data in the past was structured, during the last decade the amount of data has grown and most of it is unstructured data now. IBM has estimated that 2.3 trillion gigabytes of data are created every day and 40 zettabytes of data will be created by 2020, which is an increase of 300 times from 2005 (IBM Big Data, 2013). The **variety** describes those different types of data, because various data types are generated by industries as financial services, health care, education, high performance computing and life science institutions, social networks, sensors, smart devices, etc. Each of these data types differ from each other. This means that it is impossible to fit this kind of various data on a spreadsheet or into a database application. The **velocity** measures the frequency of the generated data that needs to be processed. As required the data could be processed in real-time or processed when it is needed. The **veracity** is required to guarantee the trustworthiness of the data, which means that the data has the quality to enable the right action when it is needed.

With the previous analysis it is obvious that Big Data plays a crucial role in today's society. Due to the different formats and size of the unstructured data, the traditional storage infrastructures could not achieve the desired Quality of Service (QoS) and could lead to data unavailability, compliance issues, and increased storage expenses. A solution, that addresses the data-management problems is the data-centric architecture, on which the data-centric systems are designed. The philosophy behind DCS is simple, as the data size grows, the cost of moving data becomes prohibitive. Therefore the data-centric systems offer the opportunity to move the computing to the data instead of vice versa. The key of this system design is to separate data from behavior. These systems are designed to organize the interactions between applications in terms of the stateful data, instead of the operations to be performed. As the volume and velocity of unstructured data increase all the time, new data management challenges appear, such as providing service dependability for applications running on the Cloud. According to the author (Kounev et al., 2012) dependability can be defined as the ability of a system to provide dependable services in terms of availability, responsiveness, and reliability. Although dependability has many definitions, in this chapter we will try to depict the dependability in a containerized, component-based system, data-centric system.

While many petascale computing applications are addressed by supercomputers, there is a vast potential for wider utilization of computing infrastructures by using the form of cloud federations. Cloud federations may relate to both computing and data aspects. This is an emerging area that has not been sufficiently explored in the past and may provide a variety of benefits to data-centric applications.

The remaining section of this chapter are structured as follows. Section 2 describes the component-based software engineering methodology, the architectural approach for data management, and container interoperability. Section 3 points out the key concepts and relations in dependability, where the dependability attributes, means, and threats are described. Section 4 describes the serving of virtual machine and container images to applications. Section 5 elaborates the QoS management in the software engineering process. Section 6 concludes the chapter.

## 2 COMPONENT-BASED SOFTWARE ENGINEERING

In order to be able to build dependable systems and applications we must rely on a well-defined methodology that would help analyze the requirements and the trade-offs in relation to dependability. Therefore it is necessary to relate the development of dependable data-centric systems and applications to modern software engineering practices in all their phases, such as requirements analysis, component development, workflow management, testing, deployment, monitoring, and maintenance.

With the constant growth of software's complexity and size, the traditional software development approaches have become ineffective in terms of productivity and cost. Component-Based Software Engineering (CBSE) has emerged to overcome these problems by using selected components and integrating them into one well-defined software architecture, where each component should be presented as a functionality independent from other components of the whole system. As a result of CBSE implementation, during the software engineering process the developer selects and combines appropriate components instead of designing and developing the components themselves—they are found in various repositories of Open-Source software. The components can be heterogeneous, written in different programming languages, and integrated in an architecture, where they communicate with each other using well-defined interfaces.

## 2.1 LIFE-CYCLE

Every software development methodology addresses a specific life cycle of the software. Although life cycles of different methodologies might be very different, they could all be described by a set of phases that are common for all of them. These phases represent major product lifecycle periods and they are related to the state of the product. The existing CBSE development lifecycle has separated component development from system development (Sharp and Ryan, 2009). Although the component development process is in many aspects similar to system development there are some notable differences, for instance components are intended for reuse in many different products, many of which have yet to be designed.

The component lifecycle development, shown in Fig. 1, can be described in seven phases (Sharp and Ryan, 2009):

1. Requirements phase: During this phase the requirement specification for the system and the application development are decided. Also during the requirements phase the availability of the components is calculated.
2. Analysis and Design phase: This phase starts with a system analysis and a design providing overall architecture. Furthermore, this phase develops a detailed design of the system. During this phase it is decided which components will be used in the development. Because the design phase impacts the QoS, during this phase the necessary QoS level to be achieved is determined.
3. Implementation phase: Within this phase the components are selected according to their functional properties. They are verified and tested independently and together as well. Some component adaptations may be necessary to ensure compatibility.
4. Integration phase: Although integration is partially fulfilled during the implementation phase, the final integration of all components within the architecture occurs during this phase.
5. Test phase: This phase is necessary to ensure component quality within the given architecture, because the component could have been developed for another architecture with different requirements. The components are usually tested individually, after integrating in assembly and after the full integration in the system.
6. Release phase: This phase includes preparing the software for delivery and installation.
7. Maintenance phase: As for all software products, it is necessary to provide maintenance with the replacement of components. The approach in CBSE is to provide maintenance by replacing the old components with new components or by adding new components into the system.

Cloud computing technologies can be used to increase the ubiquitous access to services, anywhere at anytime. Therefore Cloud architecture needs to be designed to provide the desired QoS. The author

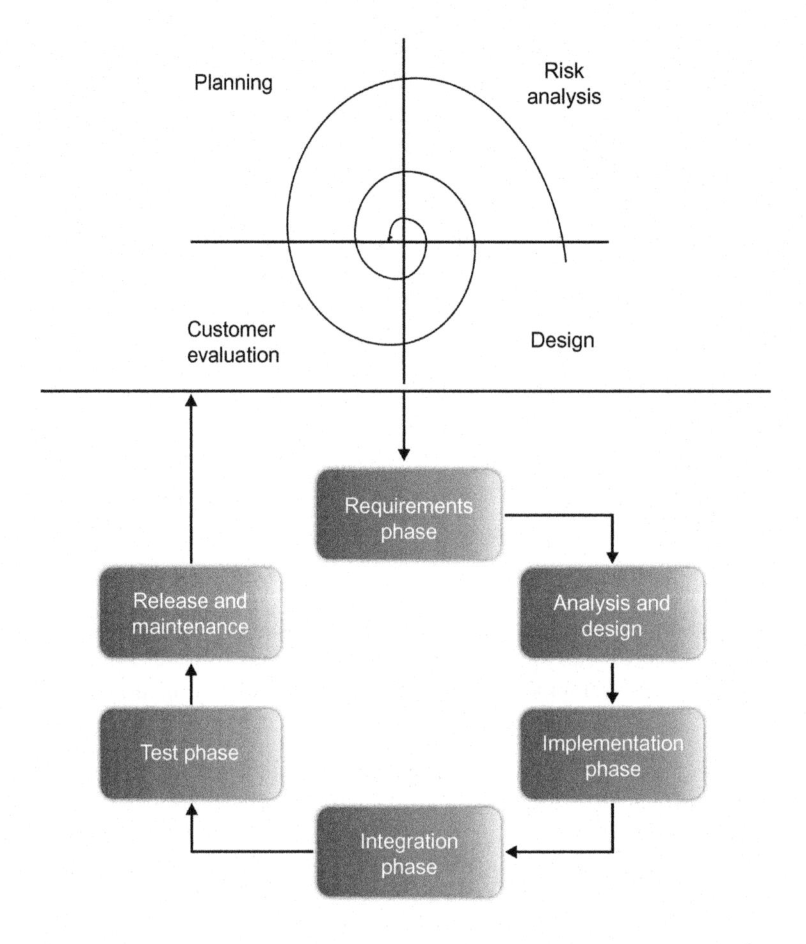


**FIG. 1**

Component-based software development lifecycle.

(Ramachandran, 2011) explains that it is essential to design the Cloud applications as Web service components based on well-proven software processes, design methods, and techniques such as the component-based software engineering techniques. One of the most common problems during Cloud application development is the SLA support. Since the SLAs vary between different service providers, this goal can only be reached using components designed for flexible interface that links to many different SLAs. This CBSE supports implementation of different design characteristics, which can be designed to support many SLAs.

## 2.2 ARCHITECTURAL APPROACHES FOR DATA MANAGEMENT

Containerization is a process of distributing and deploying applications in a portable and predictable way. The use of containers is attractive for developers and system administrators as well, because of the possibilities they offer, such as abstraction of the host system away from the containerized application, easy scalability, simple dependency management and application versioning, and lightweight execution environments (Paraiso et al., 2016). Compared to VM virtualization, containers provide a lighter execution environment, since they are isolated at the process level, sharing the host's kernel. This means that the container itself does not include a complete operating system, leading to very quick start up times and smaller transfer times of container images. Therefore an entire stack of many containers could run on top of a single instance of the host OS. Depending on the container, we can distinguish three different architectural approaches, described in the following.

### 2.2.1 Functionality and data in a container

The use of containers can make it possible to diminish the otherwise hard barrier between stateful and stateless services. The use of containers makes it possible to configure this aspect as part of the software engineering process. Fig. 2 shows a container that provides an embodiment of both functionality and data. If the service needs to be stopped and restarted somewhere else, it may be possible to use a check pointing mechanism (Stankovski et al., 2016). For example, when a service restart may be needed is there deteriorated in the network parameters, such as the latency between the client and the running service. This approach may also be useful when the service needs to be used when various events happen in the environment, for example, when a user wishes to have a videoconference, or when a user wishes to upload a file to a Cloud storage. In these cases, the geographic location of the user may influence the Quality of Experience of the applications and thus, it may be necessary to start the container in a location near to the place where the user is located.

A container with functionality and data is a container that containerizes the application code, all the libraries required for the target application and the dataset together. In general terms, the use of container technologies has many advantages over other virtualization approaches. Containers offer better portability, because they can be easily moved among any system that runs on the same host OS

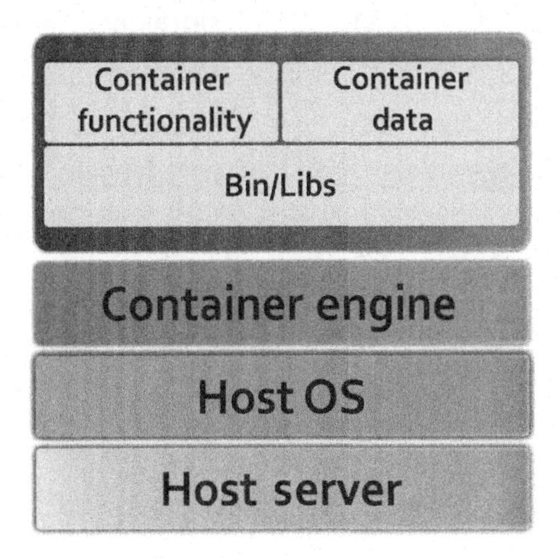


**FIG. 2**

A single container with functionality and data.

without additional code optimization. It is possible to run many containers on the same infrastructure, that is, they can easily scale horizontally. Thus the performance is improved, because there is only one OS. The main advantage of containers is that they can be created much faster than VM. This provides a more agile environment for development and a continuous integration and delivery.

However, containers are not perfect, and there are some disadvantages. With containerization there is no isolation from the host OS. Since the containers share the same OS, security threats could easily access the entire system. This security threat can be solved by combining containers and VM, where the containers are created within an OS running on a VM. With this approach the potential security breach could only affect the VM's OS, but not the physical host. Another containerization disadvantage is that the container and the base-host must use the same OS.

### 2.2.2 Clean separation between functionality and data in containers

Another approach that may be useful in relation to containers is to completely separate the functionalities from the actual data needed for their operation. This can be achieved by developing software defined systems and applications with a clear cut between the two. In such cases, when the functionality is needed, along with the data, a minimum of two containers need to be started: the first container provides the functionality, while the second container provides the data infrastructure (e.g., file system, SQL data base or knowledge base, as in the case of the ENTICE project). Fig. 3 graphically presents the possible situation.

While the time required to setup and start the needed functionality for the given user may be slightly increased (as it requires the mounting of the data container on the container that provides for the functionality and the use of a network for communication between the containers), the separation can be effectively used in situations where the privacy and security aspects are very important to the application, and single-tenant users are the architectural choice. However, this architecture may be used

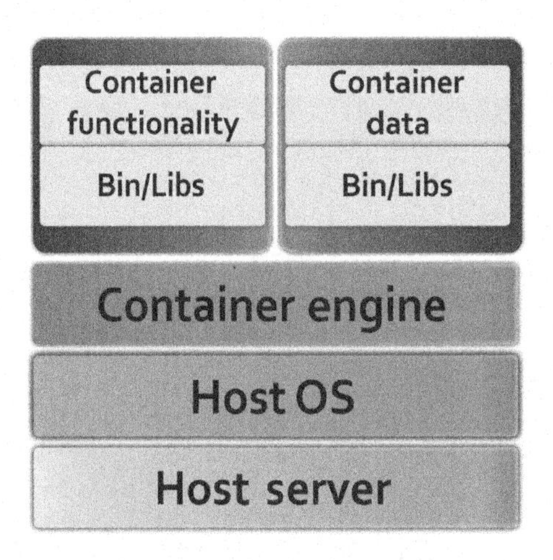


**FIG. 3**

Two linked containers provide separation of the functionality from the data.

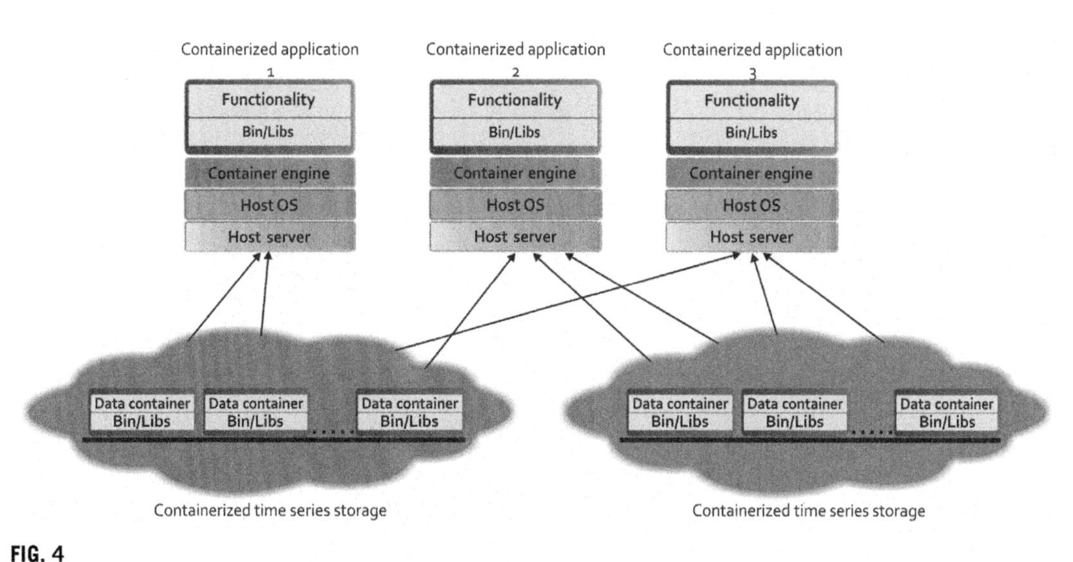


**FIG. 4**

Geographically distributed containerized data infrastructure based on a specific technology such as Cassandra.

to achieve high Quality of Service also in multitenant applications, such as those that are based on the Internet of Things (Taherizadeh et al., 2016b), where both containers containing the Web server and the SQL data base can elastically scale.

### 2.2.3 Separate distributed data infrastructure and services

The previously described approaches function well when there is no need to work with large data sets. However, long running applications produce large applications data sets, which must be stored and managed somewhere. This has led to arrangements of two parallel architectures: one for the management of data with a variety of approaches (e.g., CEPH, 2016; Amazons3, 2017; Storj.io, 2016; etc.) and another for the management of functionalities that are packed onto containers.

Such clearly separated data management infrastructures and services can be established in a virtualized format as shown in Fig. 4. An example of such a system is Cassandra (Apache Cassandra, 2016), which can operate in a distributed and containerized format, trading performance with elegance of implementation. Thus the Cloud application is developed to make use of two infrastructures, one for handling elasticity and other properties of the containerized application, and another for the application dataset, hosted on a scalable distributed storage, which can be accessible from any container on the network. In order to improve the dependability of an architecture like this, with improving the availability and reliability, could be achieved by fast reinstatation of database nodes, when the application becomes unresponsive. According to the author (Stankovski et al., 2015) this could be done by saving the container's file system with the state of the given container into disk files, copied to another server, and restarting the container without rebooting or recommencement from the beginning of the process.

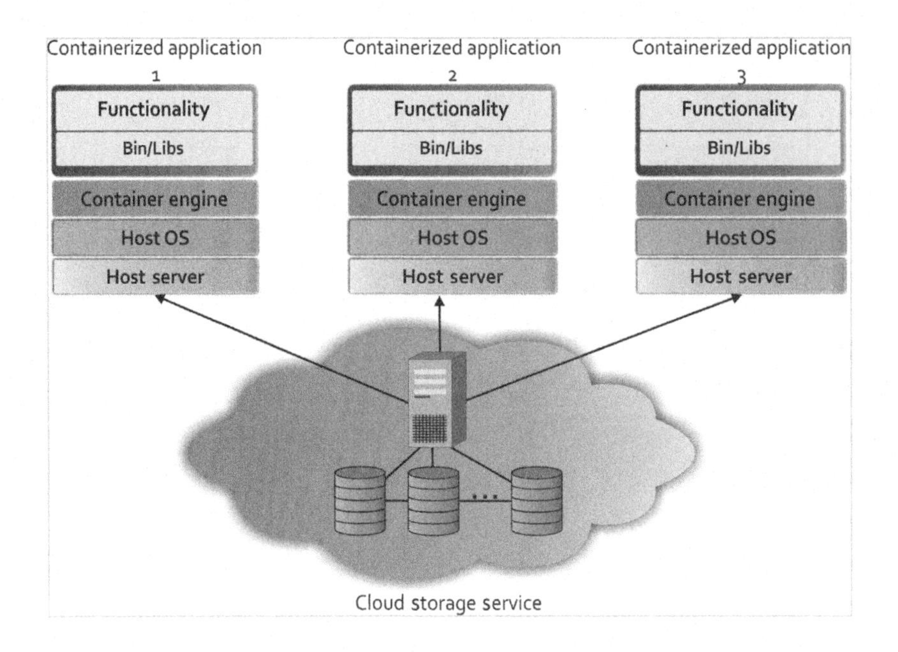


**FIG. 5**

A separate nonvirtualized data management infrastructure such as CEPH.

As previously explained, in order to improve the Quality of Service parameters (aka performability) in the case of handling great amounts of data it may be necessary to make use of data management systems, such as CEPH, which cannot be effectively virtualized (see Fig. 5).

## 2.3 EMERGING CONTAINER INTEROPERABILITY ARCHITECTURES

CNCF.io is an emerging container interoperability architecture promoted by a recently formed organization. The architecture as shown in Fig. 6 is developed to address several areas of emergent needs for Cloud applications, including the IoT and Big Data application domains. There are three main reasons for developing the CNCF architecture: container packaging and distribution, dynamic scheduling of containers, and micro-service implementation. The first part of the CNCF is designed based on technologies that build, store, and distribute container-packaged software. The second part of CNCF consists of technologies that dynamically orchestrate container images. Examples of orchestration technologies are: Kubernetes (2017) and Apache Mesos (2017). The third part of this emerging architecture consists of R&D microservices, which are designed to ensure that the deployed container image is built in a way that it can be easily found, accessed, consumed, and linked to other services. Thus this architecture can be used to build distributed systems and support the discovery and consumption of software (i.e., the overall life-cycle of Cloud applications).

As is evident from this architecture, distributed storage facilities have not been sufficiently considered, nor have they been effectively included in the CNCF architecture. The architecture specifies

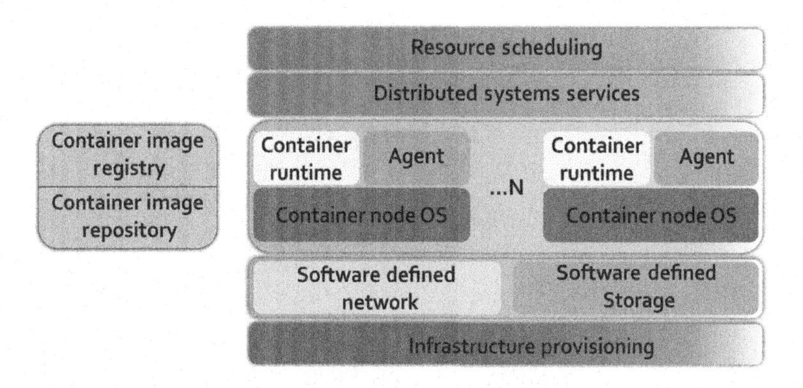


**FIG. 6**

A Cloud interoperability architecture of the Cloud Native Computing Foundation.

the need to separate the interfaces between the containers and the Software Defined Network and Software Defined Storage.

# 3 KEY CONCEPTS AND RELATIONS IN DEPENDABILITY

Based on the previously described use cases, and based on a review of the literature, it is possible to establish some key concepts and relations in **dependability**. System failures have devastating effect, resulting in many unsatisfied users affected by the failure and disruption of business continuity. The downtime results in loss of productivity, loss of revenue, bad financial performance, and damage to the user's reputation. Damage to the user's reputation could directly affect the user's confidence and credibility with costumers, banks, and business partners. Therefore, it is crucial to design a dependable data-centric system. The original definition of dependability is the ability of a system to deliver service that can justifiably be trusted. An alternate version of the definition of dependability, based on the definition of failure, is stated as the ability of a system to avoid failures that are more frequent or more severe, and outage duration that are longer than is acceptable to the user (Avizienis et al., 2004). Although many different characteristics could influence the dependability of a system, the three essential characteristics are: attributes, threats, and means.

The dependability of a system changes according to the system requirements. Thus the dependability attributes priority level may change to a greater or lesser extent. However, availability is always required, whereas reliability, safety, and security may or may not be required. Because dependability relies on many probabilistic attributes, it could never be interpreted in absolute sense. As a result to the unavoidable presence of system faults, systems could never be fully available, reliable, safe, or secure.

## 3.1 DEPENDABILITY ATTRIBUTES

The dependability attributes are the measurements that could be applied to a system to rate the dependability. Although there are many attributes according to which system dependability can be

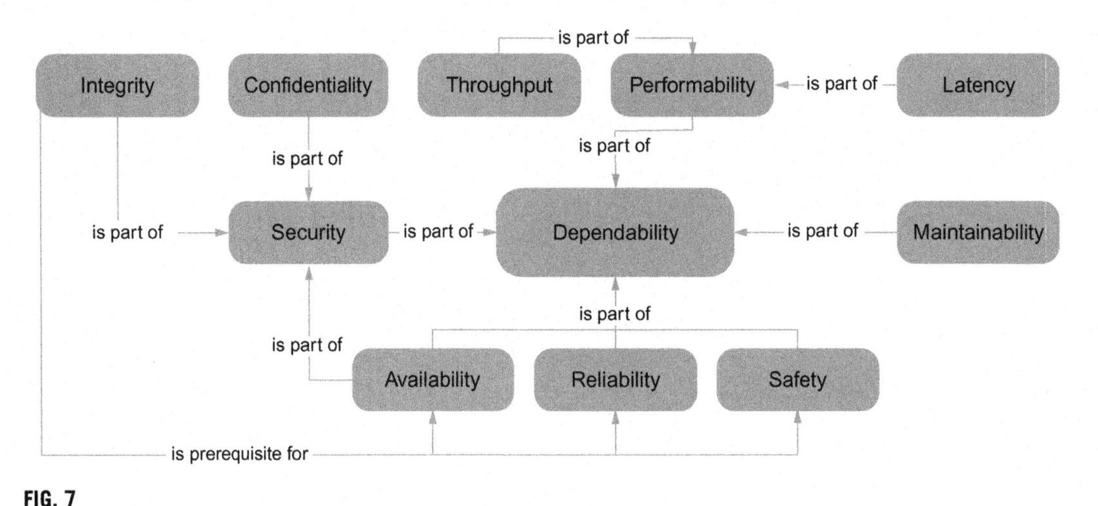


**FIG. 7**

Dependability attributes ontology.

evaluated, the generally agreed attributes of dependability are: availability, reliability, safety, and security. However, these four attributes are not always enough to build a dependable system or to satisfy the dependability requirements. Thus additional attributes are used, such as maintainability, performability, confidentiality, and integrity. Many dependability attributes rely on each other. (Fig. 7) The attributes can be quantifiable or subjective. The quantifiable attributes can be quantified by physical measurements (i.e., reliability). On the other hand the subjective attributes cannot be directly measured using metrics, but are subjective assessments which require judgmental information to be applied (i.e., safety).

### *3.1.1 Availability*

**Availability** of a system is the probability that the system is functioning correctly at some instance in time. When designing a system, not only the probability of failure is considered, but the number of failures and the time required to make repairs. In such cases the goal is to maximize the time that the system is in an operational state, performing the tasks satisfactory. These goals are manifested by the availability of the system. However, this metric is understood in broader terms compared to the definition that commercial Cloud storage providers may have. For example, Amazon treats their storage as available even if there is no answer to requests, as only two specific HTTP error responses account for error under which SLA violation applies. However, this is understandable, since when no answer is given by the storage service it might be a network issue between the requester and the storage and is thus out of the control of the storage provider. On the other hand, some other failures resulting in service not being useful could still occur on the provider's side, yet are not subject to SLA violation.

Many different planned and unplanned incidents could result in service unavailability. Planned outages include different hardware and software upgrades and maintenance processes, such as installation of new hardware, software upgrades, patches, backups, data restores, and testing. These outages are well planned and expected, but still they are the reason for service unavailability. Unplanned outages include failures as a result of many different unplanned reasons, like failure of physical or virtual components, database failures, or human mistakes. To improve the availability, risk analyses have to be performed. Risk analyses will calculate the component failure rate and the average repair time, which are calculated by mean time before failure, mean time to failure, and mean time to repair.

- Mean Time before Failure: The average time available for a system or component to perform its normal operations between failures. It is usually expressed in hours.
- Mean Time to Failure: The average time that a system or a component will function correctly before it fails. This is a statistical value and is meant to be the mean over a long period of time and a large number of units. Technically, mean time before failure should be used only in reference to a repairable item, while mean time to failure should be used for nonrepairable items. However, mean time before failure is commonly used for both repairable and nonrepairable items.
- Mean Time to Repair: The average time required to repair a failed component. While calculating, it is assumed that the fault responsible for the failure is correctly identified and the required spares and personnel are available.

When implementing availability the ping command is not enough. Since even if the storage service is answering for pings, the services backend might be down. For example, the Simple Storage Service (S3) and its widely used APIs of Amazon expose a variety of functionalities of the backend to the outside world. The technologies used to realize the backend services can be various and this is transparent to the users. The used technologies themselves can provide for various degrees of Quality of Service. Pinging the S3 service will therefore provide no insider information on the actual status of the backend services. Thus service availability cannot be deduced solely by using the ping command. To test the backend, one obvious way is to run a functionality test, that is, to try to retrieve/upload some data. Such a test has the opportunity to uncover potential problems with the backend. Thorough tests, however, are difficult to design and must be prepared for a specific usage scenario where availability is of crucial importance. In this context various lightweight adaptive monitoring techniques can be used, such as those developed in the context of the CELLAR (Giannakopoulos et al., 2014) project. The monitoring technique for availability has to be lightweight so that it does not represent significant additional load on the network, its data payload for testing should be small in size and short in time. However, this approach cannot fully address cases when the backend only starts failing when there are more significant data management needs.

### *3.1.2 Reliability*

The **reliability** attribute calculates the probability of success, whereby a system will satisfactorily performs the task it was designed, for a specific time, under well-specified environmental conditions. The reliability guarantees that no fault would prevent a continuous delivery of service, it stands for failure-free interval of operation. High-reliability is necessary when it is expected that the system will operate without interruptions and in cases when maintenance cannot be performed. Unfortunately, all systems fail, the only questions are when, and how frequently.

As the monitoring and control capabilities have been constantly advancing, and with their implementation, systems have gained the opportunity to reduce the risk of failures and improve their reliability. Higher reliability in a data-centric system can be reached by linking the system monitoring to the system servers, which would be able to quickly detect any defects and securely act toward ensuring operational continuity.

### 3.1.3 Safety

**Safety** is the ability of the system to operate without catastrophic failure, which could affect the users and the environment. Safety and reliability are related attributes. Reliability and availability are necessary for a system, but not sufficient conditions for fulfilling system safety conditions. Thus the safety attribute can be described as reliability with respect to catastrophic failures (Powell, 2012).

Data-centric systems have found their role in many crucial industries, such as medicare, defense, and nuclear power. Implementing the necessary safety standards in those industries is crucial, otherwise it could result with system failure, which could lead to human injury, loss of life, property damage, or environmental disaster. Thus the data-centric systems can also be defined as safety-critical systems. However, not all the data-centric systems operate in such environments, thus safety implementation is not always of primary importance in, for instance, systems that are used in e-commerce.

### 3.1.4 Security

**System security** offers the ability to protect itself from accidental or intentional external attack that can change the system behavior and damage data. If an attack is not prevented and it damages the system, then the reliability and safety are no longer valid. Security is not a single dependability attribute. It is a composition of confidentiality, integrity, and availability.

The **confidentiality** attribute measures if the data, such as user personal information, documents, and credit card numbers are protected from disclosure to unauthorized parties. Protecting this sensitive data is important when designing a dependable system. The **integrity** attribute controls whether or not the data has been modified by unauthorized parties. For instance, if during a bank transaction the system is not dependable enough as a result of low integrity, and the transaction data is modified from $100 to $100,000, it could be very costly to the user. The integrity is a prerequisite for availability, reliability, and safety. The implementation of low security standards during the design of the data system has direct influence on the availability of the system. This can result in system unavailability, due to denying access attacks that are common nowadays.

Higher security might be implemented by using techniques, such as when system prevents unauthorized disclosure of data, prevents unauthorized data removal, and prevents unauthorized withholding of data. Container virtualization can as well incorporate applications within encrypted workspace on one host operating system. Although many containers can share one OS host, the virtual container can be customized with policies that control the type of interactions that may happen among the applications within the virtual container. Therefore, all the interactions between applications within the container remain in the container. As a result, all of the data associated to these applications remains secure.

### 3.1.5 Maintainability

**Maintainability** is the ability of the system to go through repairs and modifications while it is up and running. Once a system fault is detected, it is desirable to be able to apply the necessary improvements as soon as possible, without having to shut down the system. Also patching update mechanisms could be implemented to improve the overall performance of the system during runtime, these mechanisms are often referred to as live upgrade or live update mechanisms.

Using the container virtualization in data-centric systems offers us high level maintainability due to the nature of containerization technologies. With proper implementation of the proposed architectural approaches for data management, described previously, the developers could achieve high maintainability.

### 3.1.6 Performability

The **performability** attribute indicates the responsiveness of a system to execute a task within a predetermined time interval. Performability can be measured using the latency or the throughput. The **throughput** indicates the expected amount of data to transfer over a time unit. The **latency** indicates the time before the first byte is retrieved from the storage.

This property is one of the most investigated properties in various Cloud computing systems and applications. Hence, various means of implementation may be possible. For example, the ongoing SWITCH project focuses on software engineering aspects of time-critical applications and places great care on performability. This project investigates the use of various protocols for communication among software components, and among the clients and the application, such as the User Datagram Protocol (UDP) or the Hypertext Transfer Protocol (HTTP). As it is obvious, the throughput parameter is hard to predict in the open Internet as the situation can dramatically change on hourly basis. In the open Internet there is no control over the network paths as well as no control over the current workload. Inside the Cloud provider various Software Defined Networking (SDN) (Kim and Feamster, 2013) techniques may be used to guarantee the Quality of Service. The SWITCH project has been investigating an edge computing approach in order to address this part of the performability problem.

The latency of the various repositories is sometimes part of the Service Level Agreements concluded between the user and the Cloud provider. For example, the latency of Amazon's Standard S3 storage when retrieving data is measured in milliseconds, while Amazons Glacier S3 storage has retrieval time of 3–5 hours on average. As a side note, Glacier S3 storage is a type of storage which is aimed at storing of backup data at lower cost, consequently having long retrieval time before the first byte is retrieved. Given the pervious example, if retrieval time is significant but not measured, high download time could give a wrong figure over network path load.

## 3.2 DEPENDABILITY MEANS

**Dependability means** describe the ways to increase the dependability of a system. When developing a dependable system, it is necessary to utilize a set of four procedures: fault prevention, fault tolerance, fault forecasting, and removal. The functionality of fault prevention and fault tolerance is to deliver a service that can be trusted, while the fault forecasting and removal goal is to deliver confidence in that dependability and security specification are adequate and that the system is ready to meet them.

### 3.2.1 Fault prevention

**Fault prevention** techniques are used to minimize the possibility of the occurrence of system faults (Pradhan, 1996). Minimizing the occurrence of system faults is possible during the specification, implementation, and fabrication stages of the design process by implementing quality control techniques. Monitoring tools, such as Nagios (2017), during those stages could be used for tracing system faults. These tools are composed of different components, which monitor the current status, collect the metrics, send data to monitoring server, analyze the reports, and notify the administrator. Nagios monitors the current state of the hosts, devices, and services and if a suspicious behavior is detected then it would notify the administrator. Another monitoring approach, proposed by Taherizadeh et al. (2016a) monitors performance from user perspective. That monitoring approach could be used to evaluate the network quality delivered to an end-user, and then it is possible to improve the overall acceptability of the service, as perceived subjectively by each user.

### 3.2.2 Fault tolerance

Fault tolerance techniques are used to prevent system faults from becoming system errors, which could lead to serious system failures. A fault tolerant system is designed to keep performing even when some of its software or hardware components break down. This type of design allows a system to continue performing at reduced capacity rather than completely shutting down after a failure.

Fault tolerant systems are divided into two groups, depending on the system's recovery behavior from errors. They can be either roll-forward systems or roll-back. The roll-back recovery uses checkpointing mechanism to return the system to some earlier correct state, after which it would continue functioning properly. The checkpointing mechanisms are designed to periodically record the system state. However, this approach gets challenging when it is used with big VMI (Goiri et al., 2010). The roll-forward recovery, on the other hand, returns to the state when the error was detected in order to correct the error and then the system continues working forward.

The system can also improve the fault tolerance using adaptive techniques. These techniques give the system the ability to adapt to the environmental changes. To ensure the system's fault tolerance, its state is monitored and reconfigured as needed.

### 3.2.3 Fault forecasting and removal

These techniques are necessary for determining and reducing the present number of faults and the consequences of the faults. **Fault forecasting** evaluates the system behavior. It estimates the amount of faults that are present in the system, the possibility of their activation in the future, and the consequences of their activation. **Fault removal** are techniques that reduce the number of faults that are present in the system according to the results of the fault forecasting techniques. This technique includes three steps: verification, diagnosis, and correction (Dubrova, 2013).

Verification assures that the system fully satisfies all the expected requirements. If it does not, then the fault is diagnosed and the suitable corrections are performed. After the correction the system is verified again to ensure that the system adheres to the given verification conditions. System maintenance in an important fault removal technique as well. Corrective maintenance is used for removing faults that have been reported for producing errors to return the system to service, while preventive maintenance is used for discover faults in the system before they could cause an error during runtime.

## 3.3 DEPENDABILITY THREATS

When developing a dependable system, one of the most important phases is calculating the dependability threats. **Dependability threats**, such as faults, errors, and failures, describe the risks that can degrade the dependability of the system infrastructure. It may be expected that our understanding of the dependability threats will increase with an increasing number of implemented data-centric systems and applications in near future.

A **fault** may be understood as a defect that occurs in some hardware or software component in the system, e.g., software bug or hard drive system fault. Such an event may definitely affect the dependability of the system. If the fault is properly detected by the monitoring system, it may be possible to mitigate the longer term effects of the fault. For example, the self-adaptation component of SWITCH may reestablish the service in another Cloud provider. This happens to be one of the main benefits of virtualization as various faults may be possible to mitigate by reestablishing the required services elsewhere.

An **error** is a system state caused by a system fault that produces incorrect result and unexpected system behavior. System **failure** is a state when the system delivers service that is different from the system specification for the specified period of time. In software defined systems and applications, both errors and failures can be addressed through the use of authentication and authorization mechanisms that can ascertain the high quality of VMI/CIs that are used. IBM has recently introduced a patent that provides encryption mechanisms for VMIs and their integral parts (Kundu and Mohindra, 2012). More modern approaches may include the use of blockchain (Shrier et al., 2016) to provide strict checking of system errors and failures and may be used to allow the use of well-tested and error-free software components.

System dependability threats depend on each other. For example, system faults are the reason for error to appear, and errors are the reasons for the failures to appear. If there is a Cloud infrastructure that suffers from a fault in some of the software components, at some moment this could produce an error and lead to unexpected system failure and could result with system unavailability.

# 4 ENTICE CASE STUDY: SERVING VIRTUAL MACHINE AND CONTAINER IMAGES TO APPLICATIONS

Several recent projects, such as mOSAIC (2017) have shown that it is possible to build Cloud application engineering platforms that can be ported to specific Cloud providers and may be used to surpass the vendor lock-in problem. In such platforms, once the Cloud application is fully engineered, it can be readily deployed across multiple Clouds.

Such portable Platforms as a Service (PaaS) have a large potential for achieving cost savings, because they offer reuse of software components in various Cloud provider settings. Thus component-based design offers great flexibility for resources management and service customization in such environments.

The ENTICE platform can be viewed as a portable PaaS, which offers the user efficient VMI/CI operations in the Cloud. Fig. 8 presents the key research areas addressed by ENTICE. It could be also considered as a repository-based system, which encapsulates a variety of subsystems (Gec et al., 2016).

1. In the first step of using ENTICE the programmers search for software components that are expressed as functionalities.
2. Certain functionalities may be developed through the means of recipes, with tools such as CHEF (2017) or PUPPET (2016).
3. The next step in the process is to synthetize a VM or container image, which is an automated process.
4. Following this, the ENTICE services provide for optimization based on the necessary functionality. This is done via the use of functionality tests.
5. The most sophisticated part of the distribution process is the multiobjective optimization, known as Pareto front, which solves the problem of conflicting QoS requirements. Instead of a single

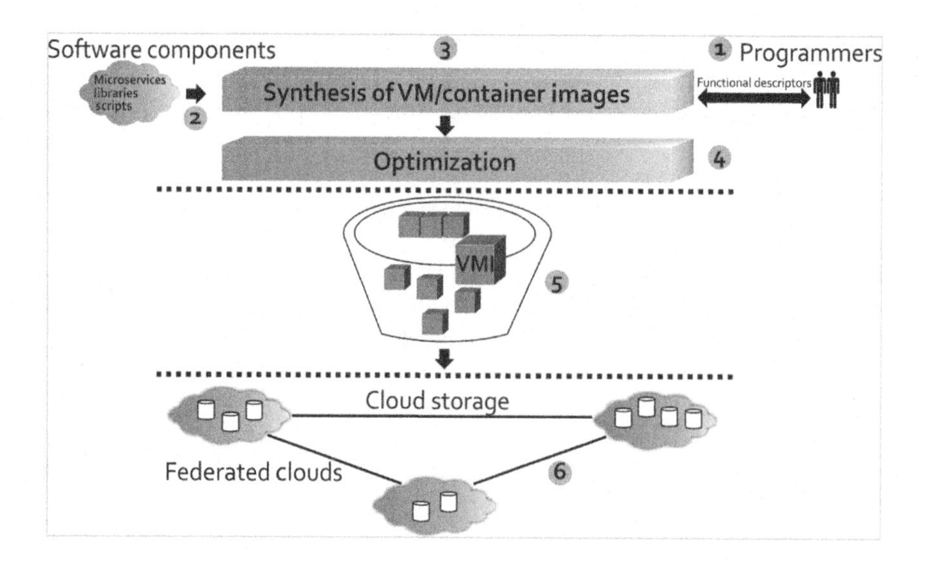


**FIG. 8**

Packaging and distribution of software components.

solution that optimizes all requirements, the multiobjective optimization results with a set of trade-off solutions. This decreases the selection of the best fitting solutions to the users' requirements. When no further optimization is possible, the ENTICE environment provides the possibility of user-friendly further manual optimization of the images. Based on the QoS requirements of the Cloud application users it is necessary to distribute the Virtual Machine and/or container images geographically on a federated Cloud repository.

6. The final stage addressed by ENTICE is the delivery and deployment of images in a designated Cloud provider where the functionality has to be provided. Improving the delivery time could result in deploying the images where their data is stored, as the data-centric approach describes.

ENTICE helps optimize not only the functioning of the federated VMI/CI storage facility, but also the obtained nonfunctional properties of the application at the runtime. An important part of this architecture is a Knowledge Base that helps gather, exchange, and manage information about the necessary QoS related to the VMI/container images repositories.

The most relevant key components of the ENTICE environment, which can be used by developers, are presented as follows.

Development tools are the instruments used to develop a Cloud application. Unfortunately, these instruments do not provide enough information for the deployment stage. As a result, creating VMI and CI images that contain specific services is a time-consuming process that requires good developer knowledge about library dependent software of the service and scalability. The ENTICE environment offers the developer fast and easy image access and repository-wide software search and discovery.

The Knowledge Base component is a service that uses triple database and mechanisms like reasoners, rules, and validations. to provide information to all other components of the architecture. The reasoning mechanisms, strategic and dynamic, are implemented according to the domain ontology.

The strategic reasoning evaluates the QoS functional properties, execution time, costs, and storage to support automatic packaging of applications. The dynamic reasoning ensures that the image packaging and preparation have been done according to the dynamic information about the federated Cloud instance.

The ENTICE user interface is a client-side application that communicates with the Knowledge Base, which is the server-side application, via APIs. Thus it offers the developer the opportunity to optimize and search VMI/CI stored on public repositories, and upload new images to the repositories or system based settings. The developer is offered the feature to generate images by using a script that includes functional requirements.

The services of the ENTICE environment process functionalities are elastic and geographically distributed. These functionalities optimize the operation based on information and knowledge on network performance through monitoring services. An ENTICE service is used to reduce the image size for every VMI/CI that contains user developed software components, in order to reduce the VMI size without affecting the running applications functionalities. These operations give the user the opportunity to save on the storage cost as soon as the software asset is deployed in the repositories.

Many public CI repositories are currently available on the Internet. By federating such repositories, a new quality may be expected. With adding new constraints, dependencies, and derived data from reasoning or user experience, new repositories can be registered in the KB.

## 5 SWITCH CASE STUDY: MANAGING QoS IN THE SOFTWARE ENGINEERING PROCESS

The Quality of Service parameters are important aspects in the software engineering process. They must be collected and recorded early in the software engineering process. This is due to the fact that current Cloud applications engineering tools are posed to radically speed up the process on one hand, but, they miss adequate tools to be able to influence the QoS that has to be obtained once the application is deployed and running in a specific Cloud provider.

In SWITCH (2017) particular care has been given to develop a SWITCH Interactive Development Environment (SIDE), which can be used to specify various QoS metrics related to individual components (e.g., databases and compute units). A semantic model such as enhanced OASIS (2013) can be used to store the necessary information. Fig. 9 presents what happens in the runtime when the QoS model is used to obtain the necessary QoS for the particular functionality.

1. The user (or a program) sends a request to use a particular functionality. For an example, the user may request to launch a video conference, which is a JITSI (2017) functionality packed into a container.
2. The SWITCH platform at runtime fetches the desired QoS model for the particular functionality.
3. Based on the desired QoS model for the application, it makes a decision on which infrastructure it is best to deploy and uses the application. It may choose from a variety of Cloud infrastructures where the software components (container images) are being delivered and deployed.
4. Following this the user can use the desired functionality at a desired high QoS. The SWITCH platform at runtime may provide for several services including monitoring and alarm-triggering. These services may cause the platform to self-adapt based on possibly violated QoS metrics.

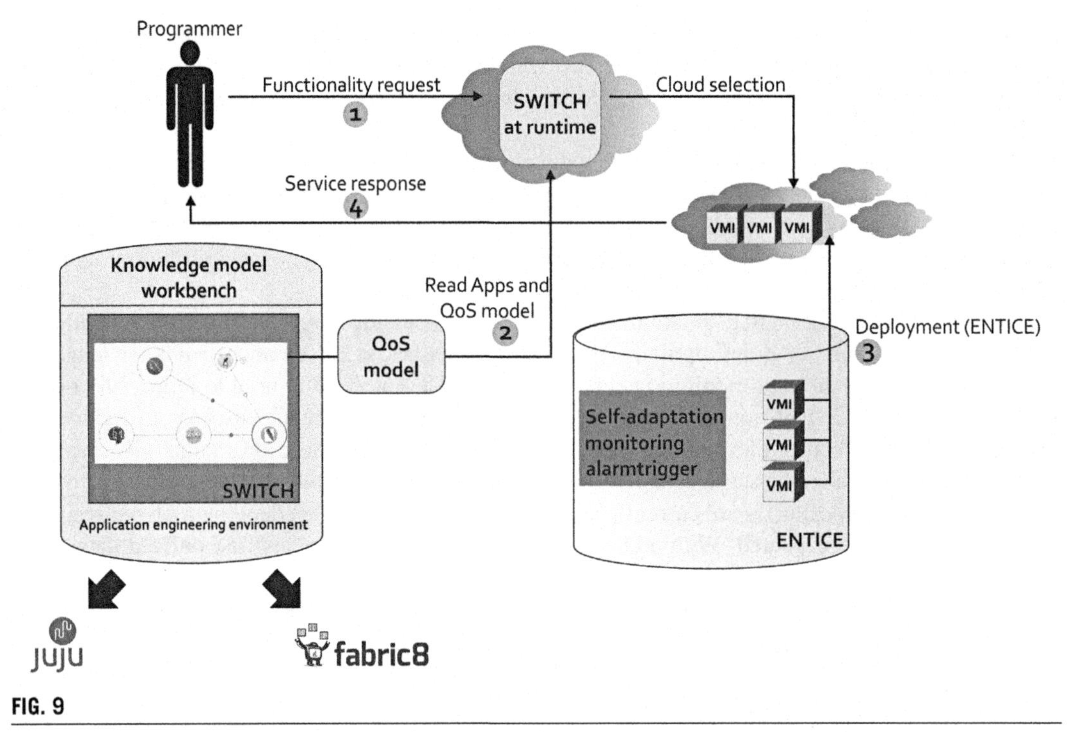


**FIG. 9**

Runtime Cloud application adaptation conforming to QoS requirements.

The key word for these types of scenarios is *multiinstance* as the application can be deployed and used at a high QoS for a single usage instance. The design of the SWITCH project offers a glimpse of the future directions in software engineering for IoT and Big Data applications that respect QoS in the overall engineering process. Both SWITCH and ENTICE environments are designed to be able to support single usage instance applications.

## 6 CONCLUSIONS

Cloud computing systems and applications already provide numerous benefits to their users. Some of these include a shorter software engineering lifecycle, independence from the Infrastructure as a Service providers, optimal usage of resources, and simplified maintenance of the applications (Rittinghouse and Ransome, 2016). All these benefits, however, cannot be obtained without having to address a certain new range of requirements. Dependability is one of the key properties for such systems and applications. In order to define the property, it was necessary to look at its relations with other needed properties, such as availability, reliability, safety, and security; and some additional attributes, such as maintainability, performability, integrity, and confidentiality.

With the emergence of the Big Data era, the Cloud architectures will need to be further transformed and adapted in order to support event-driven and single functionality usage instance scenarios. In such scenarios, dependability will have to be established on the fly.

Emerging container-based interoperability standards, notably the CNCF.io, offer new architectural approaches, where data-centric (aka edge computing) is the key concept. In such architectures, container images can be seamlessly delivered worldwide via a system of highly optimized container repositories. ENTICE (2017) is an example of an emerging architecture where the federated storage infrastructure services (for the management of container and Virtual Machine images) can be used to optimize the size of the images, their geographic distribution, and delivery and deployment times. ENTICE is therefore designed to support data-centric applications with its ability to deliver and deploy the needed functionality geographically according to the requirements of the software engineers. A technology like ENTICE may also help maintain other aspects of dependability, such as maintainability, with the possibility to deliver updated functionalities seamlessly and transparently to the end-users.

When considering the runtime environment, another technology that is supposed to improve the overall software engineering life-cycle is SWITCH (2017). A major goal in the development of the SWITCH Interactive Development Environment (SIDE) (Zhao et al., 2015) is to assure that the Quality of Service requirements are taken from the end-user early in the software design and engineering process. Once these requirements are semantically annotated and stored, they can be used later on during the use of the application to decide on which Cloud provider to deploy the application. Taking care of the needed nonfunctional properties at an early stage of the software design process is a crucial aspect that has to be taken into account when designing dependable data-centric systems.

Cloud federations (including storage) are already happening (Kimovski et al., 2016). From an applications view point, there are several possible architectures that can be used to build data-centric dependable applications. Here, three main possibilities were described: the delivery and use of data along with the functionality (i.e., in one container), the delivery of functionalities in one container and data in another, and the third option may be to use an optimized distributed storage facility, such as CEPH (2016).

Depending on the requirements, one or more of these architectures may apply. Sometimes it may be necessary to use a multiobjective optimization approach to make decisions about which of the possible application design patterns and system architectures is optimal. More efforts are needed to be able to incorporate suitable Quality of Service models in the overall process of managing the applications. Our future research will concentrate on these aspects.

## ACKNOWLEDGMENTS

The research and development reported in this chapter have received funding from the European Union's Horizon 2020 Research and Innovation Programme under grant agreements no. 643963 (SWITCH project: Software Workbench for Interactive, Time Critical and Highly self-adaptive cloud applications) and no. 644179 (ENTICE project: dEcentralized repositories for traNsparent and eficienT vIrtual maChine opErations).

## REFERENCES

Amazons3, 2017. https://aws.amazon.com/documentation/s3/. (Online; Accessed 15 May 2017).

Apache Cassandra, 2016. http://cassandra.apache.org/doc/latest/. (Online; Accessed 19 December 2016).

Apache Mesos, 2017. http://mesos.apache.org/documentation/latest/. (Online; Accessed 15 May 2017).

Avizienis, A., Laprie, J.C., Randell, B., Landwehr, C., 2004. Basic concepts and taxonomy of dependable and secure computing. IEEE Trans. Dependable Secure Comput. 1 (1), 11–33.

Bryson, S., Kenwright, D., Cox, M., Ellsworth, D., Haimes, R., 1999. Visually exploring gigabyte data sets in real time. Commun. ACM 42 (8), 82–90.

CEPH Storage, 2016. http://docs.ceph.com/docs/master/. (Online; Accessed 15 December 2016).

CHEF Automate, 2017. https://www.chef.io/chef/#chef–resources. (Online; Accessed 15 May 2017).

Dubrova, E., 2013. Fundamentals of dependability. In: Fault-Tolerant Design. Springer, New York, pp. 5–20.

ENTICE, 2017. ENTICE-decentralised repositories for transparent and efficient virtual machine operations. http://www.entice-project.eu/about/. (Online; Accessed 15 May 2017).

Gec, S., Kimovski, D., Prodan, R., Stankovski, V., 2016. Using constraint-based reasoning for multi-objective optimisation of the entice environment. In: IEEE 2016 12th International Conference on Semantics, Knowledge and Grids (SKG). IEEE, New York, pp. 17–24.

Giannakopoulos, I., Papailiou, N., Mantas, C., Konstantinou, I., Tsoumakos, D., Koziris, N., 2014. Celar: automated application elasticity platform. In: 2014 IEEE International Conference on Big Data (Big Data). IEEE, New York, pp. 23–25.

Goiri, Í., Julia, F., Guitart, J., Torres, J., 2010. Checkpoint-based fault-tolerant infrastructure for virtualized service providers. In: 2010 IEEE Network Operations and Management Symposium, NOMS 2010. IEEE, New York, pp. 455–462.

IBM Big Data, 2013. http://www.ibmbigdatahub.com/infographic/four-vs-big-data/. (Online; Accessed 12 December 2016).

JITSI, 2017. https://jitsi.org/Documentation/DeveloperDocumentation/. (Online; Accessed 15 May 2017).

Kim, H., Feamster, N., 2013. Improving network management with software defined networking. IEEE Commun. Mag. 51 (2), 114–119.

Kimovski, D., Saurabh, N., Stankovski, V., Prodan, R., 2016. Multi-objective middleware for distributed VMI repositories in federated cloud environment. SCPE 17 (4), 299–312.

Kounev, S., Reinecke, P., Brosig, F., Bradley, J.T., Joshi, K., Babka, V., Stefanek, A., Gilmore, S., 2012. Providing dependability and resilience in the cloud: challenges and opportunities. In: Resilience Assessment and Evaluation of Computing Systems. Springer, Berlin, pp. 65–81.

Kubernetes, 2017. http://kubernetes.io/docs/. (Online; Accessed 15 May 2017).

Kundu, A., Mohindra, A., 2012. Method for authenticated distribution of virtual machine images. US Patent App. 13/651,266 (Oct. 12, 2012).

mOSAIC Project, 2017. http://www.mosaic-cloud.eu/. (Online; Accessed 15 May 2017).

Nagios, 2017. https://www.nagios.org/. (Online; Accessed 15 May 2017).

OASIS, 2013. Topology and Orchestration Specification for Cloud Applications. http://docs.oasis-open.org/tosca/TOSCA/v1.0/TOSCA-v1.0.html. (Online; Accessed 19 December 2016).

Paraiso, F., Stéphanie, C., Yahya, A.D., Merle, P., 2016. Model-driven management of docker containers. In: 9th IEEE International Conference on Cloud Computing (CLOUD).

Powell, D., 2012. Delta-4: A Generic Architecture for Dependable Distributed Computing, vol. 1. Springer Science & Business Media, New York.

Pradhan, D.K., 1996. Fault-Tolerant Computer System Design. Prentice-Hall, Inc., Upper Saddle River, NJ.

PUPPET, 2016. https://docs.puppet.com/puppet/. (Online; Accessed 17 December 2016).

Ramachandran, M., 2011. Component-based development for cloud computing architectures. In: Cloud Computing for Enterprise Architectures. Springer, London, pp. 91–114.

Rittinghouse, J.W., Ransome, J.F., 2016. Cloud Computing: Implementation, Management, and Security. CRC press, Boca Raton.

Sharp, J.H., Ryan, S.D., 2009. Component-based software development: life cycles and design science-based recommendations. In: Proc. CONISAR, v2 (Washington, DC).

Shrier, D., Wu, W., Pentland, A., 2016. Blockchain & infrastructure (identity, data security). MIT, Massachusetts.

Stankovski, V., Taherizadeh, S., Taylor, I., Jones, A., Mastroianni, C., Becker, B., Suhartanto, H., 2015. Towards an environment supporting resilience, high-availability, reproducibility and reliability for cloud applications.

In: 2015 IEEE/ACM 8th International Conference on Utility and Cloud Computing (UCC). IEEE, New York, pp. 383–386.

Stankovski, V., Trnkoczy, J., Taherizadeh, S., Cigale, M., 2016. Implementing time-critical functionalities with a distributed adaptive container architecture. In: Proceedings of the 18th International Conference on Information Integration and Web-based Applications and Services.

Storj.io, 2016. http://docs.storj.io/. (Online; Accessed 19 December 2016).

SWITCH, 2017. SWITCH project (Software Workbench for Interactive, Time Critical and Highly self-adaptive Cloud applications). http://www.switchproject.eu/about/. (Online; Accessed 15 May 2017).

Taherizadeh, S., Taylor, I., Jones, A., Zhao, Z., Stankovski, V., 2016a. A network edge monitoring approach for real-time data streaming applications. In: Economics of Grids, Cloud, Systems, and Services.

Taherizadeh, S., Jones, A.C., Taylor, I., Zhao, Z., Martin, P., Stankovski, V., 2016b. Runtime network-level monitoring framework in the adaptation of distributed time-critical cloud applications. In: Proceedings of the International Conference on Parallel and Distributed Processing Techniques and Applications (PDPTA), The Steering Committee of The World Congress in Computer Science, Computer Engineering and Applied Computing (WorldComp), p. 78.

Zhao, Z., Taal, A., Jones, A., Taylor, I., Stankovski, V., Vega, I.G., Hidalgo, F.J., Suciu, G., Ulisses, A., Ferreira, P., et al., 2015. A software workbench for interactive, time critical and highly self-adaptive cloud applications (switch). In: 2015 15th IEEE/ACM International Symposium on Cluster, Cloud and Grid Computing (CCGrid), IEEE, New York, pp. 1181–1184.

CHAPTER

# RISK ASSESSMENT AND MONITORING IN INTELLIGENT DATA-CENTRIC SYSTEMS

2

**Mauro Iacono***, **Stefano Marrone***
*Università degli Studi della Campania "Luigi Vanvitelli", Caserta, Italy**

## 1 INTRODUCTION

Data-Centric Systems (DCS) are a rather new perspective opened by the convergence of two factors: the low-cost availability of unprecedented amount of data that can be harvested to produce value and the emergence of large computing systems composed of Components Off The Shelf (COTS) computing nodes. For example, inexpensive units that are coordinated to provide the means to process, by means of abstraction layers and proper software solutions, large amount of data and massively parallel computations. The effects of this convergence is currently represented by the spread of cloud and Big Data applications that are substantially a new declination of traditional computing architecture and software paradigms, probably pushed to their limits and exploited by means of proper middleware and network support. This is just the first birthcry of DCS, as they seem to propose a range of possibilities that significantly outpace current conventional large-scale computing systems: an example of what the future may bring us is given by Data-Centric Intelligent Systems that may exploit a massive base of unstructured knowledge to produce nonobvious information by using nonconventional, possibly autonomic strategies to get to their goals by capturing, classifying, analyzing, and processing data in new ways (Ranganathan, 2011).

The main characteristic of DCS is the fact that the information to be processed is such that it is not practical to move data toward computation units: this may be due to extension, costs, velocity, bandwidth, or a combination of factors. The extension of data is the main push toward the urge for new solutions: a conservative esteem of the amount of data processed or delivered by enterprise servers per year has to be measured in zettabytes, with an exponential growth (Chang and Ranganathan, 2012); the estimated total amount of online data indexed by Google has to be measured in exabyte and has increased 56 times between 2002 and 2009, in times that can be considered as the far past in this application field (Ranganathan, 2011); similar estimations are reasonably likely to be valid for the other bigger Internet players, and other companies follow similar trends. This growth significantly outperforms Moore's law as well as the current technological trends to date. The natural consequence of this data explosion is that computation has to move toward data, keeping data in place, and allocating data in the most fit way to allow an efficient computation: although mobile code already appeared for some decades in special applications, the extent of the phenomenon is now shifting mobility of computation from being a technique of limited application to becoming a new computation paradigm.

Security and Resilience in Intelligent Data-Centric Systems and Communication Networks. https://doi.org/10.1016/B978-0-12-811373-8.00002-1

Copyright © 2018 Elsevier Inc. All rights reserved.

## 1.1 CURRENT ARCHITECTURAL SOLUTIONS

Current architectural solutions for DCS are mainly based on cloud technologies that exploit specialized data center architectures. Cloud technologies enabled dramatic improvements in massively distributed computing in many directions: the cloud software stack, implementing abstractions, virtualization, and Virtual Machine (VM) mobility, allowing for enhanced performances in power consumption, sustainability, costs, reliability, and scalability, thanks to proper resource management policies. A data center is composed of a large number of similar nodes, basically with the same architecture of common computers, that are organized into racks to save space and achieve a better physical organization and interconnection. Racks are organized into corridors that form aisles. Some of the nodes are devoted to computing, some are devoted to storage. Nodes depend on power supply units and on network equipment in order to perform their duties. The network, besides its physical connections, is composed of several items with different responsibilities: switches perform local connections trying to fully exploit the available bandwidth; routers constitute the network backbone and must properly handle the traffic; firewalls protect nodes; and load balancers try to distribute requests in the best way possible to enhance performances. The spatial organization of the nodes influences the organization of the networking infrastructure.

This architecture may be used to build data-centric applications, generally founded onto a Software as a Service (SaaS) logic. The main challenge is to ensure that the overall computing infrastructure is able to fulfill the performance requirements required by the applications. For example, to ensure that large datasets are appropriately organized and distributed, and possibly relocated, and that they are available to the software, that may possibly be migrated toward data by means of VM migration mechanisms, taking advantage of the networking infrastructure.

The migration of a VM is the main tool to bring computing toward data in cloud-based DCS, as it is already available as a tool in cloud middleware for performance, power saving, and dependability reasons. Migration, is unfortunately, expensive in terms of time and in terms of bandwidth: a VM has to be launched from remote storage, or it has then stopped and moved between nodes, and to be relaunched on the target node, causing sudden and massive traffic on the networking infrastructure (Gribaudo et al., 2016c).

In order to achieve the required level of performances, proper solutions are adopted. Although modern processors can run in excess of 40–80 parallel threads, the available primary memory is generally not enough to allow and contain all the data on which intelligent data centered applications rely. As external storage plays a substantial role, and it is, at this state, the only solution for data persistence, redundant solutions are adopted to speed up performances, such as RAID (Rapid Array of Independent Disks) and JBOD (Just a Bunch Of Disks). A further step is the use of solid state disks to provide an additional level of decoupling and data caching, at the cost of complicating the overall management and tuning because of an additional indirection layer between where data are computed and where data are stored. Careful organization of the storage subsystem with proper mappings, coupled with a storage management layer capable of providing autonomic balancing, a combination of replication and coding, and good management, can provide a significant improvement for storage performance featuring the additional advantage of providing a comfortable means to also introduce mechanisms that ensure a better reliability and a better availability of data (Gribaudo et al., 2016a). Nevertheless, these mechanisms are an additional workload for the networking infrastructure.

The critical role of the network is evident.

Cloud network infrastructures adopted special architectures in order to cope with the network bottleneck and try to relax bandwidth constraints with reasonable costs. Bilal et al. (2013) presents a number of suitable solutions. The three-tier architecture is the most popular. The lowest tier is used to implement a fast interconnection between computing and storage nodes in a local area (e.g., a rack or a corridor, allowing respectively less expensive devices with lower scalability or more expensive devices with more scalability and less overall complexity). This tier, known as the Access tier, is constituted by a switch that is connected to all local nodes and to the upper tier. The Aggregation tier collects all traffic from the switches by means of another, more performant switch, creating an isle (a logical partition of the infrastructure) that it connects to the top tier. Finally, the Core tier creates the overall interconnection backbone across the infrastructure, with other, properly connected, switches. This network organization is considered to be the best compromise between scalability, performances, and costs. However, it is the part of the system that scales the least, as the number of computing and storage nodes is limited by the number of possible physical interconnections in the three tiers and their potential aggregate performances are limited by the available bandwidth. Consequently, the network infrastructure may prevent the theoretically available resources to be fully harvested to implement data centered applications. Some solutions have been proposed, based on better protocols (e.g., see Hopps, 2000), or simply by relying on more expensive technologies.

## 1.2 FUTURE ARCHITECTURAL SOLUTIONS

With all its flexibility and advantages, the cloud is still just the beginning. Cloud architectures have brought a revolution, specially by offering the macroscopic benefit of elasticity, but they are just an extreme evolution of conventional computing architectures. Roughly speaking, clouds are built by connecting COTS nodes with specialized network architectures that show their nature of bottlenecks under extreme solicitation and on a big system scale, on which proper, sophisticated middleware produce a number of abstract services oriented to the management and the execution of virtualized resources. Computing nodes are essentially conventional computers that are designed and built around the processor and that have been managed by it since the beginning of computer history. Memories are still (even if less and less) divided between primary and storage, exploiting technological advances rather than technological revolutions (solid state disks (SSD) are actually not an exception, but a proof of this fact). To further evolve our (early) DCS, new architectural solutions and new paradigms for hardware and software design are needed.

Literature defines it as an inflection point in technology (Ranganathan, 2011), and indicates a number of new technologies as key factors to build this new paradigm. The goal is a completely new point of view for system design, based on the centrality of data repositories instead of processing units, and on a blurred separation between primary memories and storage, and a reduction of the levels of intermediation between the actual physical location in which data is resident and the place in which it is processed (e.g., reducing, counterintuitively, caching). The purpose is to reduce delays, to better match the possible processing throughput with the growing rate of data by combining architectural and technological advancements to foster innovation and enable more intelligent applications. This, in turn, will exploit the richness of new data that can currently be collected by pervasive networks of nanosensors or mobile devices, avoiding the risk of being incapable of taking advantage of all available data. Together with hardware advancements, and dependent on them, software advancements should aim at harvesting all available data by a similar combination of new software architectures capable of solving the idiosyncrasies inherited by cloud software stacks from legacy software organizations, and new software solutions capable of simultaneously accessing and combining multiple data sources with structured, semistructured, and structured organization.

Some interesting new candidate technologies include a variety of proposals for nonvolatile primary memories, including new solutions such as nanostores (Ranganathan, 2011) (i.e., devices that colocate the nonvolatile memories and processors that are meant to be the basic building block of new architectures) and photonics to further decrease data transfers, or more methodological means such as a wider application of HW-SW codesign to optimize solutions. For a comprehensive and more detailed analysis the reader can refer to Chang and Ranganathan (2012).

## 2 RISK FACTORS IN DCS MANAGEMENT

The analysis of risks starts from the identification of the main risk factors about the system. Factors span over very different aspects of the management and the operation of the system, and may affect the system on different time scales. Factors that operate on a long time scale, and their consequences, are of paramount importance in shaping the management strategy, and in general have a minor impact on operations, if not by means of policy shaping; the factors that operate on short time scale are rather impacting on operations, with a level of severity that may depend on the scope and the timing. This rough, general, and nonexhaustive classification may suggest two general areas of problems that have to be considered during risk analysis: we may define the processes of assessing these two areas respectively, risk assessment in the large and in the small. Risk assessment in the large relates to the decisions about investments, research directions, and the scope of interest of the activities; it requires a vision and a solid knowledge about the external context in which the decisions have to be taken and enacted, and about the future perspectives in the context. Risk assessment in the small is instead related to more practical issues, about the system as it is, its instantaneous state, the immediate threats to which it is exposed in its context by controllable or uncontrollable factors. Risk assessment in the small may more easily be able to benefit from quantitative information about the context, a more well-defined scenario and a smaller scope, with a clear chain of consequences and well-identifiable elements to exploit or to control to minimize the risk.

### 2.1 IN THE LARGE

The main risk factors concern the aspects connected to innovation, and long-term costs of sustainability. According to literature (Chang and Ranganathan, 2012; Ranganathan, 2011), in the long run, DCS will surely undergo a radical technological switch as the paradigm is shifted. This will produce uncertainty in two dimensions of the decision space of the management: the timing with which the change has to be performed and the choice of the technological solutions. These two factors introduce potential risks that must be analyzed and contrasted by means of a proper roadmap.

Timing is a crucial factor, as this industry is very competitive: the ability to manage the best technological solution first allows for a primary role in the market. Choosing the wrong timing may lead to an economical disaster, as early adopters have no significant data to rely upon, while late adopters have to adapt and may have trouble finding their market niche, which can place their own survival in danger. The timing and the organization of the change roadmap also affects the sustainability of the innovation process, as external technological factors may significantly impact on costs, too.

Choosing the right technology is also crucial, specially for early adopters. In systems that exhibit such a level of complexity, the implications of a single choice are not easily understandable, as they may reflect both on the interactions with the rest of the system and may show up only over time. The urge to adopt new solutions can create a fast decision process that cannot be as informed as when it is limited to well-assessed alternatives with a small innovation content: while the introduction of SSD in a memory hierarchy may benefit from well-defined benchmarks and of knowledge about analogous solutions with a richness of documentation about lesson learned, a decision about switching to a nanostore centered architectural model requires investments in research, the definition of proper benchmarks, new simulation and evaluation tools, new middleware, and in general, a completely new approach to development and management. A wrong decision may lead to a disaster.

A less extreme example is given by two well-known risk factors in very large architectures: power consumption and security. While these factors may seem to be classifiable as fitting in the small, they operate on different time scales. While power management seems to be an activity that is in the scope of resource scheduling techniques, a strategy for power management may be an important competitiveness factor to stay on the market (Rossi et al., 2016). Security is another issue that needs a strategy to implement prevention, besides the solution of incidents, short time monitoring and interventions.

## 2.2 IN THE SMALL

In the small, risk factors mainly depend on the architecture of the system and on the behavior of its components (and, out of the scope of this chapter, on the human factor). A first factor is given by the scale of the system itself: even if components in isolation are characterized by very high Mean Time Between Failures (MTBF), the number of components is such that the overall frequency of failures in the system, due to isolated problems or to propagation or interactions, is not negligible, thus a proper management approach is needed. Luckily, most of the architectural solutions that are adopted to fulfill the performance requirements have a benefit of dependability as well, but the cases of repeated execution of software due to crashes and of performance degradation due to the need for reconstruction of (temporarily) lost or corrupted data into storage are very common, and must be managed autonomically by the middleware.

The scale is critical for its impact on the management of the network infrastructure as well. Besides being the less scalability-friendly component and the principal bottleneck of the system; it is also a point of failure. A thorough evaluation of the network and its behaviors, as a whole, in the tiers and in its components, is crucial to be able to face problems, and constant monitoring may also help in case of external threats (such as security attacks).

Proper power management helps to keep costs sustainable, and is a factor that protects economical margins. Power management relies on VM consolidation and migration (Ciardo et al., 2016), thus it may also produce workload for the network infrastructure; an accurate design of data distribution and of the interconnection logic of the system with respect of software needs may help in minimizing the power consumption of networking devices; and, finally, the cooling subsystem is a significant source of power consumption, thus an appropriate scheduling of VMs may have a twofold beneficial effect.

In general, these risk factors may be evaluated with quantitative approaches, as there is a richness of tools that help monitoring and modeling current architectures. A presentation of tools and approaches for performance evaluation of cloud and Big Data systems may be found in Gribaudo et al. (2016d).

An additional factor is obviously related to security issues that in turn may influence all the others. For its paramount importance, the rest of this chapter focuses on security risk assessment in the case of active threats.

## 3 TRADITIONAL INFORMATION RISK ASSESSMENT

The first research on "data-centric," systems in its broadest acceptions, refer to studies on code mobility where the innovative idea was to move code working on data rather than data: mobile software agents have constituted an important research trend in this field (Pham and Karmouch, 1998). The motivation of those works mainly relied on the efficiency improvement of transferring few code lines rather than considerable amount of data over a network. Furthermore, even if the first focus was on the scalability of applications, it was soon evident there were also benefits in terms of security of the data.

On the other hand, industries were faced with the problem of securing their data starting from those data vital for the correct functioning of the companies. Enterprise Information Security Architecture (EISA) is the set of practices applying a comprehensive and rigorous method for assuring security in the processes of an organization and among the people in a company (Shariati et al., 2011; Anderson and Choobineh, 2008). Although often strictly associated with information security technology, it was soon evident that EISA could not be managed only with technological devices and countermeasures: security must be pursued by means of holistic solutions across the sets of procedures, people, and technologies. Not all data are created equal since some of these piece of data play a more critical role in the life of the entire enterprise with respect to other parts: protecting all the data in the same manner could create ineffective solutions and/or inefficient management of the securing effort. To cope with the objective of rationalizing security efforts, the concept of risk started to be used to quantify the datas' level of criticality: the evidence can be found in the BITS schema (see Fig. 1) where Business, Information, Technology, and Security are grouped under the same umbrella.

First risk assessment procedures have been borrowed from the safety world where well-assessed methodologies, recognized by both industrial and scientific communities, were already known: there are plenty of general, as well as, sector-specific standards about safety-related risk assessment methodologies. Fig. 2 reports a classical schema as can be found in one of these standards.[1]

All risk assessment procedures base their evaluation of risk (***R***) on the formula:

$$\boldsymbol{R} = \boldsymbol{T} * \boldsymbol{D}; \tag{1}$$

where ***T*** is the probability of having an hazard and ***D*** is an estimation of loss in case of damage to the system or to stakeholders. Among safety assessment approaches, two main categories can be found: qualitative and quantitative procedures. Quantitative approaches assign a probability to ***T*** (expressed as a real number between 0 and 1) and an economical loss to ***D*** (e.g., euros, dollars); instead, qualitative procedures simply indicate some qualitative ranges (i.e., probable, impossible, frequent) for ***T***. To evaluate risk, the first are based on a loss (economical) threshold after which the hazard is not tolerable, the second on a risk matrix that tells, for every couple attack-damage, the actions to be pursued. Even if quantitative approaches are surely more accurate and can be the base for solid risk assessment, qualitative approaches are easier to apply and more readily used in industrial settings (and are sometimes the only viable alternative, when there is lack of quantitative information and big uncertainty, such as for the "in the large" perspective).

The huge advancements of ICT have determined in these years a rapid growth of connectivity not only in enterprises but also in products: previously off-line products and companies are now facing the challenges of data security. This is the case with the Internet of Things (IoT) and with the Industry 4.0

[1] http://www.rssb.co.uk/improving-industry-performance/management-of-change.

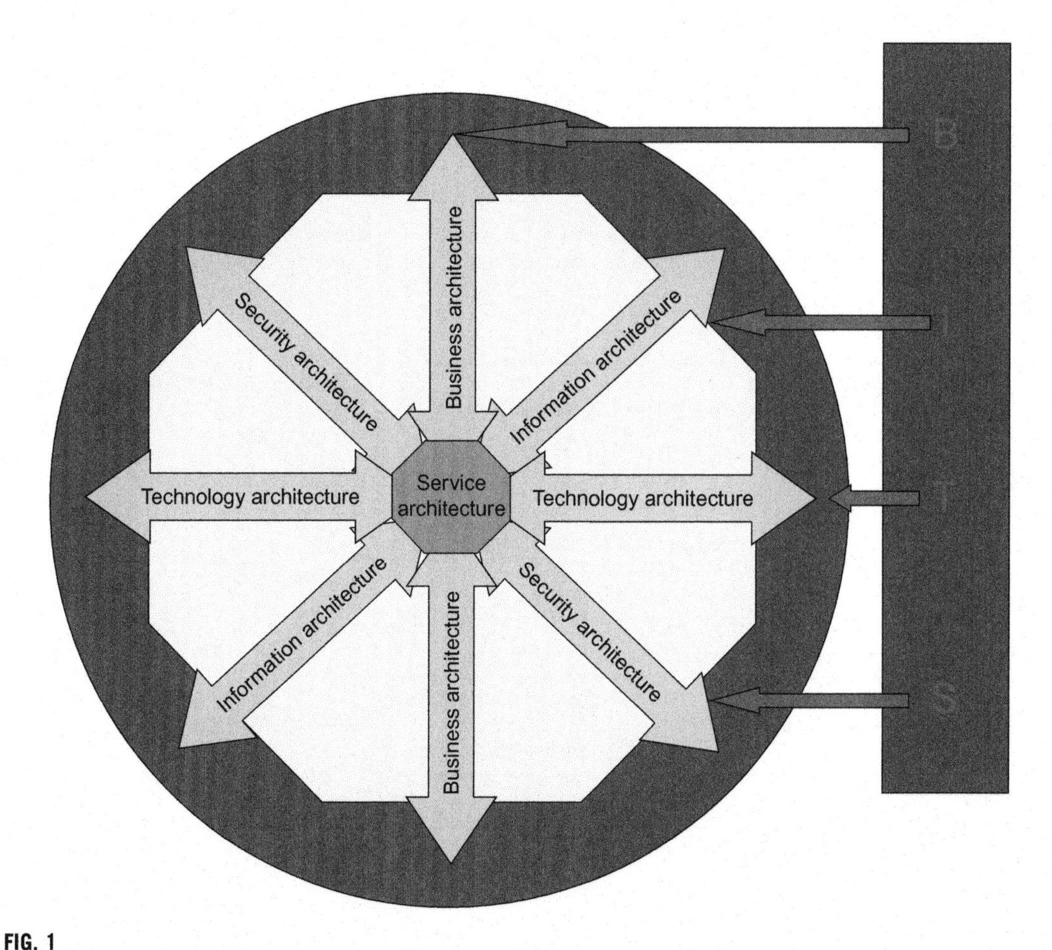


**FIG. 1**

Schema of the BITS approach.

paradigm, which are changing how data are generated, stored, and processed. IoT, despite its name, deeply deals with the everyday activities of society. Cisco states that IoT will generate more that 40 zettabytes per year by 2018, that is to say about 5 Terabyte per person per year. It is not difficult to understand that such data will surely contain personal and sensitive data: guaranteeing data privacy will be a must in future society (Cisco, 2016).

As cybercrime has become a worldwide plague, security-related companies have invested in methods to assess security-related risk of their products. For security, it is more accurate to refer to an extension of the previous formula:

$$\boldsymbol{R} = \boldsymbol{T} * \boldsymbol{V} * \boldsymbol{D}; \tag{2}$$

where $\boldsymbol{V}$ represents the vulnerability of the system: the probability that an attack could lead to a compromised system.

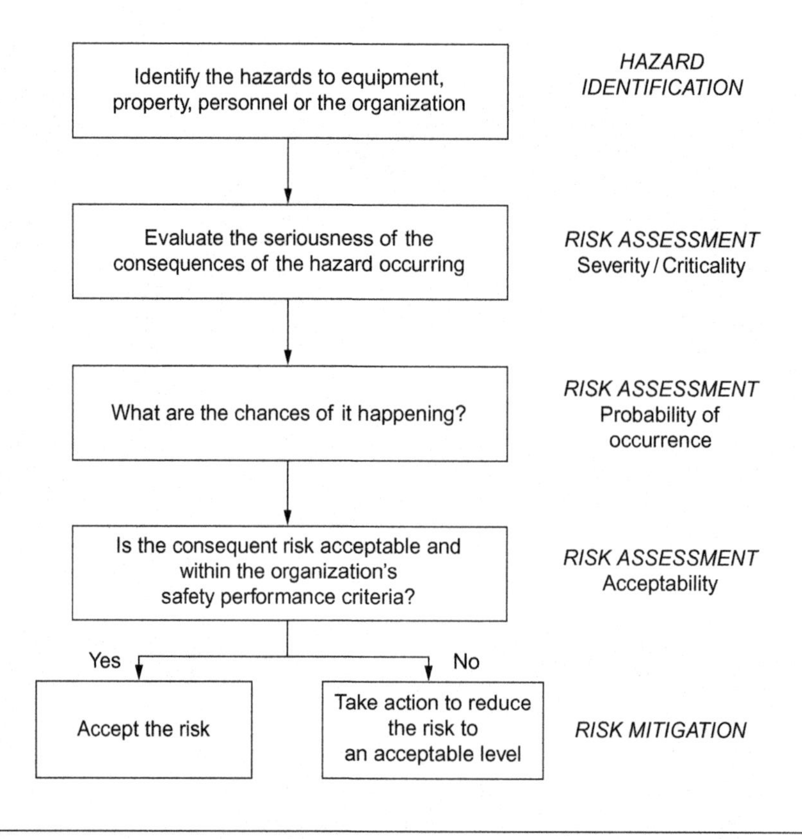


**FIG. 2**

Schema of risk assessment process.

Connected artifacts will also control many of our lives: smart cars and autonomous vehicles are just a few examples of Cyber-Physical Systems (CPS), computer-based systems that will deeply interact with our lives, and will control and determine our safety. According to TrendMicro (2015) as well as other organizations, even if cyberattacks are mainly directed against corporate and private sectors, the attacks against critical infrastructures are growing rapidly, since systems that have traditionally been closed are now open to the Internet and are exposed to malware, Denial of Service (DoS) attacks, and other threats. This is a problem of both safety and security; in the future, several researchers expect these two concerns will be merged into one (Bloomfield et al., 2013).

Finding a proper risk assessment method that integrates safety and security aspects is not easy and there are no known methods that have been assessed. Research is working on this concern and some research findings are already available. In Macher et al. (2016) a first attempt to integrate both these concerns into a joint risk assessment method is made in the automotive domain: this domain is a pioneer in the field due to the massive introduction, not only of the electronics, but also of the great level of connectivity cars are achieving. Other valuable resources are Ward et al. (2013) and Raspotnig et al. (2013). More in general, different international conferences and workshops are looking to these new concern, as are mature conferences in the field of dependability (e.g., DSN, EDCC, SSCS, S4CIP).

The final destination of data is the cloud: as interconnection to the Internet and broadband connections are becoming more and more pervasive, accessing data by storing to and manipulating them from cloud systems is becoming a standard practice. Since a misuse of cloud computing techniques can open severe security vulnerabilities for data, also cloud systems must be designed and assessed according to risk assessment procedures. Such procedures replicate the best performing solutions for safety and security, which are just customized according to specific threats. The European Union Agency for Network and Information Security (ENISA) published a cloud computing risk procedures report in 2009 that highlight such risks and addressed each threat with proper countermeasures (ENISA, 2009); in 2016 the Cloud Security Alliance (CSA) updated this study going deeper in technical security issues (CSA, 2016).

## 4 A RISK ASSESSMENT APPROACH FOR CPS

As stated in Section 3, one of the main pitfalls of existing risk procedures is the exclusive reliance on qualitative evaluations: Fig. 3 shows the schema at the base of our approach. Since the main aim is to provide a comprehensive model-based approach to security systems design and evaluation,

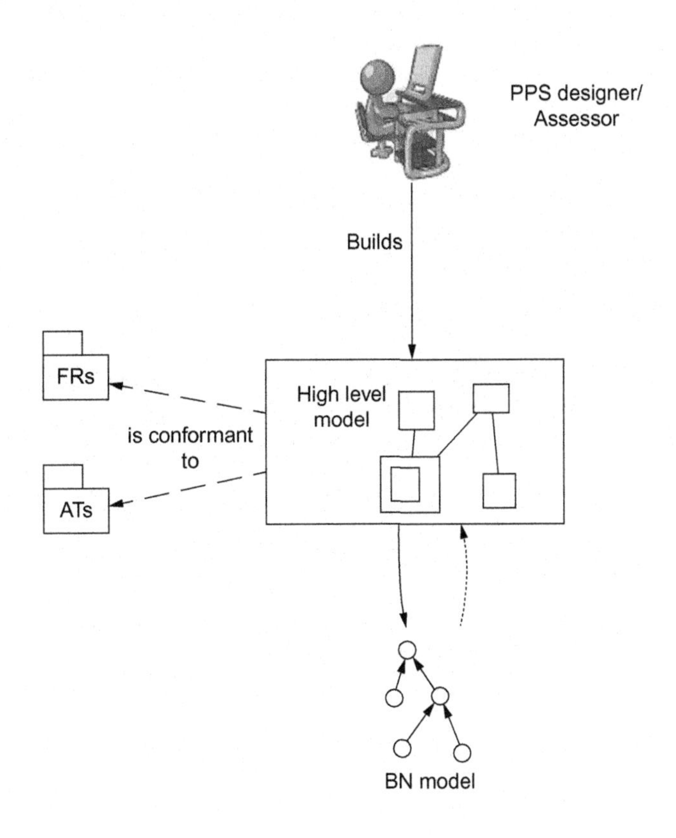


**FIG. 3**

Risk modeling and analysis approach.

designers and assessors manipulate formal models of the system focusing on different aspects: threats and hazards, system responses, and further counteractions. Designers are addressed to model such aspects according to some formalisms as Attack Trees (AT) and Fuzzy Rules (FR).

At this point, two sets of automatic model-to-model transformations are considered to generate more analyzable models from these two: AT to Bayesian Networks (BN) and FR to BN. Both of them are in charge of generating a formal model for the evaluation of vulnerability. The two transformations return the results of vulnerability analysis and the solution of the decision problem and integrate them for the user.

## 5 THE PROPOSED APPROACH

Fig. 4 depicts the risk assessment process proposed in this work. The approach works against natural hazards as well as security threats, hence, it can be used in joint risk assessment for both safety and security concerns.

In the first step, a risk analyst models both the action and the counteraction behaviors of the adversary. We generally denote as actions the behaviors of an adversary that accesses the system; as the method can be applied to different cases, the following concerns can be reported to this situation: malicious actions of attackers, misbehaviors of authorized users, failure of external services the system relies on, as well as errors of internal system components. On the contrary, the counteractions are the adversary's responses to some system defenses. One of the original features of the proposed approach is to demonstrate the benefits of the use of a model composed by two heterogeneous submodels in a multiformalism approach (Gribaudo and Iacono, 2014): the Action Model and the Counteraction Model respectively capturing actions and counteractions.

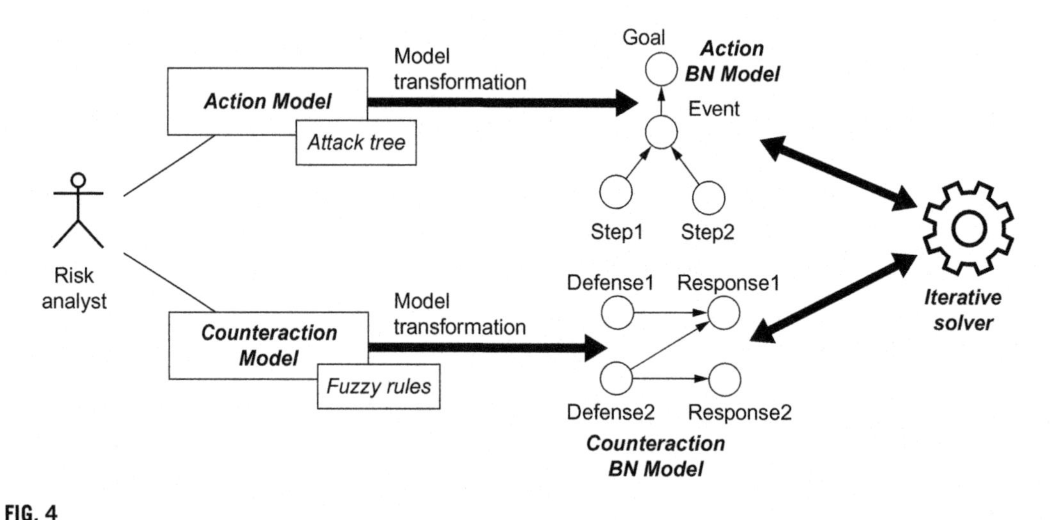


**FIG. 4**

Overview of the proposed approach.

The Action Model combines the effects of the considered threats toward the compromised system state (e.g., exploitation of the system, disruption of the service). In order to capture this information, proper languages must be chosen in an effective and efficient way. AT is one of the most suited formalisms since it can be used to model malicious behaviors and, since it is an extension of the most assessed Fault Tree formalism, it fits the modeling of natural hazards.

On the other hand, the Counteraction Model focuses on the likelihood of some responses to defender strategies. The model may be seen as a set of rules describing what an adversary does when it sees that some paths and possibilities are not viable in the execution of an attack. We explore the use of FR as a way of describe and formalize such rules; this formalism requires, in our case, a translation toward another language that enables more flexible analyses.

BN is a powerful formalism that demonstrates its ability to describe reasoning under uncertain conditions in several domains, such as artificial intelligence (Charniak, 1991), system dependability (Bobbio et al., 2001; Bernardi et al., 2013), and software and system security (Marrone et al., 2013; Xie et al., 2010). Several analysis techniques using efficient algorithms can be applied to BN that do not require state space-based solutions; there are many available tools supporting such analysis as well.

The core of this approach is constituted by the translation of both Action Model and Counteraction Model into two BN models, respectively the Action BN Model and the Counteraction BN Model. This translation is performed by means of model transformations (Gribaudo and Iacono, 2014). Once the two original models have been translated into the same language, an Iterative Solver can alternatively analyze the two models. This tool is in charge of analyzing the first BN model and report the results as input for the analysis of the second model, then it reports these outputs as inputs for a re-analysis of the first model, iterating until a termination condition is reached.

## 5.1 ON THE STRUCTURE OF THE MODELS

Figs. 5 and 6 show the structure of both the Action Model and the Counteraction Model. These structures are described in terms of their interfaces.

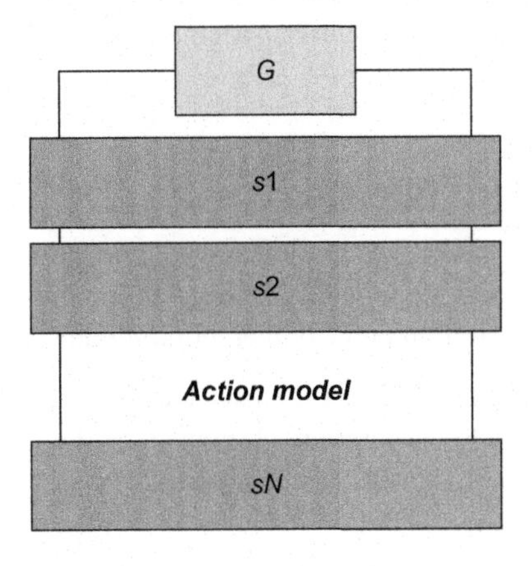


**FIG. 5**

Structure of the Action Model.

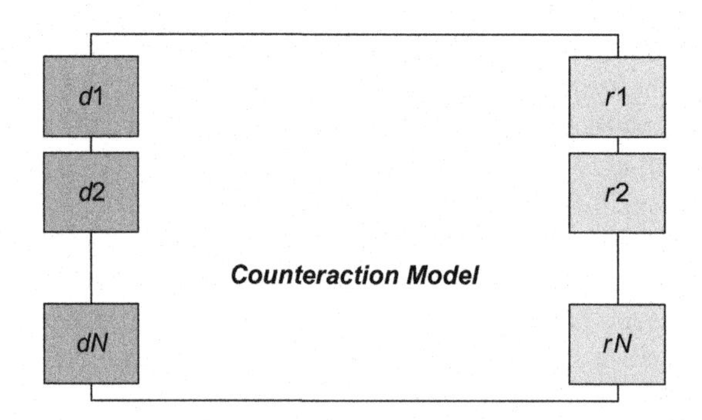


**FIG. 6**

Structure of the Counteraction Model.

In greater detail, the Action Model (Fig. 5) presents these sets of interfaces: (1) $\mathcal{S} = \{s_1, \ldots, s_N\}$, which represent the set of the basic actions that can be performed by the adversary, (2) $\mathcal{G} = \{g_1, \ldots, g_K\}$, that is the set of sensitive targets the adversary can exploit with its attack steps. The way the basic attack steps contribute to determine the likelihood of exploiting the goals (see Section 6). The Counteraction Model (Fig. 6) presents two sets of interfaces: (1) $\mathcal{D} = \{d_1, \ldots, d_N\}$ represents the possibility that an adversary can find one or more of the possible actions it can make already blocked by the defender; (2) $\mathcal{R} = \{r_1, \ldots, r_N\}$, which represents the actions the adversary performs as a consequence of the strategy of the defender. $\mathcal{S}$, $\mathcal{D}$, and $\mathcal{R}$ have the same cardinality. Furthermore, let $\mathcal{V}$ be $\mathcal{D} \cup \mathcal{R} \cup \mathcal{S}$.

All the interfaces are supposed to be of the Boolean type: we indicate the Boolean set with $\mathcal{B}$. This notwithstanding, they have different meanings:

- $x \in \mathcal{G}$ is `true` if the related target is compromised, otherwise it is `false`;
- $x \in \mathcal{S} \cup \mathcal{R}$ is `true` if the related attack basic step occurs, otherwise it is `false`;
- $x \in \mathcal{D}$ is `true` if the related defending action has been set by the defender, otherwise it is `false`;

## 5.2 ON THE ITERATIVE SOLUTION PROCESS

The aim of the iterative solution process is to jointly analyze both Action and Counteraction models to compute the defender's and adversary's strategies. The iterative solution process is supported by two composition mechanisms between Action and Counteraction models as in Fig. 7:

- *iteration*: it occurs inside a single iteration. It maps $s_j$ on $d_j$;
- *interiterations*: it occurs between two iterations. It maps $r_j$ of the Counteraction Model at $i$th iteration on $s_j$ of the Action Model at $(i+1)$th iteration.

The algorithm applied by the iterative solution process is described in the pseudo-code reported in Listing 1.

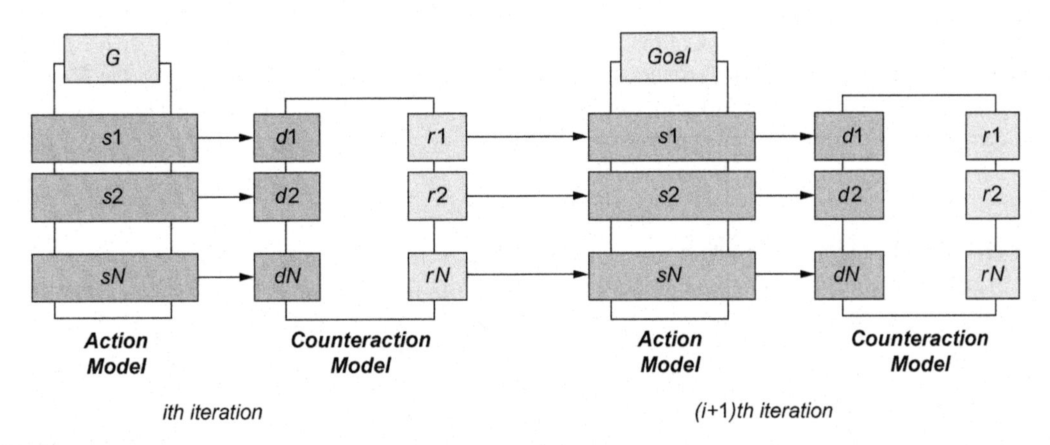


**FIG. 7**

Composition of the models during the solution iterations.

**LISTING 1**

Iterative Solver pseudo-code

```
void isolver(int maxIter, double riskVL) {
  int iterCounter = 0;
  double risk = 0;

  int exitCondition = 0;
  while(exitCondition == 0) {
    risk = Action.computeGoals();
    iterCounter++;
    exitCondition = (iterCounter == maxIter) || (risk < riskVL);

    if (exitCondition == 0) {
      Counteraction.defenses = Action.defStrategy();
      Action.steps = Counteraction.attStrategy();
    }
  }
}
```

The function accepts two parameters: `maxIter`, that is the maximum number of iterations allowed, and `riskVL`, that represents the maximum allowed risk level that a defense strategy could tolerate. In this chapter it is computed by the sum of the cost of each compromised target $C_i$ weighted by its vulnerability $P_i$:

$$R = \sum_{i=1}^{K} C_i \cdot P_i \quad (3)$$

For the purpose of this chapter, vulnerability is the probability that the infrastructure does not resist to a given attack. The two models are represented in the algorithm by two objects, `Action` and `Counteraction`. After a proper initialization, the first step of the algorithm (inside the loop) is the computation of the Goals by means of (`Action.computeGoals()`), which updates the vulnerabilities of each goal and the total risk index of the infrastructure. Then if none of the target (*maxIter* or *riskVL*) has been reached, the new defense strategy is computed against the adversary strategy (i.e., the $\mathcal{R}$ of the previous iteration) by means of `Action.defStrategy()`; the new counter-strategy of the adversary is then computed by calling the `Counteraction.attStrategy()` procedure.

## 6 ENABLING FORMALISMS

This section describes how the Action and Counteraction models can be defined and translated into BN. Moreover, Section 6.3 details how the iterative solver works at the BN level.

### 6.1 THE ACTION MODEL

AT and its variants constitute a powerful set of formalisms that both academic and industrial communities use to model intrusion scenarios. In this paper, we refer to the specific variant of AT described in Gribaudo et al. (2015) where a model-to-model transformation describing how to generate a BN model from an AT is described. Consequently, in the following, Action Models respect both the formalism presented in Gribaudo et al. (2015) and the model structure introduced in Section 3. In this framework, the elements of the $\mathcal{S}$ of the Action Model are modeled as the Basic Events of the AT, and the $\mathcal{G}$ of the Action Model are instead the Top Events of the AT. In this chapter, we fully use the approach in Gribaudo et al. (2015) where multiple AT may be combined and analyzed by translating them into BN models. Then the modeler can populate the AT models by describing how basic threats combine in order to compromise the final targets. This task is accomplished by adding proper events collected in $\mathcal{E}$ to the model (i.e., internal AT nodes).

The model transformation from AT to BN is not fully described here; the translation is based on a similar one from FT to BN (Bobbio et al., 2001) and is fully described in Gribaudo et al. (2015). In brief, each AT is translated into a BN node; the structure of the AT induces a similar structure into the BN model: attack logical operators (OR, AND, KooN) are modeled in BN by means of CPTs. Then the BN generated by each AT are merged by following the rules in Gribaudo et al. (2015).

All these BN variables belong to $\mathbb{B}$; despite the fact that they differ in the semantics of these values:

- BN nodes generated by $\mathcal{G}$ mean `true` if the representing target is compromised, otherwise it is `false`;
- BN nodes generated by $\mathcal{E}$ mean `true` if the representing attacking (internal) event occurs, otherwise it is `false`;
- BN nodes generated by $\mathcal{S}$ mean `true` if the representing attack basic step occurs, otherwise it is `false`;

Fig. 8 reports an example of such mapping in which a two goals Action Model (Fig. 8A) is translated into a single optimized BN model (Fig. 8B).

### 6.2 THE COUNTERACTION MODEL

Since the Counteraction Model is defined to represent the behavior of the adversary when it perceives that the system/infrastructure administrator is adopting some kind of defending strategy, it seems

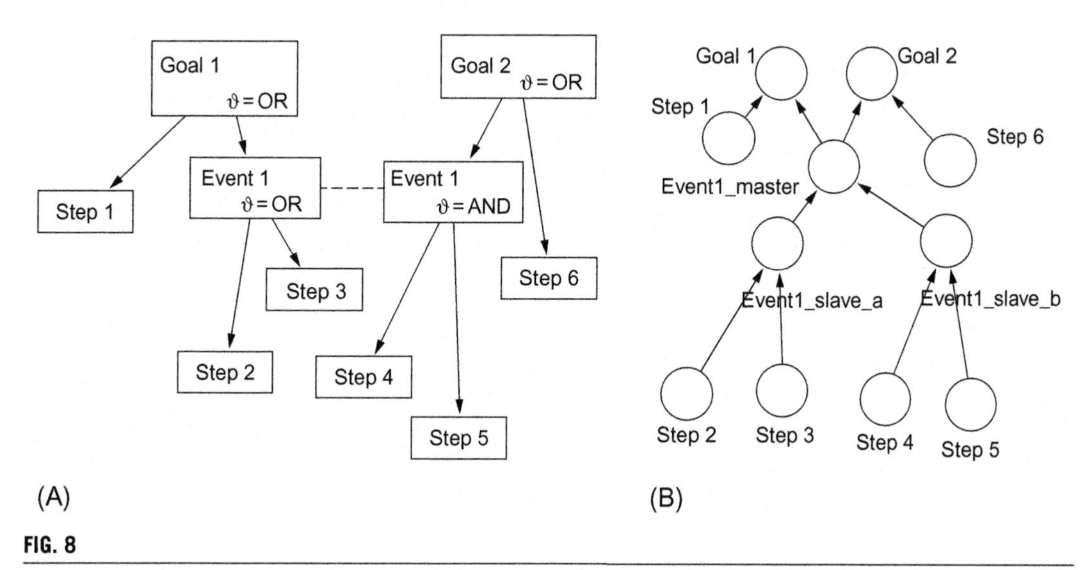


**FIG. 8**

Example of AT-to-BN mapping: original combined AT model (A), BN model (B).

natural to model it by following a rule-set schema. In this chapter, we mainly explore the effects of the uncertainty aspect behind the decisions of the adversary, by adopting a description based on fuzzy logic (Kusko, 1993), namely FR. A fuzzy description of the modus operandi of the adversary allows modeling his probabilistic behaviors, overcoming the limitations to use deterministic crisp logic to model this kind of concept.

One of the most important concept of fuzzy logic is the notion of membership function:

$$\mu_F : \mathcal{V} \longrightarrow 0 \ldots 1 \subset \mathbb{R} \text{ with } F \in \mathbb{B} \tag{4}$$

Due to the probabilistic nature of the variables, we may define this function as

$$\mu_F(x) = \Pr(x == F) \tag{5}$$

Moreover, a set of fuzzy operators is introduced by borrowing from traditional Boolean logic:

- $\oplus : \mathcal{D} \times \mathcal{D} \longrightarrow \mathcal{D}$ means the fuzzy OR;
- $\odot : \mathcal{D} \times \mathcal{D} \longrightarrow \mathcal{D}$ means the fuzzy AND;
- $\neg : \mathcal{D} \longrightarrow \mathcal{D}$ means the fuzzy NOT;

The Counteraction Model is a set of rules following this schema:

$$\textbf{\textit{IF}}\ \textit{IF\_CLAUSE}\ \textbf{\textit{THEN}}\ \textit{THEN\_CLAUSE}; \tag{6}$$

Each rule is composed of two parts: an IF_CLAUSE and a THEN_CLAUSE. The IF_CLAUSE is a fuzzy predicate using Boolean constants, elements of $\mathcal{D}$ and the fuzzy operators defined above (e.g., $(d_i == \textit{TRUE}) \odot (d_j == \textit{FALSE})$ where $i, j \in \{1, \ldots N\}$. The THEN_CLAUSE is constituted by a list of counteractions triggered by the (fuzzy) truth of the IF_CLAUSE; it contains a set of assignments of Boolean values to variables in $\mathcal{R}$ (e.g., $(r_m = \textit{TRUE}), (r_n = \textit{TRUE})$ where $m, n \in \{1, \ldots N\}$

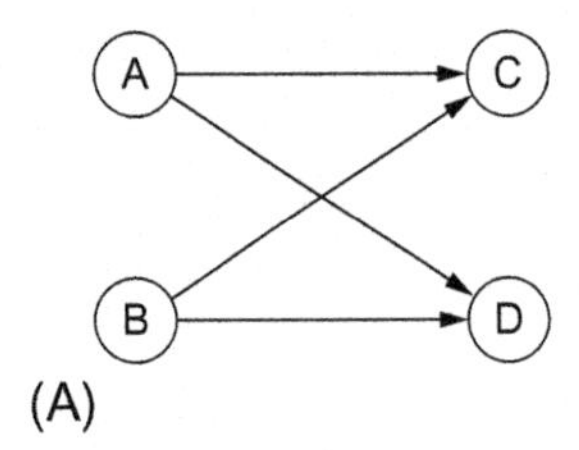


| A | B | C = true | C = false | D = true | D = false |
|---|---|---|---|---|---|
| false | false | 1 | 0 | 0 | 1 |
| false | true | 0.5 | 0.5 | 0.5 | 0.5 |
| true | false | 1 | 0 | 0 | 1 |
| true | true | 1 | 0 | 0 | 1 |

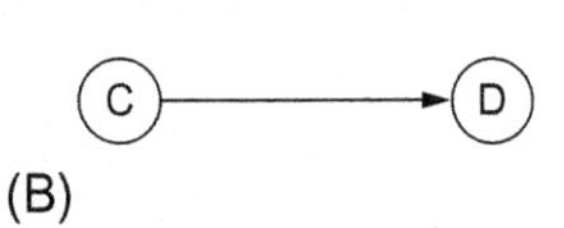


| C | D = true | D = false |
|---|---|---|
| false | 0.5 | 0.5 |
| true | 0 | 1 |

**FIG. 9**

Example of FR-to-BN mapping: BN model and CPT tables of the Eq. (7) (A), BN model and CPT table of the Eq. (8) (B).

As for the Action BN Model, BN nodes belong to $\mathbb{B}$; in particular:

- a BN node generated by $\mathcal{D}$ is `true` if the representing defending action has been set by the defender, otherwise it is `false`;
- a BN node generated by $\mathcal{R}$ is `true` if the representing attack step occurs, otherwise it is `false`;

The transformation between FR and BN works as follows. Each rule generates a BN model where:

- each variable in the `IF_CLAUSE` or in the `THEN_CLAUSE` generates a BN node;
- an arc is drawn from each variable in the `IF_CLAUSE` to each variable in the `THEN_CLAUSE`;
- for each BN node generated by a variable in the `THEN_CLAUSE`, a CPT is set implementing in a deterministic way the logic predicate in the `IF_CLAUSE`.

By setting proper probability distributions on the BN nodes translating variables in the `IF_CLAUSE`, and by querying the values on BN nodes related to `THEN_CLAUSE` variables, the mechanisms of the fuzzy inference may be replicated in by exploiting BN.

To give an example, let a Counteraction Model be constituted by the following rules:

$$\textbf{\textit{IF}}\ (A\ ==\ \textit{true}) \oplus (B\ ==\ \textit{false})\ \textbf{\textit{THEN}}\ (C\ =\ \textit{true}),\ (D\ =\ \textit{false}); \tag{7}$$

$$\textbf{\textit{IF}}\ \neg(C\ ==\ \textit{false})\ \textbf{\textit{THEN}}\ (D\ =\ \textit{false}); \tag{8}$$

Two Counteraction BN Models match these equations: they are represented in Fig. 9.

## 6.3 COMBINING AND ANALYZING THE BN MODELS

As abstractly introduced in Section 3, the analysis process works as follows. As first, the Action BN Model is solved by computing the probability of the nodes representing the goals without any observation of BN nodes representing steps. The evaluation is accomplished by using estimated a priori probability distributions of the possible moves of the adversary.

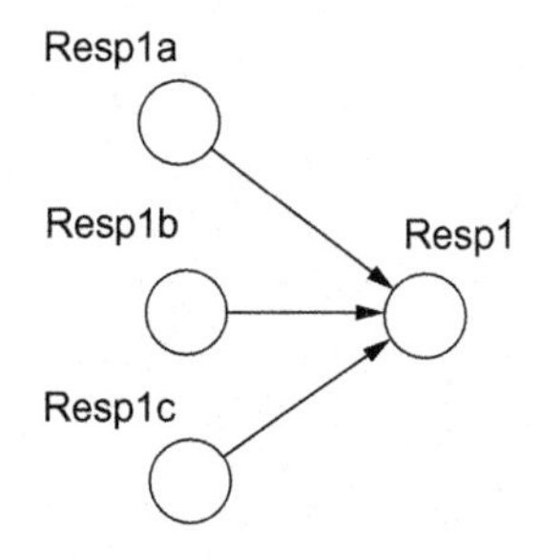


| Resp1a | Resp1b | Resp1c | Resp1 = true | Resp1 = false |
|---|---|---|---|---|
| false | false | false | 0 | 1 |
| false | false | true | 0.33 | 0.67 |
| false | true | false | 0.33 | 0.67 |
| false | true | true | 0.67 | 0.33 |
| true | false | false | 0.33 | 0.67 |
| true | false | true | 0.67 | 0.33 |
| true | true | false | 0.67 | 0.33 |
| true | true | true | 1 | 0 |

**FIG. 10**

BN linear consensus mechanism.

Now, the most critical goal is chosen as a base for further analysis.[2] Let us now observe how this goal has been reached. By calculating the expectation probabilities, it is possible to evaluate the most likely cause (i.e., the attack step that is more likely to be successful) for getting to the critical goal. Hence, the first move of the defender is to avoid that the adversary may accomplish this attack step.

At this point, the Counteraction model is considered. If the defender sets his/her move by defending $s_i$, the $d_i$ is set to true and $d_j$ to false $\forall j \neq i$. The Counteraction BN Model is considered by setting observations according to the values of $\mathcal{D}$, and it is used to compute the probability values of the nodes in $\mathcal{R}$. The values of these variables represent the set of counteractions the adversary takes into considerations knowing the moves of the defender.

The second iteration starts by copying (as observations) the values of the responses to the nodes representing the considered attack steps. Now, a second evaluation of the criticality of goals may be done and a new defending strategy may be elaborated. However, only strategies that improve the overall risk of the infrastructure are considered. The process ends by reaching a maximum number of iterations as well as by lowering the level of the risk under a tolerable threshold.

The possibility of having a node of $\mathcal{R}$ present in the THEN_CLAUSE of more that one rules is a crucial issue of this approach. Since these rules can produce different values of this variable, a consensus mechanism is needed as the one proposed in Fig. 10.

## 7 RELATED WORKS

A comprehensive presentation of the typical solutions currently adopted in the data centers to support data-centric computing applications can be found in Han et al. (2016). The paper presents the specifications of major server vendors' products and the most popular simulators that are used to evaluate the characteristics and the performances of data centers, in order to provide the means for a comparison and support the design and management of large infrastructures. A cloud architecture for data-centric applications, based on a Software as a Service approach, is presented in Mandal et al.

[2] In the hypothesis that all the assets have the same economic value, the probability of being exploited represents a criticality measure.

(2013). The paper provides a number of useful references that describe the structure of other solutions, and proposes a new solution that enhances customizability, scalability and multitenancy to provide better performances. Vishnu et al. (2010) provides an insight on the main problems that impact on fault tolerance in large DCS and also includes a list of interesting related papers. In Gribaudo et al. (2016a) and Gribaudo et al. (2016b) the problem of efficiently solving the fault tolerance problem in storage subsystems for large architectures is analyzed, and a modeling approach is presented suitable for performance prediction. A systematic review of the problems in data distribution service middleware is provided in Ömer KÖksal and Tekinerdogan (2017).

Topics in evaluating the performance and dependability of large-scale computer systems are presented (Castiglione et al., 2014, 2015; Barbierato et al., 2016) in which also the problems in modeling deriving from the scale are presented. Xu et al. (2014), Yan et al. (2012), Distefano and Puliafito (2014), and Lian et al. (2005) analyze the storage subsystem and related problems from the point of view of optimal management, performance, and dependability. Specifically, Lian et al. (2005) analyzes the problem of properly distributing replicated data across a storage subsystem, and the consequences in terms of efficiency and performances; an interesting solution that exploits nanodatacenters to implement a massively distributed backup storage system, is presented in Simon et al. (2014) and may be of interest for the readers, as it discusses cost issues and gives some useful analytical model in terms of distributions. Finally, useful lesson learned are reported in Oral et al. (2014) about large-scale data-centric parallel file system deployment and operation.

As VMs have a principal role in current architectures, both for software relocation and cloud resource management to complete the framework, we briefly suggest some readings that helped in building the background for this chapter. Menasce' (2005) provides a performance-oriented introduction to the topic, including a comprehensive bibliography, while Gribaudo et al. (2012), Huber et al. (2010), Watson et al. (2010), Benevenuto et al. (2006), Vasile et al. (2015), and Sfrent and Pop (2015) describe the impact of different complexity factors on performance and, in general, on the issues that have to be considered for consistent management.

The issues that characterize cloud networking infrastructure are well-described in Singh et al. (2015), which analyzes the solutions adopted by Google to deal with the scale problem (Zafar et al., 2016; Azizi et al., 2016). The three tiers network architecture presented in this chapter is not the only proposal in literature: the other most adopted solutions are Clos, fat tree (Al-Fares et al., 2008) and DCell (Bilal et al., 2013; Couto et al., 2015). Other interesting ones are VL2 (Greenberg et al., 2009) and CONGA (Alizadeh et al., 2014; Luo et al., 2014) describes a specifical approach for DCS. Additional useful readings are Fiore et al. (2014) about data dependent network structures, Palmieri et al. (2016) and Spoto et al. (2014) about distributed data centers, and Saleh (2010) about implementing data-centric networks on the optical network virtualization platform. Performance analysis in loud network infrastructures are mainly performed by means of simulation tools on the base of monitoring data. Bilal et al. (2013) and Couto et al. (2015) offer a good comparison of existing methods, while Fiandrino et al. (2015) and Ruiu et al. (2016) also consider power consumption issues, and Gribaudo et al. (2016c) provides an analytical approach.

The main references about DCS and their future development used for this chapter are Chang and Ranganathan (2012) and Ranganathan (2011). In Chang and Ranganathan (2012) the authors deal with the problem of shaping future data centers around data resources to adapt them to DCN. They analyze current trends and technological proposals in order to derive a comprehensive scenario, and present an interesting evolutive taxonomy of recent system designs for data-centric applications in terms of architectural solutions. Chang and Ranganathan (2012) also provides a significant list of further references. Ranganathan (2011) specially focuses on the transition from microprocessors to nanostores, which the authors advocate as the future building blocks of DCS oriented data centers. The paper presents a useful taxonomy of data-centric workloads to derive the needs of future applications and present a number of hypotheses. In particular, the authors stress the implications of data centric workloads for system architecture as a driver for innovation in design, and detail a complex architecture based on nanostores, following a critical approach. In Bourhis et al. (2016) an analysis of complex data-centric application is performed, in terms of motivation, possible alternatives, and design choices. The authors deal with data processing problems by exploring the topic by using models to ensure a formal approach and to try and get to a solid description that is able to encompass all the main related problems. Chen et al. (2013) and Okamoto et al. (2014) specifically deal with data-centric networking architecture. From the point of view of the DCS management, the readers will find useful hints related to middleware topics in Ravindran (2015) and Ravindran and Iannelli (2014) that present two approaches for Quality of Service management. Saleh (2010) and Saleh et al. (2009, 2013) deal with data-centric Web services.

## 8 CONCLUSIONS

Risk analysis and risk management are current practices for every manager. In a field that naturally evolves at a very fast pace, the body of knowledge needed to support an informed decision is vast wide and, at the same time, continuously variable, especially when the interest is on the most advanced applications of the discipline. When the scenarios in which decisions have to be effective inherit a natural complexity from the inner nature of the problem, the availability of proper conceptual and practical tools is of paramount importance. If, additionally, the application field is innovation bound, the only solution to the decision problem is to try and be as aware as possible of what is driving the change and how new technologies will actually impact on the future (i.e., whether or not they cause a paradigm shift).

In this chapter, we presented the current and future context in which Data-Centered Systems operate and are likely to dramatically evolve. We presented the main risk factors that should be taken into consideration when evaluating risk. To provide a practical application, we have chosen a significant risk factor, defined security with its implications, and showed how general risk analysis principles may be specialized for a single risk factor and how a quantitative tool can be developed by reusing existing, well-assessed tools by adding knowledge and following a rigorous conceptual process.

## ACRONYMS

**AT** Attack Trees
**BITS** Business, Information, Technology and Security
**BN** Bayesian Networks
**COTS** Components Off The Shelf
**CPS** Cyber-Physical System
**CSA** Cloud Security Alliance
**DCS** Data-Centered System

**EISA** Enterprise Information Security Architecture
**ENISA** European Union Agency for Network and Information Security
**FR** Fuzzy Rules
**FT** Fault Trees
**IoT** Internet of Things
**JBOD** Just a Bunch of Disks
**MTBF** Mean Time Between Failures
**RAID** Rapid Array of Inexpensive Disks
**SaaS** Software as a Service
**SSD** solid state disk
**VM** Virtual Machine

## GLOSSARY

**Action Model** a model describing the evolution of an attacker behavior.
**Adversary** in this chapter, an attacker who has malicious intentions against an infrastructure or an abstract entity in case of natural events/disaster.
**Counteraction model** a model describing the evolution of a defender behavior.
**Formalism** a modeling language.
**Multiformalism model** a model that is properly composed of (explicit or implicit) submodels described by using different formalisms.

## REFERENCES

Al-Fares, M., Loukissas, A., Vahdat, A., 2008. A scalable, commodity data center network architecture. SIGCOMM Comput. Commun. Rev. 38 (4), 63–74.

Alizadeh, M., Edsall, T., Dharmapurikar, S., Vaidyanathan, R., Chu, K., Fingerhut, A., Lam, V.T., Matus, F., Pan, R., Yadav, N., Varghese, G., 2014. Conga: distributed congestion-aware load balancing for datacenters. SIGCOMM Comput. Commun. Rev. 44 (4), 503–514.

Anderson, E.E., Choobineh, J., 2008. Enterprise information security strategies. Comput. Secur. 27 (1–2), 22–29. doi:10.1016/j.cose.2008.03.002.

Azizi, S., Hashemi, N., Khonsari, A., 2016. A flexible and high-performance data center network topology. J. Supercomput. 1–20. doi:10.1007/s11227-016-1836-2.

Barbierato, E., Gribaudo, M., Iacono, M., 2016. Modeling and evaluating the effects of Big Data storage resource allocation in global scale cloud architectures. Int. J. Data Warehouse. Min. 12 (2), 1–20.

Benevenuto, F., Fernandes, C., Santos, M., Almeida, V.A.F., Almeida, J.M., Janakiraman, G.J., Santos, J.R., 2006. Performance models for virtualized applications. In: Min, G., Martino, B.D., Yang, L.T., Guo, M., Rnger, G. (Eds.), ISPA Workshops, Lecture Notes in Computer Science, vol. 4331. Springer, Berlin, Heidelberg, pp. 427–439.

Bernardi, S., Flammini, F., Marrone, S., Mazzocca, N., Merseguer, J., Nardone, R., Vittorini, V., 2013. Enabling the usage of UML in the verification of railway systems: the DAM-Rail approach. Reliab. Eng. Syst. Saf. 120, 112–126.

Bilal, K., Khan, S.U., Zhang, L., Li, H., Hayat, K., Madani, S.A., Min-Allah, N., Wang, L., Chen, D., Iqbal, M.I., Xu, C., Zomaya, A.Y., 2013. Quantitative comparisons of the state-of-the-art data center architectures. Concurrency Computat. Pract. Exper. 25 (12), 1771–1783.

Bloomfield, R., Netkachova, K., Stroud, R., 2013. Security-Informed Safety: If It's Not Secure, It's Not Safe. Springer Berlin Heidelberg, Berlin, Heidelberg, pp. 17–32. doi:10.1007/978-3-642-40894-6_2.

Bobbio, A., Portinale, L., Minichino, M., Ciancamerla, E., 2001. Improving the analysis of dependable systems by mapping fault trees into Bayesian networks. Reliab. Eng. Syst. Saf. 71 (3), 249–260. doi:10.1016/S0951-8320(00)00077-6.

Bourhis, P., Deutch, D., Moskovitch, Y., 2016. Analyzing data-centric applications: why, what-if, and how-to. In: 2016 IEEE 32nd International Conference on Data Engineering (ICDE), pp. 779–790. doi:10.1109/ICDE.2016.7498289.

Castiglione, A., Gribaudo, M., Iacono, M., Palmieri, F., 2014. Exploiting mean field analysis to model performances of big data architectures. Futur. Gener. Comput. Syst. 37, 203–211.

Castiglione, A., Gribaudo, M., Iacono, M., Palmieri, F., 2015. Modeling performances of concurrent big data applications. Softw. Pract. Exp. 45 (8), 1127–1144.

Chang, J., Ranganathan, P., 2012. (Re)designing data-centric data centers. IEEE Micro 32, 66–70. doi:10.1109/MM.2012.3.

Charniak, E., 1991. Bayesian networks without tears. AI Mag. 12 (4), 50–63.

Chen, J., Zhang, H., Zhou, H., 2013. Topology-based data dissemination approaches for large scale data centric networking architecture. China Commun. 10 (9), 80–96. doi:10.1109/CC.2013.6623506.

Ciardo, G., Gribaudo, M., Iacono, M., Miner, A., Piazzolla, P., 2016. Power consumption analysis of replicated virtual applications in heterogeneous architectures. In: Caporarello, L., Cesaroni, F., Giesecke, R., Missikoff, M. (Eds.), Digitally Supported Innovation: A Multi-Disciplinary View on Enterprise, Public Sector and User Innovation. Springer International Publishing, Cham, pp. 285–297.

Cisco, 2016. Cisco global cloud index: forecast and methodology, 2015–2020. https://www.cisco.com/c/dam/en/us/solutions/collateral/service-provider/global-cloud-index-gci/white-paper-c11-738085.pdf.

Couto, R.D.S., Secci, S., Campista, M.E.M., Costa, L.H.M.K., 2015. Reliability and survivability analysis of data center network topologies. CoRR abs/1510.02735.

CSA, 2016. Big Data—Security and Privacy Handbook. https://downloads.cloudsecurityalliance.org/assets/research/big-data/BigData_Security_and_Privacy_Handbook.pdf.

Distefano, S., Puliafito, A., 2014. Information Dependability in Distributed Systems: The Dependable Distributed Storage System, pp. 3–18.

ENISA, 2009. Cloud Computing Risk Assessment. https://www.enisa.europa.eu/publications/cloud-computing-risk-assessment.

Fiandrino, C., Kliazovich, D., Bouvry, P., Zomaya, A., 2015. Performance and energy efficiency metrics for communication systems of cloud computing data centers. IEEE Trans. Cloud Comput. PP (99), 1–1.

Fiore, U., Palmieri, F., Castiglione, A., De Santis, A., 2014. A cluster-based data-centric model for network-aware task scheduling in distributed systems. Int. J. Parallel Prog. 42 (5), 755–775.

Greenberg, A., Hamilton, J.R., Jain, N., Kandula, S., Kim, C., Lahiri, P., Maltz, D.A., Patel, P., Sengupta, S., 2009. Vl2: a scalable and flexible data center network. SIGCOMM Comput. Commun. Rev. 39 (4), 51–62.

Gribaudo, M., Iacono, M., 2014. An introduction to multiformalism modeling. In: Gribaudo, M., Iacono, M. (Eds.), Theory and Application of Multi-Formalism Modeling. IGI Global, Hershey, pp. 1–16.

Gribaudo, M., Piazzolla, P., Serazzi, G., 2012. Consolidation and replication of VMS matching performance objectives. In: Analytical and Stochastic Modeling Techniques and Applications, Lecture Notes in Computer Science, vol. 7314. Springer Berlin Heidelberg, Berlin, Heidelberg, pp. 106–120.

Gribaudo, M., Iacono, M., Marrone, S., 2015. Exploiting Bayesian networks for the analysis of combined attack trees. Electron. Notes Theor. Comput. Sci. 310, 91–111. doi:10.1016/j.entcs.2014.12.014. Proceedings of the Seventh International Workshop on the Practical Application of Stochastic Modelling (PASM).

Gribaudo, M., Iacono, M., Manini, D., 2016a. Improving reliability and performances in large scale distributed applications with erasure codes and replication. Futur. Gener. Comput. Syst. 56, 773–782.

Gribaudo, M., Iacono, M., Manini, D., 2016b, Modeling replication and erasure coding in large scale distributed storage systems based on CEPH. In: Proc. of XII conference of the Italian chapter of AIS, Lecture Notes in Information Systems and Organisation. Springer Berlin/Heidelberg, Berlin, Heidelberg.

Gribaudo, M., Iacono, M., Manini, D., 2016c, Three layers network influence on cloud data center performances. In: Proceedings—30th European Conference on Modelling and Simulation, ECMS 2016, pp. 621–627.

Gribaudo, M., Iacono, M., Palmieri, F., 2016d, Performance modeling of big data oriented architectures. In: Kolodziej, J., Pop, F., Di Martino, B. (Eds.), Resource Management for Big Data Platforms and Applications, Computer Communications and Networks. Springer International Publishing, Cham, pp. 3–34.

Han, M., Kim, M., Park, C., Na, Y., Kim, S.W., 2016. Server system modeling for data-centric computing: in terms of server specifications, benchmarks, and simulators. In: 2016 International Conference on Electronics, Information, and Communications (ICEIC), pp. 1–4. doi:10.1109/ELINFOCOM.2016.7562993.

Hopps, C., 2000. Analysis of an Equal-Cost Multi-Path Algorithm. RFC Editor, United States.

Huber, N., Von Quast, M., Brosig, F., Kounev, S., 2010. Analysis of the performance-influencing factors of virtualization platforms. In: Proceedings of the 2010 International Conference on the Move to Meaningful Internet Systems: Part II, OTM'10. Springer-Verlag, Berlin, Heidelberg, pp. 811–828.

Kusko, B., 1993. Fuzzy Thinking: The New Science of Fuzzy Logic. Harper Collins, London.

Lian, Q., Chen, W., Zhang, Z., 2005. On the impact of replica placement to the reliability of distributed brick storage systems. In: 25th IEEE International Conference on Distributed Computing Systems, 2005. ICDCS 2005. Proceedings, pp. 187–196. doi:10.1109/ICDCS.2005.56.

Luo, H., Zhang, H., Zukerman, M., Qiao, C., 2014. An incrementally deployable network architecture to support both data-centric and host-centric services. IEEE Netw. 28 (4), 58–65. doi:10.1109/MNET.2014.6863133.

Macher, G., Armengaud, E., Brenner, E., Kreiner, C., 2016. Threat and risk assessment methodologies in the automotive domain. Procedia Compu. Sci. 83, 1288–1294. doi:10.1016/j.procs.2016.04.268. The 7th International Conference on Ambient Systems, Networks and Technologies (ANT 2016)/The 6th International Conference on Sustainable Energy Information Technology (SEIT-2016)/Affiliated Workshops, http://www.sciencedirect.com/science/article/pii/S1877050916303015.

Mandal, A.K., Changder, S., Sarkar, A., Debnath, N.C., 2013. A novel and flexible cloud architecture for data-centric applications. In: 2013 IEEE International Conference on Industrial Technology (ICIT), pp. 1834–1839. doi:10.1109/ICIT.2013.6505955.

Marrone, S., Nardone, R., Tedesco, A., D'Amore, P., Vittorini, V., Setola, R., De Cillis, F., Mazzocca, N., 2013. Vulnerability modeling and analysis for critical infrastructure protection applications. Int. J. Crit. Infrastruct. Prot. 6 (3–4), 217–227.

Menasce', D.A., 2005. Virtualization: concepts, applications, and performance modeling. In: Proc. of the Computer Measurement Groups 2005 International Conference.

Ömer KÖksal, Tekinerdogan, B., 2017. Obstacles in data distribution service middleware: a systematic review. Futur. Gener. Comput. Syst. 68, 191–210. doi:10.1016/j.future.2016.09.020.

Okamoto, S., Zhang, S., Yamanaka, N., 2014. Energy efficient data-centric network on the optical network virtualization platform. In: 2014 12th International Conference on Optical Internet 2014 (COIN), pp. 1–2. doi:10.1109/COIN.2014.6950540.

Oral, S., Simmons, J., Hill, J., Leverman, D., Wang, F., Ezell, M., Miller, R., Fuller, D., Gunasekaran, R., Kim, Y., Gupta, S., Vazhkudai, D.T.S.S., Rogers, J.H., Dillow, D., Shipman, G.M., Bland, A.S., 2014. Best practices and lessons learned from deploying and operating large-scale data-centric parallel file systems. In: SC14: International Conference for High Performance Computing, Networking, Storage and Analysis, pp. 217–228. doi:10.1109/SC.2014.23.

Palmieri, F., Fiore, U., Ricciardi, S., Castiglione, A., 2016. Grasp-based resource re-optimization for effective big data access in federated clouds. Futur. Gener. Comput. Syst. 54, 168–179.

Pham, V., Karmouch, A., 1998. Mobile software agents: an overview. IEEE Commun. Mag. 36 (7), 26–37. doi: 10.1109/35.689628.

Ranganathan, P., 2011. From microprocessors to nanostores: rethinking data-centric systems. Computer 44, 39–48. doi:10.1109/MC.2011.18.

Raspotnig, C., Katta, V., Karpati, P., Opdahl, A., 2013. Enhancing chassis: a method for combining safety and security. In: Proceedings—2013 International Conference on Availability, Reliability and Security, ARES 2013, pp. 766–773. doi:10.1109/ARES.2013.102.

Ravindran, K., 2015. Agent-based QOS negotiation in data-centric clouds. In: 2015 IEEE 4th International Conference on Cloud Networking (CloudNet), pp. 331–334. doi:10.1109/CloudNet.2015.7335333.

Ravindran, K., Iannelli, M., 2014. SLA evaluation in cloud-based data-centric distributed services. In: 2014 23rd International Conference on Computer Communication and Networks (ICCCN), pp. 1–8. doi:10.1109/ICCCN.2014.6911762.

Rossi, G.L.D., Iacono, M., Marin, A., 2016. Evaluating the Impact of EDOS Attacks to Cloud Facilities. ACM, New York, NY, USA.

Ruiu, P., Bianco, A., Fiandrino, C., Giaccone, P., Kliazovich, D., 2016. Power comparison of cloud data center architectures. In: Proceedings of the 2016 IEEE International Conference on Communications (ICC).

Saleh, I., 2010a, Specification and verification of data-centric web services. In: 2010 6th World Congress on Services, pp. 132–135.

Saleh, I., Kulczycki, G., Blake, M.B., 2009. Demystifying data-centric web services. IEEE Internet Comput. 13 (5), 86–90. doi:10.1109/MIC.2009.106.

Saleh, I., Kulczycki, G., Blake, M.B., Wei, Y., 2013. Formal methods for data-centric web services: from model to implementation. In: 2013 IEEE 20th International Conference on Web Services, pp. 332–339. doi:10.1109/ICWS.2013.52.

Sfrent, A., Pop, F., 2015. Asymptotic scheduling for many task computing in big data platforms. Inform. Sci. 319, 71–91.

Shariati, M., Bahmani, F., Shams, F., 2011. Enterprise information security, a review of architectures and frameworks from interoperability perspective. Procedia Comput. Sci. 3, 537–543. doi:10.1016/j.procs.2010.12.089.

Simon, V., Monnet, S., Feuillet, M., Robert, P., Sens, P., 2014, May. SPLAD: scattering and placing data replicas to enhance long-term durability. Rapport de recherche RR-8533. INRIA, 23pp. http://hal.inria.fr/hal-00988374.

Singh, A., Ong, J., Agarwal, A., Anderson, G., Armistead, A., Bannon, R., Boving, S., Desai, G., Felderman, B., Germano, P., Kanagala, A., Provost, J., Simmons, J., Tanda, E., Wanderer, J., Hölzle, U., Stuart, S., Vahdat, A., 2015, August. Jupiter rising: a decade of CLOS topologies and centralized control in Google's datacenter network. SIGCOMM Comput. Commun. Rev. 45 (4), 183–197.

Spoto, S., Gribaudo, M., Manini, D., 2014. Performance evaluation of peering-agreements among autonomous systems subject to peer-to-peer traffic. Perform. Eval. 77, 1–20.

TrendMicro, 2015. Report on cybersecurity and critical infrastructure in the Americas. https://www.trendmicro.com/cloud-content/us/pdfs/security-intelligence/reports/critical-infrastructures-west-hemisphere.pdf.

Vasile, M.A., Pop, F., Tutueanu, R.I., Cristea, V., KoÅĆodziej, J., 2015. Resource-aware hybrid scheduling algorithm in heterogeneous distributed computing. Futur. Gener. Comput. Syst. 51, 61–71.

Vishnu, A., Dam, H.V., Jong, W.D., Balaji, P., Song, S., 2010. Fault-tolerant communication runtime support for data-centric programming models. In: 2010 International Conference on High Performance Computing, pp. 1–9. doi:10.1109/HIPC.2010.5713195.

Ward, D., Ibarra, I., Ruddle, A., 2013. Threat analysis and risk assessment in automotive cyber security. SAE Int. J. Passenger Cars Electron. Electr. Syst. 6 (2). https://www.scopus.com/inward/record.uri?eid=2-s2.0-84878771276&partnerID=40&md5=06754d2815cb6e3aaa12ab34e8dec4ec.

Watson, B.J., Marwah, M., Gmach, D., Chen, Y., Arlitt, M., Wang, Z., 2010. Probabilistic performance modeling of virtualized resource allocation. In: Proceedings of the 7th International Conference on Autonomic Computing, ICAC '10. ACM, New York, NY, USA, pp. 99–108.

Xie, P., Li, J., Ou, X., Liu, P., Levy, R., 2010. Using Bayesian networks for cyber security analysis. In: Setola, R., Geretshuber, S. (Eds.), Proceedings of the 40th IEEE/IFIP International Conference of Dependable Systems and Networks, pp. 211–220.

Xu, L., Cipar, J., Krevat, E., Tumanov, A., Gupta, N., Kozuch, M.A., Ganger, G.R., 2014. Agility and performance in elastic distributed storage. Trans. Storage 10 (4), 16:1–16:27. doi:10.1145/2668129.

Yan, F., Riska, A., Smirni, E., 2012. Fast eventual consistency with performance guarantees for distributed storage. In: 2012 32nd International Conference on Distributed Computing Systems Workshops (ICDCSW), pp. 23–28. doi:10.1109/ICDCSW.2012.21.

Zafar, S., Bashir, A., Chaudhry, S.A., 2016. On implementation of DCTCP on three-tier and fat-tree data center network topologies. SpringerPlus 5 (1), 766. doi:10.1186/s40064-016-2454-4.

CHAPTER

# THE CYBER SECURITY CHALLENGES IN THE IoT ERA

# 3

**Antonio Scarfò***

*Next-Era Prime Spa, Naples, Italy**

## 1 INTRODUCTION

We are in the Internet of Things era, an era that will change the world as we know it. These eminent studies put this phenomena in perspective:

- Cisco Systems (Evans, 2011) says that by the year 2020 there will be 50 billion connected devices in place;
- Gartner (Lovelock, 2016) says that 6.4 Billion devices are going to be in place by the 2016;
- BI's report (Business Intelligence, 2015) claims that, by 2019, the Internet of Things will result in $1.7 trillion in value added to the global economy;
- McKinsey Global Institute (2015) estimates that the Internet of Things will have an economic impact of $11.1 trillion by 2025.

So, it seems that this phenomenon will be a real breakthrough with countless consequences on business, societies, and people's way of life.

At the same time, the concerns related to security challenges that the Internet of Things is bringing on the cyber security arena are one of the biggest barriers that could slow down the adoption of IoT solutions from business and governments perspectives.

Indeed, a report by the Internet of Things Institute (2016) says that the two main reasons why businesses are not embracing IoT yet are:

- Data Privacy: (40%) due to the huge potential harvest of data that increase dramatically security breaches costs;
- Security Concerns: (40%) due to the potential tricky concern that an IoT project can implicate in a "wild-west" like atmosphere;

Still, these reasons do not seem to be affecting the rise of the IoT wave that is estimated to reach 43% CAGR in 2019 by the same institute. In fact, this rush is one of the main problems of the IoT phenomena.

**Security and Resilience in Intelligent Data-Centric Systems and Communication Networks. https://doi.org/10.1016/B978-0-12-811373-8.00003-3**

Copyright © 2018 Elsevier Inc. All rights reserved.

# 2 THE CYBER SECURITY SCENARIO

In this chapter, we are going to provide the main facts about the cyber security current scenario in order to define the state-of-the-art as a starting point to catch the correlations with IoT areas of development.

## 2.1 WHERE ATTACKS HAPPEN, ATTACKS TARGETS AND PATTERNS

In order to have a global and real-life scenario of cyberattacks, we are going to take into consideration two reports about data breaches and information security incidents coming from the Ponemon Institute LLC (2016), sponsored by IBM, and the 2016 DBIR by Verizon (2016). The main insights come from the following questions:

Who are the targets? In term of industries, we can find three main clusters from statistics of confirmed data loss attacks:

- +35% Finance;
- 5%–12%, Accommodation, Information, Public, Retail, Healthcare;
- less than 2.5%, all others industries: Entertainment, Manufacturing, Transportation, Mining, Real Estate, Constructions, and Agriculture.

If we consider also those attacks that did not result in a breach, the public has more than 73% and all the others less than 5%.

Who are the attackers? The Ponemon Institute LLC (2016) claims that about 50% of attacks are malicious whereas all the others come from system glitch and human errors. Also, more than 80% attacks come from external sources.

Why do they attack? Verizon (2016) says that about 89% of malicious attacks are caused by financial gains (80%) and espionage objectives.

How do they attack? we can see three clusters of ways: Hacking and Malware are the most used, then there are social and finally Errors, Misuse, Physical, Environmental (Ponemon Institute LLC, 2016).

In terms of means used, some of them are increasing:

- Malware: C2, Export data, Spyware/Keyloggers;
- Hacking: Use of stolen credentials, use of backdoor or C2;
- Social: Phishing;

and some others that are decreasing:

- Hacking: Brute force, RAM;
- Malware: Backdoor.

Further, the following picture shows the pattern of attacks of confirmed data breaches by industry. Web app attacks are a clear growing trend (Fig. 1).

## 2.2 ATTACK COSTS

Now that, we have taken a look at attacks' means, patterns, and targets, let's look at how much they cost.

According to a Ponemon Institute report (Ponemon Institute LLC, 2016), it is possible to arrange the data gathered from the following perspectives:

| Industries | Crime ware | Espionage | DoS | Every thing else | Stole assets | Errors | Card skimmers | POS | Privileges misuse | Web apps |
|---|---|---|---|---|---|---|---|---|---|---|
| Accommodation | <1% | <1% | 20% | 1% | 1% | 1% | <1% | 74% | 2% | 1% |
| Administrative | | | 56% | 4% | | 2% | | 4% | 22% | 11% |
| Edu | 2% | 2% | 81% | 2% | 3% | 4% | | | 1% | 5% |
| Entertainment | | | 99% | | <1% | | | 1% | | 1% |
| Finance | 2% | <1% | 34% | 5% | <1% | 1% | 6% | <1% | 3% | 48% |
| Healthcare | 4% | 2% | | 11% | 32% | 18% | | 5% | 23% | 4% |
| Information | 4% | 3% | 46% | 21% | <1% | 11% | | <1% | 2% | 12% |
| Manufacturing | 5% | 16% | 33% | 33% | | 1% | | 1% | 6% | 6% |
| Professional | 1% | 2% | 90% | 2% | 1% | 1% | | | 2% | 1% |
| Public | 16% | <1% | 1% | 17% | 20% | 24% | | <1% | 22% | <1% |
| Retail | 1% | <1% | 45% | 2% | | 1% | 3% | 32% | 1% | 13% |
| Transportation | 10% | 16% | 26% | | | 6% | | | 6% | 35% |

**FIG. 1**

Patterns of attack by industries (Verizon, 2016).

**Table 1 Data Breach Cost By countries**

| $ Per Capita (Average Data Breach Cost) | | Total Cost M$ | |
|---|---|---|---|
| US | 221↑ | US | 7.01↑ |
| DE | 213↑ | DE | 5.01↑ |
| CA | 211↑ | CA | 4.98↑ |
| FR | 196↑ | FR | 4.72↑ |
| UK | 159↓ | AB | 4.61↑ |
| IT | 156↑ | UK | 3.95↑ |
| JP | 142↑ | JP | 3.30↑ |
| AB | 140↑ | IT | 3.26↑ |
| AU | 131↓ | AU | 2.44↓ |
| SA | 101 | BZ | 1.92↑ |
| BZ | 100↑ | SA | 1.87 |
| ID | 61↑ | ID | 1.60↑ |

- *By countries*: Table 1 shows the cost of data breaches per capita and the resulting total cost of organization. The United States, Germany, and Canada have the highest cost per capita. India, South Africa, and Brazil have the lowest cost per capita. The United States also has the highest total cost, followed by Germany and Canada. India, South Africa, and Brazil confirm the last three positions about total costs as well. All countries have a growing trend of costs in the last three years, mostly Brazil, except for the United Kingdom and Australia. Australia is the only country showing a decreasing trend of total cost.

**Table 2 Data Breach Cost per Capita by Industries**

| | $ |
|---|---|
| Healthcare | 335 |
| Education | 246 |
| Financial | 221 |
| Services | 208 |
| Life science | 195 |
| Retail | 172 |
| Communications | 164 |
| Industrial | 156 |
| Energy | 148 |
| Technology | 145 |
| Hospitality | 139 |
| Consumer | 133 |
| Media | 131 |
| Transportation | 129 |
| Research | 112 |
| Public | 80 |

**Table 3 Root Causes Analysis**

| Root | Incidence | $ Cost | MTTI | MTTC |
|---|---|---|---|---|
| Malicious or criminal attack | 48% | 170 | 229 | 82 |
| System glitch | 27% | 138 | 189 | 67 |
| Human errors | 25% | 133 | 162 | 59 |

- *By Industry*: Table 2 shows the cost of breach pro-capita per industries, the clear outcome is that the more regulated sectors, like healthcare, suffer the higher costs.
- *By root*: the report highlights the effect of the three main causes of breaches. Malicious and criminal attacks are about the 50% of the attacks, which confirms that the trend of external and malicious attacks is growing. Also, the criminal attacks cost more than others. Finally, there are reported two key parameters to measure the effectiveness of organizations to manage attacks: incident response (Mean Time to Identify MTTI) and containment processes (Mean Time to Contain MTTC). Table 3 shows the MTTI/MTTC, measured in days, by the main three root causes of attacks.

  There is a significant link between the total breach costs and MTTI, which is $3.2 million with MTTI less than 100 days and $4.38 million with a more than 100 days MTTI. Of course, the same link between breaches costs and MTTC, with an MTTC less than 30 days total data breach cost is $3.18 million whereas, with a MTTC more than 39 days the total data breach cost is $4.35 million. These data are coherent with the needed to have a response process in place able to identifies and remediates attacks.

**Table 4 Factors which affect the cost of breaches**

| Factor | Value |
|---|---|
| Incident response team | 16 |
| Extensive use of encryption | 13 |
| Employee training | 9 |
| Participation in threat sharing | 9 |
| BCM involvement | 9 |
| Extensive use of DLP | 8 |
| CISO appointed | 7 |
| Board-level involvement | 6 |
| Data classification schema | 5 |
| Insurance protection | 5 |
| Provision of ID protection | −3 |
| Consultants engaged | −5 |
| Lost or stolen devices | −5 |
| Rush to notify | −6 |
| Extensive cloud migration | −12 |
| Third party involvement | −14 |

- *By cost factors*: another interesting statistic concerns the factors that affect the costs of breaches. Table 4 shows those factors that affect positively and negatively the breach costs (the higher the better):

## 3 HOW IoT WILL AFFECT THE CYBER SECURITY SCENARIO

In order to understand the potential impact of IoT on cyber security, we will take a look at the most promising areas of development, then we will take a look at the main sources of threats coming from these IoT development areas.

### 3.1 IoT DEVELOPMENT AREAS, WHERE IoT IS EXPECTED TO GROW

Interesting insights about IoT areas of development come from the analysis provided by McKinsey Global Institute (2015). The McKinsey Global Institute foresees a potential economic value of IoT technologies between $3.9 and $11 Trillion by 2025, split in nine settings, as follows:

The values listed in Table 5 are ordered by their economic impact, so that we can deduce that:

- in the best case, Factories, Cities, and Human should get the maximum value from IoT;
- in the worst case, Factories, Cities, and Outside should get the maximum value from IoT;
- the percentage of the total IoT value for factories and retail do not depend on forecast variability whereas the other settings are strongly dependent on it, mostly Human and Outside.

**Table 5 The Economic Impact of IoT on Areas of Activities According to Mckinsey**

| Areas of Activity | Economic Impact | % of the Whole Economic Impact | |
|---|---|---|---|
| | | Min | Max |
| Factories | Operations optimization, predictive maintenance, inventory optimization, health and safety | 30.87 | 33.24 |
| Cities | Public safety and health, traffic control, resource management | 23.72 | 14.91 |
| Human | Monitoring and managing illness, improving wellness | 4.39 | 14.29 |
| Retail | Automated checkout, layout optimization, smart CRM, in-store personalized promotions, inventory shrinkage prevention | 10.46 | 10.42 |
| Worksites | Operations optimization, equipment maintenance, health and safety, IoT enabled R&D | 4.08 | 8.36 |
| Outside | Logistics routing, autonomous cars and trucks, navigation | 14.29 | 7.64 |
| Vehicles | Condition-based maintenance, reduced Insurance | 5.36 | 6.65 |
| Home | Energy management, safety and security, chore automation, usage-based design of appliances | 5.10 | 3.14 |
| Offices | Organizational redesign and worker monitoring, augmented reality for training, energy monitoring, building security | 1.79 | 1.35 |

Other interesting insights, for our aims, are related to IoT barriers and enablers, McKinsey says in its report:

- Technologies: low cost sensors, low cost batteries, low cost data communication, cloud storage—applications and analytics;
- Interoperability among devices and system;
- Privacy and confidentiality related to data gathered by sensors and analyzed by application property of individuals and/or organizations;
- Security, protection for organizations, individuals, and all the stakeholders involved in the IoT arena.

The factors above should be characteristics of an IoT valuable project. We can add some elements to the McKinsey ones. For instance, sensors need to be long-lasting in those cases where electricity is not readily available. Also, data communication, new protocols, architectures, organizations, and mechanisms (Palmieri, 2013, 2017; D'Angelo et al., 2015) are emerging in order to address the IoT requirements for flexibility, robustness, exploration, narrow bandwidth, scalability, and low-energy consumption.

## 3.2 IoT'S IMPACT ON DIGITALIZATION

The more digitalization spreads, the more cyber security takes a key role. As the data and processes are digitalized, existing processes are improved and new ways to do business and human activities are enabled. This is a process that is going to change the whole society and ways of living. Of course, the more digital processes are utilized, the more problems become digital. This even includes criminality. We cannot hesitate to define digitalization as a step of human evolution, much like the Industrial Revolution has been.

In that evolution, the IoT is a fundamental step forward. Indeed, thanks to the IoT the human race will digitalize a huge amount of processes not yet or not yet fully digitalized and will enable new business life style models. And this is another step forward in the pulverization of IT, which started with mainframes, then proceeded though PC, laptop, and smartphone, and now approaching sensors.

The IoT development requires the evolution of some technologies, which become "enablers," meaning that without these technologies IoT applications are not implementable or not advantageous as explained above. That means that a new trend of technologies development is in place:

- sensors;
- communication and transportations;
- data protocols;
- device management;
- semantics;
- applications and analytics; and
- security.

So, the IoT is a challenge for cyber security and in the next points we try to summarize what kind of threats IoT will bring on the security arena.

- Culture: the implementation of IoT solutions involves also those companies that have been using legacy and closed solutions till now. These companies are not used to developing cyber security proven solutions and, worse, the legacy solutions already in place are cyber-insecure by design. People and organizations who are used to working with these legacy and closed solutions have not developed the culture of security.
- Evolution: metering and remote control are not new concepts, the opportunity to optimize supply chains has lead companies to use these kinds of technologies for a long time before people began to talk about IoT. Simply, a lot of things are already installed but not yet connected to the Internet. So, we have to think that in a first phase of IoT development many running legacy systems will be integrated with new IoT technologies. We can imagine that these legacy systems, as above mentioned, are not ready to be connected in the Internet world from cyber security perspective.
- Interoperability: the winning factors of an IoT project is to have devices and systems broadly connected as well as data shared among different organizations. The process of increasing exposure, through which the edge of the organizations become more and more grey and always more overlapped with one of customers and partners, reduces the ability of control and extends the surface of a cyber security attack. For instance, could you have an attack from your home plant? If we think about the BYOD diffusion, it is easy to understand that a smartphone can be used like a bridge to connect personal areas and working area. So, the surface of attack for hackers is suddenly extended to personal IoT applications like sensors for a home plant. Furthermore, smartphones and cellular communication technologies, if properly exploited, can become both the enabler and the target for wide-scale DoS attacks (Lee and Kim, 2015).

- Early adopters (pioneers): market competition, government to drive the innovation and sound excitement lead a rush of IoT implementations. In the rush, people often lose focus on some aspects not directly related to the final objectives but equally important, like cyber security. IoT is not an exception, indeed actually IoT is a field where fast prototyping is key because valuable use cases are not so clear yet. Often the IoT early adopters are not so careful about security.
- Technology evolutions (innovators): as mentioned before, IoT needs the development of new enabler technologies that will make it possible and favorable. New frameworks, protocols, and application have to be developed and mature. So, as well as for early adopters, the evolution of new technologies must undergo a phase of immaturity that involves also the cyber security aspects.
- Cloud: it is a fundamental enabling factor for IoT, without cloud services most potential users will perceive IoT a unfavorable or unaffordable. So, cloud security threats will be part of IoT security threats.
- Use cases: one peculiarity, which is also a barrier to adoption, of IoT is that favorable use cases are not yet well known. In this exploration phase, unclear use cases contribute to increased security threats due the continuous adjustment of the solutions.
- Devices: they represent a key challenge in the IoT scenario (Lee and Kim, 2015). Of course, all previous points are applicable to device development and the utilization of new devices. In this phase of development, we often ask for some basic features to be provided by each device, such as being inexpensive, intelligent enough, and energy efficient. We also want our devices to behave in a way that is energy aware (in order to avoid exposing additional surface to attacks, Palmieri et al., 2011). In addition, devices should be easily upgradable, configurable, controllable, and require a minimum bandwidth.
- Digital virtualization: people, processes, and things are becoming more and more digital. The virtual and physical worlds are overlapping due the increasing digitalization of things that increasingly have a digital identity and digital capabilities. As a consequence, all aspects regarding cyber security and physical safety should be managed as one whole system, in an integrated manner, while reshaping the cyber security mission. Indeed, cyber security breaches could more and more affect people and the environment. That is shown by Gartner (Perkins and Byrnes, 2015) in the following new model aimed to take in consideration safety and reliability in the digital virtual world (IT, OT, IoT) (Fig. 2).

In order to show the potential cyber security vulnerabilities in the IoT world, just looking at devices, an HP analysis (Hewlett Packard Enterprise, 2015a,b) shows a terrible state regarding IoT threats:

- 90% of devices collect at least one piece of personal information;
- 70% of devices use unencrypted network services;
- 70% of devices, in cloud and mobile applications, enable attackers to identify valid user accounts using the account enumeration;
- 80% of devices, in cloud and mobile applications, fail to set and maintain safe passwords (complex and long enough); and
- 60% of devices, which provides user interface, are vulnerable to a range of issues such as persistent XSS and weak credentials.

HP, as well as other vendors and research organizations, has started to categorize IoT security threats (Lee and Kim, 2015). We see that the basic principles of cyber security do not change at all, what is

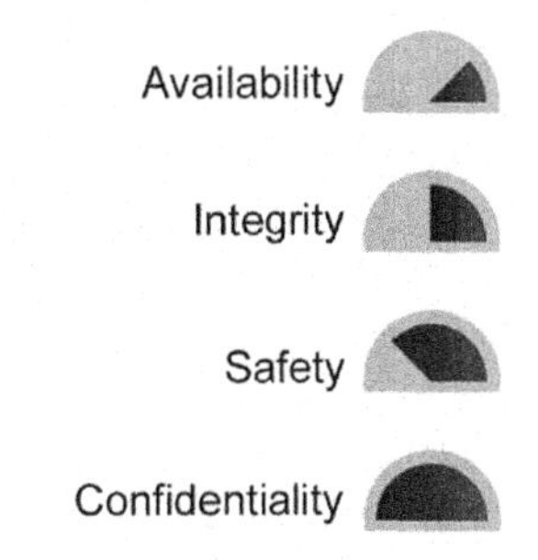


**FIG. 2**

Security elements in the IoT era (Perkins and Byrnes, 2015).

different is that there are some specific challenges coming from new devices, new IT frameworks, new processes, and business paradigms.

About classification, we would mention the OWASP Internet of Things Project (OWASP, 2017), its purpose is "to help manufacturers, developers, and consumers better understand the security issues associated with the Internet of Things, and to enable users in any context to make better security decisions when building, deploying, or assessing IoT technologies." This is a live project, based on a community, which has been split in five main areas:

- The attack surface, which list the points of attack like memory, authentication, and ecosystem;
- The vulnerabilities related to each attack surface;
- Firmware, which is security test guidance for firmware vulnerabilities; and
- SCADA/ICS, a topic related to weakness of SCADA and ICS applications.

There are also vertical projects about line guides or surveys useful to set best practices to develop secure devices and applications.

## 3.3 IoT DIGITALIZATION STRATEGIES

Of course, not all organizations potentially involved in IoT will face this evolution in the same way. According to Gartner (Perkins and Byrnes, 2015) there will be four different approaches categorized by the degree of openness and by the distribution of value between core and edge of the organization:

- *Guard the Jewels*, these companies, which are in the majority, maintain a traditional approach based on core services and they stay conservative for confidentiality, integrity, and availability. They will have an opaque approach to the data transparency. The access privileges to data and assets will remain a critical factor. These organizations will not change the cyber security approach.
- *Share the Wealth*, these companies are core centric but are experimenting the pressure to be more transparent. Usually, they are social media providers or massive datacenters with cheap services, which have to rely on patents in order to become more transparent while maintaining their values.

- *Expand the Empire*, these companies are moving their values toward the edge through the adoption of cloud services or through the relaxing of privileges of access to low-risk assets. These organizations need to align safety and cyber security. They can use a wrapped-like approach for their assets at the edge, and the main problem they are facing is the rush for the delivery of service at their edge.
- *Lead the Revolution*, these companies embrace the change and move values closer to the edge where digital meets physical. They need to integrate the management of cyber security and safety (IT, OT, physical, IOT). at the same time they open the access to their data. So, traditional security technologies have to be combined with the asset management, processes, people and the enterprise mobility management. Also, they need to adopt analytics from core to the edge to manage the complexity.

So, Gartner gives us an idea of the trajectories that companies could undertake, an aspect that appears clearly is that some of them will have to face the convergence between the digital and physical worlds, with the need to evolve their cyber security models and their approaches about risk management.

## 4 THE PROMISING WORLD OF INDUSTRIAL CONTROL SYSTEMS (ICS)

Factories individuate one the areas most affected by the IoT revolution and, in turn, by digital and physical convergence. Industrial control systems are used to adopt quite closed control systems or Operational Technologies (OT) and the IoT adoption opens new vectors of attacks. This perspective of evolution leads business risk managers to take care of the Integrity, Availability, and Confidently of data, but also of the Safety of people and of environments involved in operations processes (Perkins and Byrnes, 2015). Attacks not only affect data but also the physical security of people and of the environment, as shown by attacks like the German steel mill (Zetter, 2015) and Stuxnet (Langner, 2011).

Relevant part of this OT world are the huge amount of SCADA and ICS systems, which are quite closed systems devoted to control operative industrial processes.

Following, we will give an overview of vulnerabilities of this set of applications (Dell Inc., 2016).

First of all, the attacks against ICS components and SCADA systems have increased in recent years and the main causes of attacks clearly show that these systems are not accustomed to dealing with cyber security development philosophy and its best practices. The main causes of vulnerabilities are:

- Memory buffer bounds
- No input validation
- Information exposure
- Resource management errors
- Permissions and privileges control
- Credentials management
- Crypto issues

| Siemens | WinCC | Siemens SIMATIC | S70 300 | SINUMERIK | Tecnomatlx factory link | TIA Portal | | | |
|---|---|---|---|---|---|---|---|---|---|
| Schneider Electric | InTeractve Graphical SCADA System | Wonderware | Modicon Quantum | VAMPSET | Modicon M340 | BMX NOE 0110 | Clear SCADA | Accutech Manager | Magelis XBT HMI 6001 |
| Advantech | Advantech Studio | Advantech WebAccess | ADAMVew | | | | | | |
| CoDeSys | CoDeSys | | | | | | | | |
| DATAC | DATAC RealWin | | | | | | | | |
| Measuresot | Measuresoft ScadaPro | | | | | | | | |
| Ecava | IntegraXor | | | | | | | | |
| Elipse Software | Elipse E3 | | | | | | | | |
| Yokogava | Centum CS 3000 | | | | | | | | |
| AzeoTech | DAQFactory | | | | | | | | |
| MatrikonOPC | MatrikonOPC | | | | | | | | |
| Sinapsis | eSolar Light | | | | | | | | |
| General Electric | D20 PLC | Proficy | | | | | | | |

**FIG. 3**

Vendors versus vulnerabilities relations (Recorded Future, 2016), the blackest, the worst.

Using NIST CVE database is possible find over 70k vulnerabilities, of which 408 related to the main vendors. The picture below shows the relation vendor versus vulnerabilities (Recorded Future, 2016) (Fig. 3).

Mostly attacks are destructive and extortionist in nature like in the case of Sony (Kaspersky Lab, 2016).

The key point of the research is that ICS exploits are a limited domain which is dominated by few researchers. So, attacks are mostly done via well-known technologies like desktop software, web infrastructures, and financial infrastructures because ICS are more and more intertwined with business and internet systems.

Also, the availability of ICS systems in the Internet is becoming a serious threat because these systems are usually designed assuming that their networks would be completely isolated. They are cyber-insecure by design (Fig. 4).

Simply using Shodan we can find more than 200k ICSs, accessible on the Internet, running on 180k hosts in 170 countries. The 88.8% of the protocols used are open and insecure by design (Kaspersky Lab, 2016), they run on 91.6% of hosts. Narrowing the analysis, there are 1433 large organizations with 17,042 ICSs software running on 13,698 hosts. We are talking about organizations managing critical services like electricity, aerospace, transportation (including airports), oil and gas, metallurgy, chemical, agriculture, automotive, utilities, drinks and food manufacturing, construction, liquid storage tanks, smart cities, which are located mostly in The United States (21.9%), France (9.7%), and Italy (8%). It is likely that the most widespread systems are used by large organizations into the enterprise segment, which are from the following vendors Moxa (5057 services—29.7%), Siemens (3559—20.9%), Rockwell Automation (2383—13.9%), and Schneider Electrics (2107—12.4%).

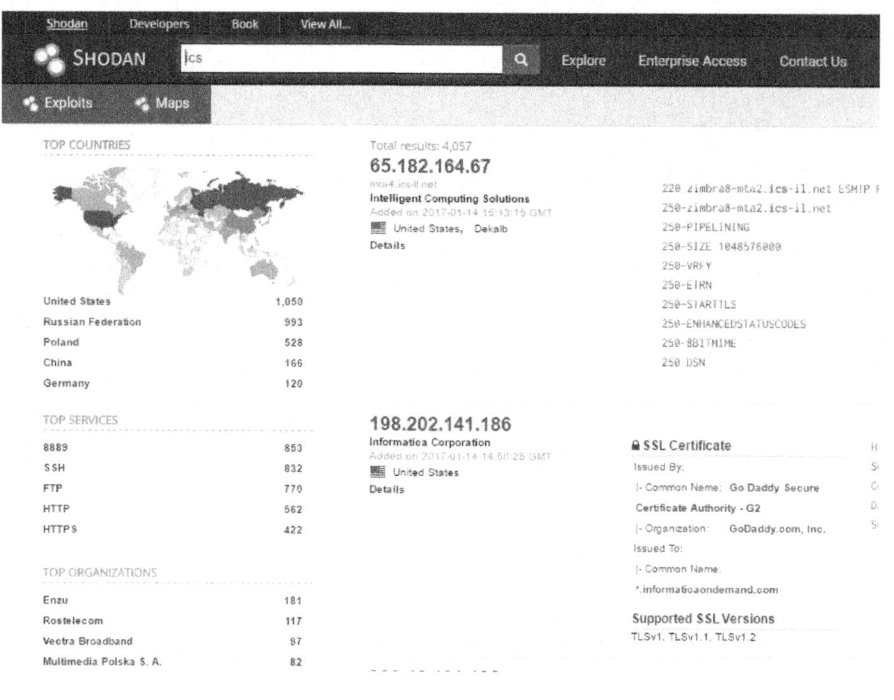


**FIG. 4**

Shodan ICS (Screenshot from Shodan, August 2016).

In terms of ICS components' vulnerabilities, 13,033 have been found on 11,882 hosts, which correspond to 6.3% of all hosts with externally available components detected. These vulnerabilities are ranked high risky by CVSS base scores (Fig. 5).

Combining ICS components vulnerabilities and insecure protocols, the total number of externally available and vulnerable ICS hosts are 172,982 (92%). In the most cases (93%), hosts contain medium risk vulnerabilities whereas 7% of the vulnerable hosts have critical vulnerabilities.

Of course, the results are likely underestimated in respect to the real number of ICS systems associated with significant risks. And also medium risks could cause significant damages, for instance, the exploitation of Siemens S7 protocol security flaws (estimated as having medium-risk), could lead to unauthorized reflashings of Siemens PLCs, hampering the corresponding technological process.

Cyber security of ICS systems can dramatically affect the physical security of the environment and the population. The risk increases significantly for small and medium companies and for consumers that are not equipped to face cyber security issues. Even if large enterprises and governments are able to understand the risks correlated to ICS systems, to fight them is not simple because ICS components are considered like black boxes, which makes it hard to modify them. So, a change in culture is required to avoid assuming that cyber security is a security axiom for ICS components.

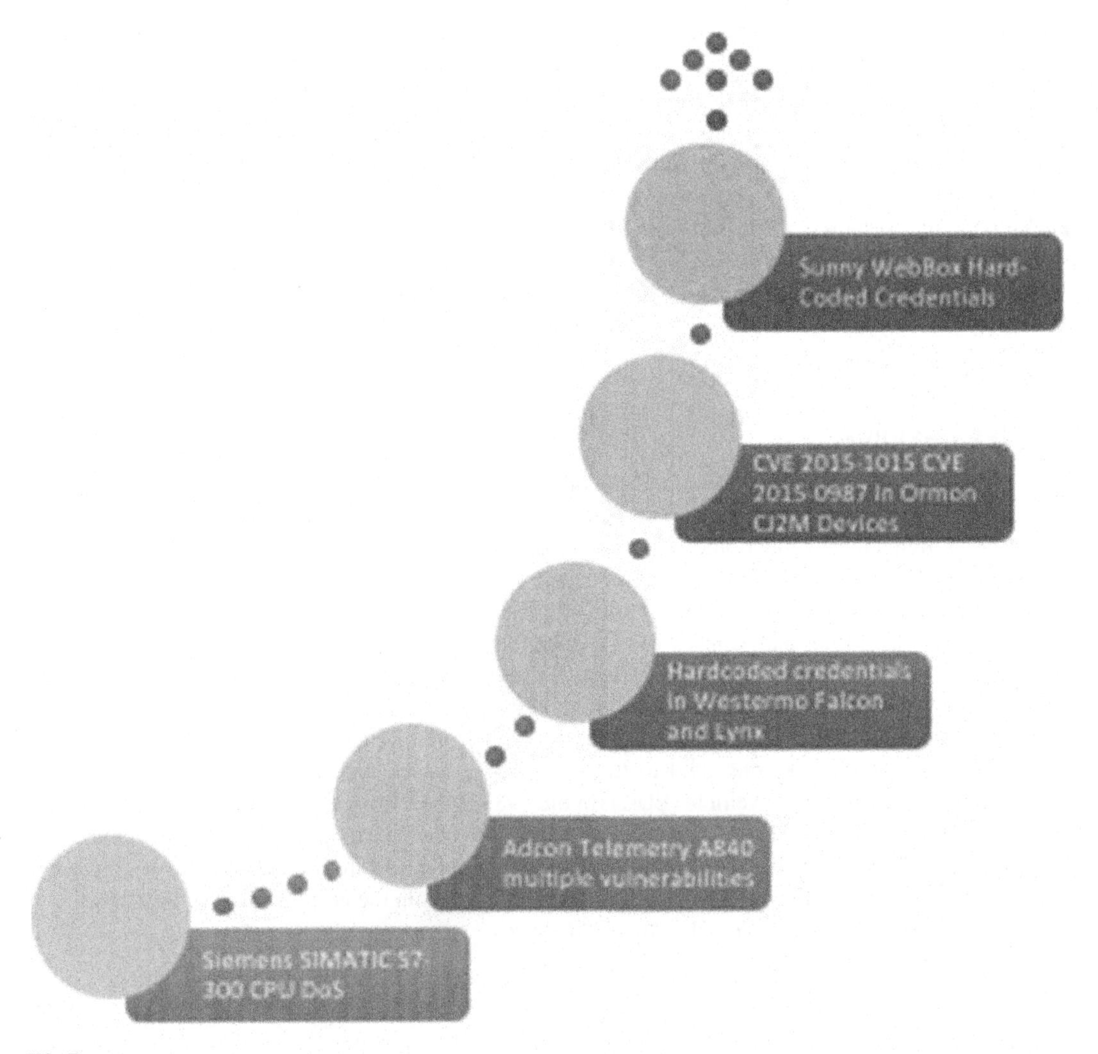


**FIG. 5**

The vulnerabilities of ICS components. Industrial control systems and their online availability, Karspersky 2016.

## 5 A DIFFERENT APPROACH IN THE AUTOMOTIVE SECTOR

In this part of the article we are going to take a look at a different approach to the digitalization in automotive sector as part of the industrial sector. We present two analyses proposed by Miller and Valasek (2015) about Jeep, and Mahaffey (2015) about Tesla.

The main insights of the analysis confirms that some of the elements we have mentioned about organizational culture and the evolution of legacy systems lead to critical vulnerabilities. Tesla has a strong secure architecture because Tesla designs their cars with the mindset of connected things and digitalization, embracing a brand new approach for a brand new product. On the other side, Jeep is adapting existing architecture to new market requirements concerning digital functionalities, which results in a dangerously attackable architecture.

We can see in these examples how the IoT opens, for the industrial sector, a wide attack surface due to the new possibilities that criminals and malicious attackers can exploit. This new surface will increase the amount of attacks and their costs. Also, we should take into serious consideration that attacks to the industrial sector can seriously affect people's safety, reshaping the concept of the cost of cyber security.

So, the necessity of a set of best practices for sectors like the automotive industry is becoming more and more critical. Experiences and data sharing are basic to accelerate the creation of a book of best practices and guidelines.

## 5.1 JEEP CHEROKEE

The analysis from Miller and Valasek (2015) about Jeep Cherokee was done in 2015. Jeep, as well as other car models on the market, is trying to gain a competitive advantages from digitalization and IoT innovations. Often, the new functionalities, several of which are related to grant a better drivers' safety, are created on the base of the original car project and architecture then, somewhat, they are part of the migration process already mentioned in the article.

In their analysis, Miller and Valasek tried to take the car control through the information system, which makes the car connected. First, they identified the architecture of the information systems and, in particular, of the network system which is based on CAN buses (Fig. 6).

What is interesting about the architecture that our attackers have discovered, is that the "infotainment" system (the Radio) is connected to the CAN buses, which are the core of the network that connects all control units of the vehicle (EUC). In the 1983 Robert Bosh started the development of the CAN architecture, which was released in 1986 by SAE. Now we have to highlight that the architecture shown above and the Chips that are used to implement that architecture are the same in several cars. So, we could extend the following analysis to several car models in the current automotive market.

We can notice that Cherokee Jeep has two CAN buses:

- CAN-C, which is an high speed bus connecting vital systems like brakes, the engine, airbags, steering control, and transmission control.
- CAN-IHS, which is a low speed bus which connects comfort system elements like radio and climate control.

Miller and Valasek have identified the attack surface, all points accessible from outside, and what, consequently, internal systems were possible to attack through what kind of vulnerabilities. The results are listed in Fig. 7.

Among the systems that equipped Cherokee Jeeps, there was the Uconnect systems which was the sole source of infotainment, Wi-Fi connectivity, navigation, apps, and cellular communications. Uconnect contains microcontrollers and software components able to communicate with other modules using Controlled Area Network, CAN-IHS, and CAN-C buses.

The attackers were able to force the WPA2 password of the WiFi connection and scan the services open on WiFi. Particularly, port 6667 was the access to a D-Bus over IP which is accessible through interprocess communication (IPC) and remote procedure call (RPC). Exploiting this vulnerability is

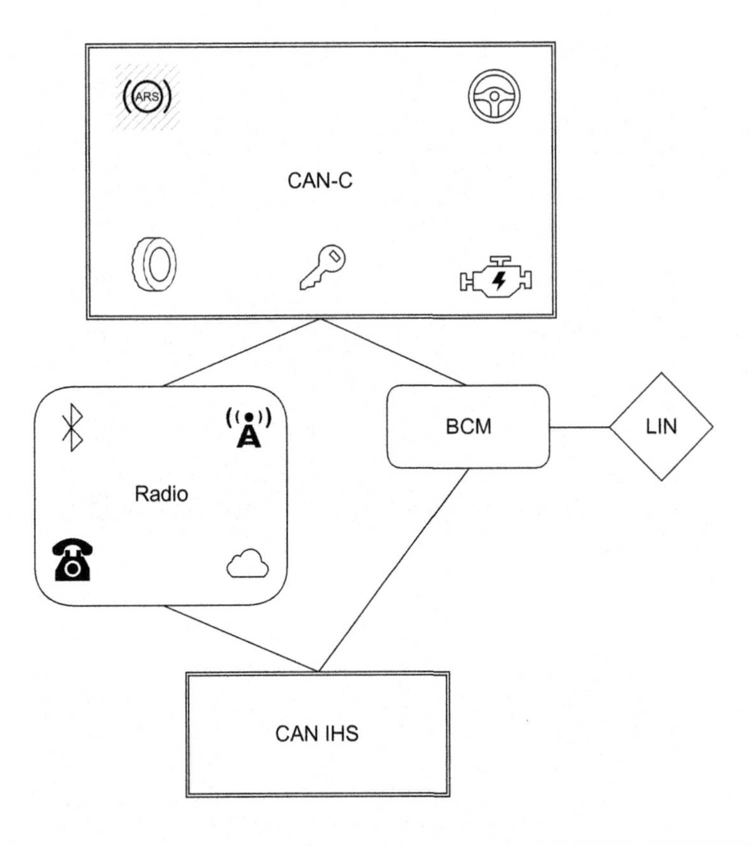


**FIG. 6**

Jeep's network (Miller and Valasek, 2015).

| Entry point | ECU | Bus |
|---|---|---|
| RKE | RFHM | CAN C |
| TPMS | RFHM | CNA C |
| Bluetooth | Radio | CAN C, CAN IHS |
| FM/AM/XM | Radio | CAN C, CAN IHS |
| Cellular | Radio | CAN C, CAN IHS |
| Internet / Apps | Radio | CAN C, CAN IHS |

**FIG. 7**

Jeep's network infotainment connections (Miller and Valasek, 2015).

possible get the control of head unit of Uconnect and access to all EUCs of the vehicle. That why D-Bus is not thought to manage attacks.

The attackers were even able to change the picture of the head unit's display, replacing it with their own image (Fig. 8).

**FIG. 8**

Attackers' image.

*From Charlie Miller, Chris Valasek "Remote Exploitation of an Unaltered Passenger Vehicle", June 2015.*

Using that approach, they first got control of the radio system, then they got the control of the GPS, and finally they were able to send commands through the CAN buses in order to control, practically, everything. The remote attack of Charlie Miller and Chris Valasek could be applied to several of FCAs car models that they estimate to be about 1.4 million units.

## 5.2 TESLA S MODEL

A test very similar to the Charlie Miller and Chris Valasek's one, but having as a target a Tesla S Model, has been performed by Mahaffey (2015). Tesla has a completely new information system, designed to be connected to the Internet by design. Indeed, attacking a Tesla S model information system, like Mahaffey says, "*had more in common with an advanced enterprise threat, where the attacker needed to move laterally into multiple systems, than a simple embedded system hack.*" The software development team has created a strong security architecture thinking of the car like an digital interconnected system, like ICT systems designer do.

The Tesla network architecture connects the two infotainment systems, named CID & ID, through a LAN, which is, in turn, connected to the CAN bus through a gateway. As usual, the CAN bus interconnects the vehicle controllers devices: the vital systems (Fig. 9).

The attack was stepped in several stages to find the right attack vectors:

- Exploration for attack surface (e.g., physical connections);
- Penetration test on attacks surface identified: browser, Bluetooth, USB, memory cards, WiFi, unknown connectors;

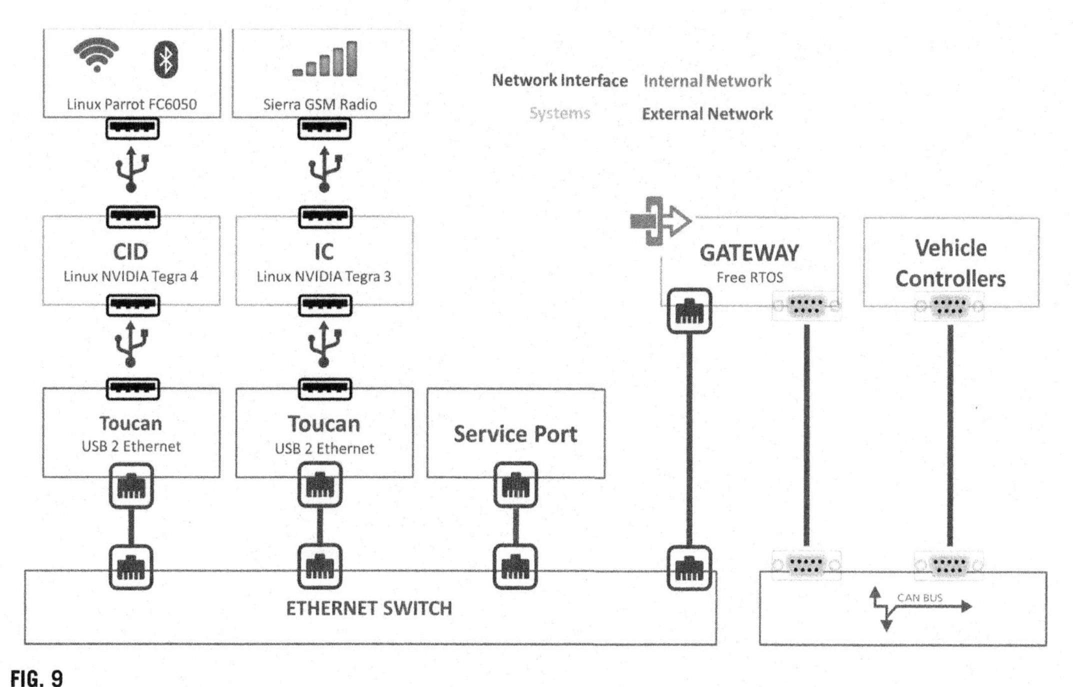


**FIG. 9**

Tesla's network connections diagrams (Mahaffey, 2015).

Particularly, the unknown connector was an Ethernet interface with a nonstandard connector like shown in the following picture (Fig. 10).

Using that nonstandard Ethernet connector, it was possible to access the services exposed in the infotainment system (following step)

- Internal services exposed by infotainment system (30)

The number of vulnerabilities found from the above steps of the analysis are listed follow:

- Vulnerability #1, Browser WebKit-based
- Vulnerability #2, Insecure DNS Proxy
- Vulnerability #3, Insecure HTTP Service
- Vulnerability #4, CID and the IC were running X11 without any form of access control
- Vulnerability #5, open download http address
- Vulnerability #6, shadow file containing passwords, very weak passwords of accounts of IC

In order to get to the root status of CID and IC systems, the attackers used the update process which made it possible to download a firmware form a Tesla's Internet servers. This firmware included the password file mentioned in Vulnerability #6, which has been cracked and, as a consequence, the attackers were able to gain the root role for the IC system. Using a token scheme to access CID, which

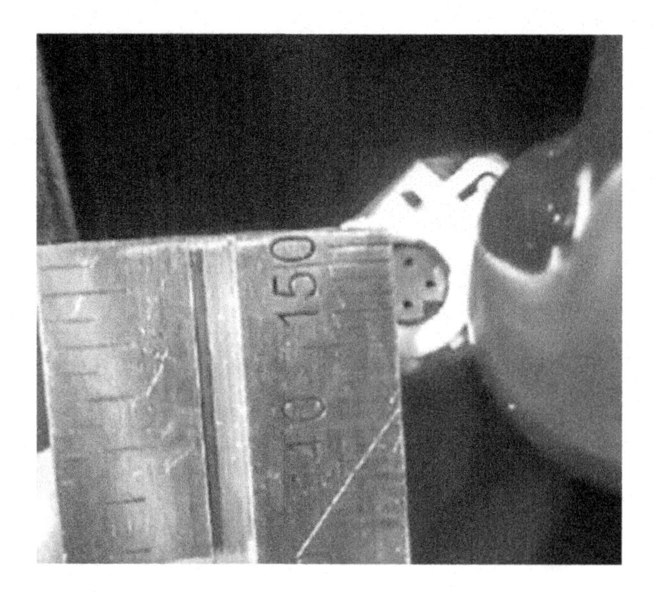

**FIG. 10**

Nonstandard Ethernet connector (Mahaffey, 2015).

was discovered upon a further analysis of the firmware file, the attackers became the root user of CID (main) and IC infotainment systems.

From here it was also possible to access X11 services and, for instance, disrupt the instrument display while driving.

Then, as a root user, our attackers noticed that CID talks with CAN systems but there was no raw CAN frames, instead there was a Gateway that allows the CID to ask for a specific set of permitted actions (VAPI, Vehicle API). This is a classical separation that we can find in the field of networking between two zones with a different level of security. It is created to make the whole system more resilient.

Anyway, the "legitimate" commands that CID can issues are not critical for the safety of the car and for its drivers/passengers. Except for the ability to "power off" the car, which means to stop the vehicle, which could seem dangerous, but was conducted in a very safe way for the driver in the experiment.

Fig. 11 shows how it has been possible to access the IC and CID systems and what kind of critical damages were possible to provoke. Anyway, the gateway was a key element to secure the resilience of the Vehicle and mainly of its passengers.

Tesla is addressing all vulnerabilities found by Mahaffey, nothing is critical indeed the software team has created a good security architecture that can be used as a template for other car producers. Some remarkable elements addressed form the findings of Mahaffey's analysis are:

- Over the air update process;
- Isolation between infotainment system and vehicle system.

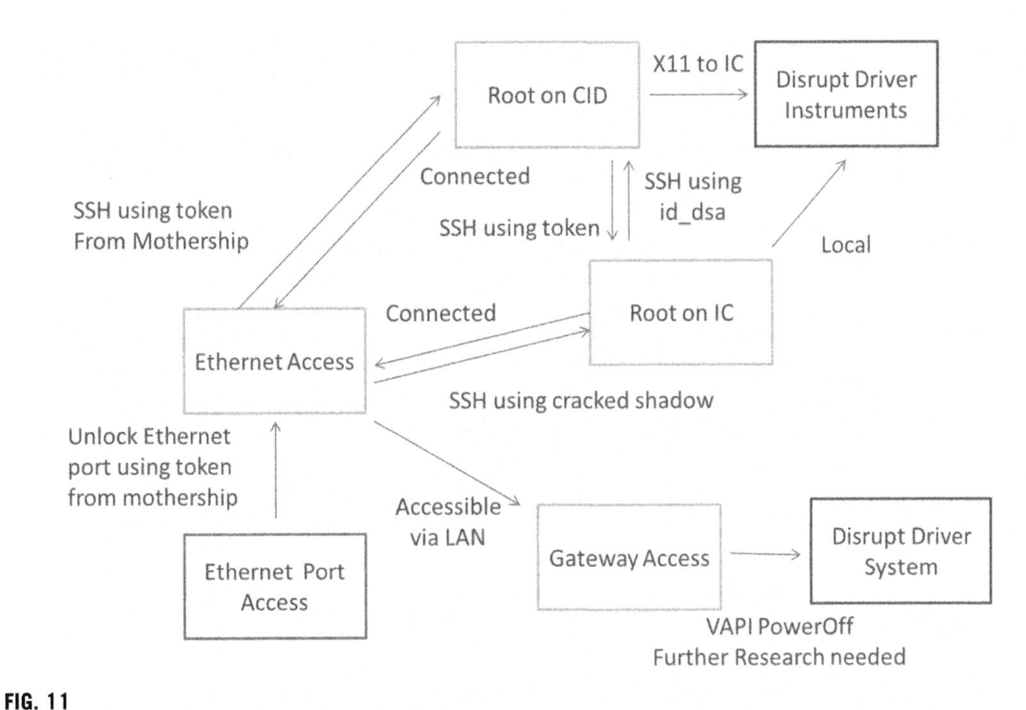


**FIG. 11**

Lateral attacks approach for Tesla S, the kill-chain (Mahaffey, 2015).

# 6 ARTIFICIAL INTELLIGENCE

The increasing number of surface attacks, the huge amount of data knowledge to analyze in order to defend digitized systems, and the high level of dynamicity of attack vectors require more and more holistic and adaptive systems, which are able to react and even predict automatically the attacks and to manage knowledge in the right timeframe.

The current algorithms applied in the cyber secure domain, which are unable to learn and adapt automatically their logic to the evolving attack vectors, represent a limitation in the cyber security battle.

From that perspective, AI represents a very promising field of research for its capability to rapidly analyze vast amounts of unstructured data, to be adaptive, and to be predictive. We can mention several examples of emerging application of cyber security based on Artificial Intelligence, before it is useful to sort Artificial Intelligence methodologies and their areas of applications (Patil, 2016).

- Neural Nets: Neural Net's history began in 1958 with the invention of perceptron by Roseblatt, which is a men-made nerve cell. By combining different typology of perceptrons, it is possible to get a net able to learn and solve issues, similar to how the human brain does. This way is very effective in solving problems related to patterns recognition, for classification, and for select actions to respond to cyber attacks. Indeed Neural nets are very effective means of performing massive parallel learning and decision-making. In cyber security they are important for detection, particularly for DoS, worm, spam, zombie. Also they are applicable for malware classification and rhetorical investigation. A valid example is the use of Neural nets in Misuse Detection which means the capability to identify network attack by comparing current and usual activities against the expected actions of an attacker (Cannady, 1998).

- Expert Systems: Expert Systems are probably the most famous and widely-used tool in Artificial Intelligence. They are "associate-skilled systems" used to find answers in some domains. Expert systems are made up of a mental object, which represents the knowledge in a specific domain, and an illation engine, which is used to generate answers and generate further information from the knowledge base. Expert Systems are very useful in simulation and in creating calculations, for instance, Expert Systems are generally used in diagnosis, finances, and computer networks. They are the basis for the acquisition of information. Therefore, an incomplete set of basic information implies a weak system, unable to produce answers. In the cyber security domain, Expert Systems are used for security architecture design, providing hints for optimal usage of resources. Also Expert Systems are potentially able to identify attacks and the ways to solve them (Rani and Goel, 2015).
- Intelligent Agents: Intelligent Agents are software solutions that can perform intelligent actions (rational behavior in economy) upon sensoring the environment that they are devoted to control and/or based upon given or learned knowledge (Fig. 12).

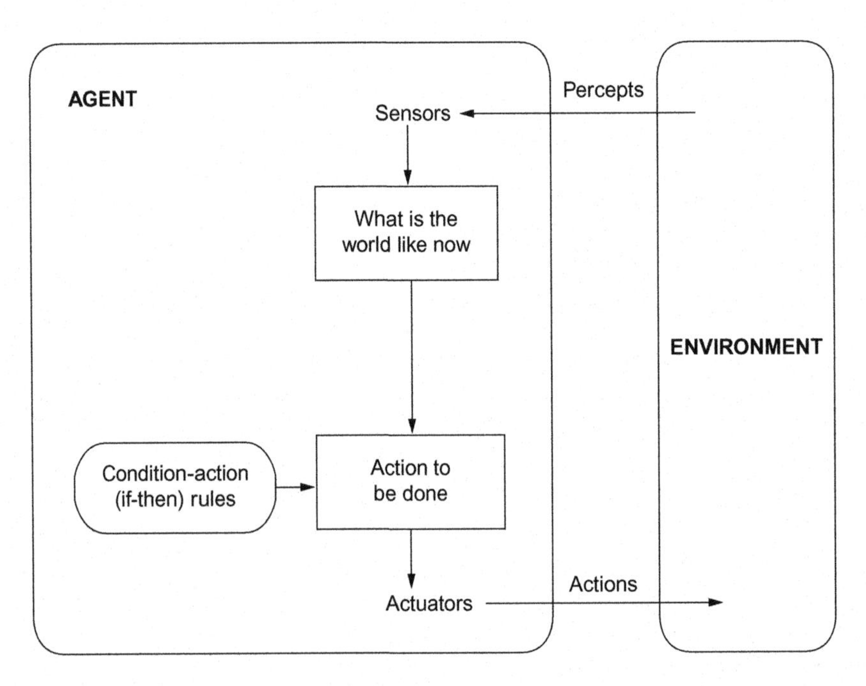


**FIG. 12**

Intelligent Agent paradigm by Wikipedia (August 2016).

In the cyber security domain, Intelligent Agents are used successfully, in cooperative fashion, against DDoS attacks. Also, hybrid models, involving Neural nets and multiagents, have been projected for intrusion detection applications and they are very promising.

- Learning: Learning is related to all processes aimed to create or extend a system of knowledge. Machine Learning is a domain of Artificial Intelligence that includes processes to automatically infer new data and skills and make predictions. Learning can be extremely complex or quite simple. To understand how complex it could be, think about how it can be hard to represent behaviors or ideas. There are three basic models to implement machine learning:
  - Supervised learning: Some examples of inputs/(desired) outputs, are given, the goal is to learn a rule that maps inputs to outputs.
  - Unsupervised learning: In this case, finding structure in input data is the goal of the algorithm; it is useful to discover hidden patterns in data or it could be a mean toward other aims.
  - Reinforcement learning: In this case, software interacts with a dynamic environment; its goal is get some objectives. The software learns upon feedback aimed to reward or punish how the learning system navigates its problem space.

  The applications of Machine Learning are in the areas of spam filtering, detection intruders, and detection of insiders that trying of data breach.
- Search Methods: A Search Method is a technique to address goal-based problem-solving. It is a central topic in Artificial Intelligence and there are several methods of search developed in Artificial Intelligence research that are commonly used in many areas of applications. Think about Google, it is a mean that we use all the time that is based on the search methodologies from Artificial Intelligence. The problem is related to finding a candidate that matches the requirements. There are many methodologies and basically they are related to a finding a path in a graph. Methods like $\alpha\beta$-search or minax are used in games are useful in decision-making for cyber security. Indeed it helps in reduce the amount of choices in the competition of two challengers.

We have listed some interesting ways to implement Artificial Intelligence applications with the goal of showing how Artificial Intelligence could be a strong candidate to face cyber security attackers. The key point is that Artificial Intelligence has promising capabilities to use large amount of unstructured data and evolve automatically and rapidly in order to keep the pace with the mutation of threat vectors.

In April 2016, MIT announced a system which predicts 85% of cyberattacks using inputs from human experts. The system is AI2 (Veeramachaneni and Arnaldo, 2016); it has achieved triple the accuracy when compared to the previous benchmarks and the false positives have been reduced by 5%. Basically, it is based on an unsupervised machine-learning approach, presenting its findings to human experts and learning from their feedback. Actually, AI2 superposes a supervised machine learning model to an unsupervised one (actually it uses three unsupervised methods), in order to not overwhelming its human overlords.

## 7 CONCLUSIONS

Our journey in cyber security evolution, leaded by the growth of the IoT adoption, has showed us some critical elements that should be seriously be taken into consideration in the coming years.

We have presented a picture of threat scenarios, by highlighting the main targets, the fundamental menaces, and attacker profiles and the reasons why they attack, together with the attack strategies. Outcomes of this first set of analyses are that Finance is the market that is more frequently attacked with success, whereas public administration is the most attacked overall. The main motivations of attacks are financial gain and espionage. Finally, we have some growing vectors of attack, like the Web, malware, hacking, and social.

Also, we have reported statistics about attacks from a costs perspective and we have found that the most regulated sectors, like Healthcare, Finance, and, surprisingly, Education are the sectors with the main costs of data breach. The same applies for amount of attacks, malicious and criminal attacks have the main costs per breach. By combining these two first analyses it is possible conclude that companies that are involved in finance and healthcare should prioritize the management of cyber security in order to minimize a successful data breach and, continuously improving MTTI and MTTC in order to minimize the costs of a data breach. Even if there are famous attacks like German steel mill and Stuxnet, other sectors, like manufacturing, agriculture, mining, transportation, and energy, seem to be relatively affected by cyber security phenomena. These sectors have some elements in common, they are relatively connected to the Internet and there is limited knowledge about their control systems, which are also called "obscure." Also, attacks on the control systems of these industries could affect the physical safety of people and the environment.

Now, if we focus the analysis on the most promising areas of development of the IoT, we notice that, most of the expectations, in terms of value and implementations, are precisely in these industrial sectors, which will have to face a dramatic improvement in terms of digitalization.

SCADA and ICS are very good examples. IoT promises miracles about the value coming from OT processes managed by SCADA/ICS, but we have shown that attacks to these systems are increasing, and mostly are coming from the internetworking because SCADA/ICS were designed like semi-isolated systems. Also, there is a remarkable lack of culture about credentials management, policies, systems design, cryptography, and information exposure. Finally, these systems are attackable through classical (aged) IT systems (e.g., browser and desktop vulnerabilities), which are more and more intertwined with SCADA/ICS as part of the Internet business. This is a lateral attack approach.

The difficulty of migration toward digital and connected systems, is also well explained in the comparison between penetration test for a Tesla and a Jeep. The attack, performed laterally through infotainment systems, shows how Tesla has a resilient brand new architecture with respect to Jeep, which is essentially a migration project.

At the very end, we show that:

- IoT/digitalization is an open door for attackers, mainly to some industry sectors less evolved in terms of organization, architectures, and culture.
- Even if there are sectors already used to dealing with cyber security, they will need to evolve their model of risk management due to the new model of business interaction and collaboration. This is a real challenge for those organizations that are shifting their values to the edge and open to share information.
- IoT Digitalization means also that the safety of people and the environment is part of cyber security risk management.
- It seems clear that sharing cross industrial information is a basic step to improve the resilience of the operations, so there are emerging open projects to share information and best practices.

Organizations should prioritize the improvement of their culture about cyber security, technologies, and processes depending of their strategic trajectories.

Organizations that are embracing the evolution, shifting their values from the core to the edge and embracing data openness, have to deal mostly with people and their culture. Technologies are basic factors to enable the optimization of processes or to support new business models, but preparing people to deal with the evolution, such as with new processes and technologies, is the first step. Also, about the prioritization model:

- Culture: Those organizations that Gartner calls *Lead the Revolution*, have to work on processes and then on enabler technologies in order to capture the value of the new models they want embrace. But more important than processes and technologies is the basic necessity to work on people in order to prepare them to be part of this continue evolution. Lead the Revolution organizations that will be impacted by IoT Digitalization, but currently are not so used to deal with cyber security like the ones we have seen in industrial sector using ICS/SCADA, should prioritize the culture aspect.
- Technologies: Those organizations are more closed, focused on protect their assets—*Guard the Jewels and Expand the Empire* in the Gartner language—likely should be focused on technologies evolution to keep themselves on the edge of the protection. Indeed, we expect that those companies will hardly modify their processes in place and this approach reduce the needs of culture evolution.
- Processes: Organizations that are embracing a more open approach to the external environment, like those that Gartner calls *Share the Wealth*, should prioritize the processes aspect because of the change in the ways of dealing with the partners customers. Of course, technologies are also basic because the technologies can enable the change of processes.

Future works: in this fast-evolving environment, where technologies are enabling new business models and new way to implement operations, Artificial Intelligence systems appears very promising due to the ability to evolve themselves autonomously, the capabilities to parallelize computing, and the ability to analyze the growing amount of unstructured data. Projects like the ones developed by Allsecurity (cloud-based bug-finding system) based on a symbolic execution engine and a directed fuzzer, Deep Armor by SparkCognition, which uses cognitive algorithms to find and remove malicious files, Darktrace, which tries to automate the defense using its self-learning algorithms, and Antigena technologies that replicates human antibodies to identify and neutralize cyber threats automatically, are evidence that there is an emerging thinking about the new frontiers to face cyber security threats. So, it is worth to investigate AI as a potential next step in cyber security evolution.

## REFERENCES


Business Intelligence, 2015, November. The Internet of Things Report.

Cannady, J., 1998. Artificial Neural Networks for Misuse Detection. School of Computer and Information Sciences Nova Southeastern University.

D'Angelo, G., Palmieri, F., Ficco, M., Rampone, S., 2015. An uncertainty-managing batch relevance-based approach to network anomaly detection. Appl. Soft Comput. J. 36 (17), 408–418.

Dell Inc., 2016. Dell Security, Annual Threat Report.

Evans, D., 2011, April. The Internet of Things, How the Next Evolution of the Internet is Changing Everything.

Hewlett Packard Enterprise, 2015b. Securing the Internet. December, 2015.

Hewlett Packard Enterprise, 2015a. Internet of Things Research study. December 2015.
Internet of Things Institute, 2016, April. Top 10 Reasons People Aren't Embracing the IoY, Yet.
Kaspersky Lab, 2016. Industrial Control Systems and Their Online Availability.
Langner, R., 2011. Stuxnet: dissecting a cyberwarfare weapon. IEEE Secur. Priv. 9 (3), 49–51.
Lee, Y., Kim, D., 2015. Threats analysis, requirements and considerations for secure internet of things. Int. J. Smart Home 9 (12).
Lovelock, J., 2016. The Internet of Things is Shifting Hackers' Targets. March 11, 2016
Mahaffey, K., 2015. Hacking a Tesla Models: What We Found and What We Learned. August 7, 2015.
McKinsey Global Institute, 2015, June. The Internet of Things: Mapping the Value Beyond the Hyep.
Miller, C., Valasek, C., 2015, June. Remote Exploitation of an Unaltered Passenger Vehicle.
OWASP, 2017. The Open Web Application Security Project. www.owasp.org.
Palmieri, F., 2013. Scalable service discovery in ubiquitous and pervasive computing architectures: a percolation-driven approach. Futur. Gener. Comput. Syst. 29 (3), 693–703.
Palmieri, F., 2017. Bayesian resource discovery in infrastructure-less networks. Inform. Sci., 376, 95–109.
Palmieri, F., Ricciardi, S., Fiore, U., 2011. Evaluating network-based DoS attacks under the energy consumption perspective: new security issues in the coming green ICT area. In: Proceedings—2011 International Conference on Broadband and Wireless Computing, Communication and Applications, BWCCA 2011, pp. 374–379.
Patil, P., 2016. Artificial intelligence cyber security. Int. J. Res. Comput. Appl. Robot. 4 (5), 1–5.
Perkins, E., Byrnes, F.C., 2015, July. Cybersecurity Scenario 2020 Phase 2: Guardians for Big Change.
Ponemon Institute LLC, 2016, June. The Cost of Data Breach Study: Global Analysis.
Rani, C., Goel, S., 2015. CSAAES: an expert system for cyber security attack awareness. In: 2015 International Conference on Computing, Communication & Automation (ICCCA).
Recorded Future, 2016. Threat Intelligence Report.
Veeramachaneni, K., Arnaldo, I., 2016. Ai2: training a big data machine to defend. In: IEEE 2nd International Conference on Big Data Security on Cloud (BigDataSecurity), IEEE International Conference on High Performance and Smart Computing (HPSC), and IEEE International Conference on Intelligent Data and Security (IDS).
Verizon Report DBIR, 2016. 2016 Data Breach Investigation Report.
Zetter, K., 2015, August. A Cyberattack has Caused Confirmed Physical Damage for the Second Time Ever. Wired.

## FURTHER READING

Merlo, A., Migliardi, M., Gobbo, N., et al., 2014. A denial of service attack to UMTS networks using SIM-less devices. IEEE Trans. Dependable Secure Comput. 11 (3), 280–291, Art. no. 2315198.

CHAPTER

# IoT AND SENSOR NETWORKS SECURITY

4

**Gianfranco Cerullo*, Giovanni Mazzeo*, Gaetano Papale*, Bruno Ragucci*, Luigi Sgaglione***

*University of Naples "Parthenope", Naples, Italy**

## 1 INTRODUCTION

Nowadays, over seven billion users in the world are connected to the Internet for sending and reading emails, browsing Web pages, accessing e-commerce services, playing games, and sharing experience on social media. The wide-scale diffusion of the Internet has been the driving force of an emerging trend, the use of such global communication infrastructure to enable machines and smart objects to communicate, cooperate, and make decisions on real-world situations. This promising paradigm is known as the "Internet of Things" (IoT) and its evolution goes hand-in-hand with the progress of the supporting technologies addressed to this new vision of the wireless communication scenario, such as RFID, NFC, wearable sensors, Wireless Sensor Networks (WSNs), actuators, and Machine-to-Machine(M2M) devices. The term IoT was used for the first time in the late 1990s by the entrepreneur Kevin Ashton, one of the founders of the Auto-ID Center at MIT, referring to the linking of objects to the Internet through RFID tags (Madakam et al., 2015). A comprehensive definition of the IoT has been provided by a recommendation of International Telecommunication Union—Telecommunication Standardization Bureau (ITU-T). It describes the IoT as a global infrastructure for the information society, enabling advanced services by interconnecting (physical and virtual) things based on existing and evolving information and communication technologies (ITU-T, 2012). Even today, despite the ITU-T Recommendation, the scientific community, standardization entities, and businesses do not in agree on the real meaning of the "Internet of Things" term. The reason under this divergence is that the IoT combines in a single paradigm three different visions, namely Things-oriented, Internet-oriented and Semantic-oriented (Fig. 1). The Things-oriented vision points out that every object can be monitored using pervasive technologies. The Internet-oriented vision emphasizes the ability of the objects to interact among them through the network, making them "smart." The last vision (i.e., Semantic-oriented) is related to the huge amount of heterogeneous data collected by several available sensors. This data needs to be processed to overcome the interoperability issues and the use of semantic technologies is necessary. Consequently, the researchers have a different perspective of the IoT depending on the vision they approach this technology from, their objectives, and knowledge. Also, the fusion in a single paradigm of more visions leads to face various challenges, including the

Security and Resilience in Intelligent Data-Centric Systems and Communication Networks. https://doi.org/10.1016/B978-0-12-811373-8.00004-5

Copyright © 2018 Elsevier Inc. All rights reserved.

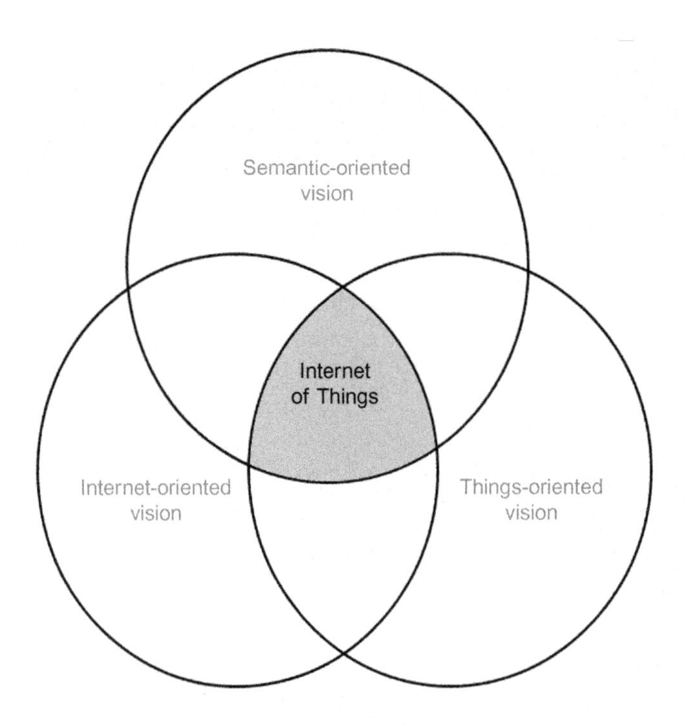


**FIG. 1**

IoT as convergence of three different visions.

need of sensors with high computational features and low energy consumption, the development of methods to prevent attacks against IoT platforms, and the protection of user privacy. The paper aims at providing the reader with a global view of the Internet of Things, showing the advantages and the risks that the adoption of this technology brings with it. The work is organized as follows. In Section 2, the key elements and an overall architecture of the IoT are presented. Section 3 describes the main application domains and examples of real use of the IoT paradigm. The IoT security and privacy risks are the subject of Sections 4 and 5. Device constraints and attacks against one of the key IoT-enable technology, such as the Wireless Sensor Networks, are shown in Sections 6 and 7. Finally, Section 8 provides a conclusion and future trends.

## 2 IoT ELEMENTS AND ARCHITECTURE

This section will introduce the main elements needed to deliver the functionality of the IoT and will describe a five-layer architecture that summarizes the operations performed by this technology.

### 2.1 IoT ELEMENTS

The functionality of the IoT can be explained through the following crucial elements: Identification, Sensing, Communication, Computation, and Services.

1. *Identification*
   In an IoT application the need of interconnection and iteration among objects, including physical devices, requires development of technologies and techniques for uniquely identifying various things over the network. Identification aims at providing a clear and unambiguous identity for each device within the network. To this end, a wide set of identification methods are available, and they can be classified in three main classes:

   - **Object Identifiers** are used to uniquely identify physical and virtual devices. Electronic Product Codes (EPC), Universal Product Codes (UPC), Universally Unique Identifiers (UUID), and ubiquitous Codes (uCode) belong to this category.
   - **Communication Identifiers** are used to uniquely identify devices during the communication with other devices, including Internet-based communications. Classic examples of communication identifiers are the IPv4 and IPv6 addresses.
   - **Application Identifiers** are methods to represent Application-level IDs. In the scope of the IoT, the identification of applications and services involves Uniform Resource Locator (URL) and Uniform Resource Identifier (URI).

2. *Sensing*
   The IoT sensing element deals with two main tasks:

   - acquiring and gathering environmental data from the objects within the network;
   - sending aforementioned data to a database, a cloud, or a data warehouse.

   The processing and analyzing of this information will allow sensors to make decisions and specific action depending on the service. For instance, there are several mobile applications that enable users to monitor and control a great number of smart devices managing every aspect at their home (i.e., comfort, security, and energy consumption).
3. *Communication*
   In an IoT application the sensors, or more in general the nodes, must interact and communicate over a medium (i.e., the radio frequency channel) affected by noise, losses, and security issues. The role of communication technologies is to enable the connection and communication of heterogeneous devices to provide a reliable, fast, and secure smart service. A list of communication technologies adopted in the IoT scenario includes: Radio Frequency Identification (RFID), Near Field Communication (NFC), Wi-Fi, Bluetooth and its low-energy version, and ZigBee. For a more comprehensive description of the features of these communication technologies (i.e., operating band, throughput, range of transmission, and encryption/security mechanisms), refer to Section 5.
4. *Computation*
   The computational capability of the IoT is represented by the processing units like CPUs, microcontrollers, SOCs, and software applications. On the market, several hardware platforms were designed to work with IoT applications. Some examples of hardware platforms IoT-dedicated are: Friendly ARM, Intel Galileo, Arduino, and Raspberry PI. Similarly, many software suites were developed to provide IoT functionality. In this context, Operating Systems are fundamental since they run for the entire lifecycle of a device. In particular, Real Time Operating Systems (RTOS) are valuable candidates in the scope of IoT. Another important role for the computational task of IoT is performed by the Cloud platforms. These ones provide a set of services such as

features to facilitate the transfer of data from a device to the cloud, for real-time processing of the data, and for the user to benefit of the information extracted from the collected data.

5. *Services*
   A great number of applications that can use the IoT multiplies the possible services targeted by this paradigm. A classification of the IoT services in four different categories is presented in Gigli and Koo (2011) and Xiaojiang et al. (2010). The services included in the first class, known as *Identity-related services*, are used from all IoT applications. In fact, they aim at identifying the real-world objects to bring them to the virtual world. Therefore an Identity-related service consists of the things equipped with a means of identification (i.e., RFIDs, NFC), and a device that reads the identity of the equipped object. *Information Aggregation services* is concerned with the process of acquired data from the heterogeneous sensors located into the environment. Parsing and normalization are typical operations performed on the acquiring raw data. Moreover, these services are responsible for transmitting and reporting data through the network infrastructure to the IoT application. The *Collaborative-aware services* leverage on the aggregated data obtained from the previous services to make decisions and perform reactions. The effectiveness of these services relies both on the reliability/speed of IoT infrastructure and the computational capability of the devices. The last class of IoT services is called *Ubiquitous services*. Their goal can be summarized by three words: anytime, anyone, anywhere. An ubiquitous service must provide a Collaborative-Aware service "anytime" that it is needed, to "anyone" that requests it, and "anywhere" on the IoT infrastructure. In the IoT applications, reaching the provided level of ubiquitous services is a great challenge due to the number of issues to be addressed. In particular, the presence of different technologies and protocols represents a barrier to the ubiquity.

## 2.2 IoT ARCHITECTURE

The main task of the IoT is to interconnect a plethora of heterogeneous objects through the Internet. These objects will produce as much larger traffic on the Network as more stored data is needed. Therefore an effective architecture for the IoT needs to address many aspects like scalability, interoperability, QoS, privacy, and security. Today, even though several projects and researches have tried to design a common architecture for IoT, a reference model has not been reached yet. Tree-layers, five-layers, Middleware, and SOA-based architecture are only some of the models proposed in Wu et al. (2010), Anithaa et al. (2016), and Chaqfeh and Mohamed (2012). In Fig. 2, a general architecture for the IoT is shown. It is structured in five layers: Objects Layer, Object Abstraction Layer, Service Management Layer, Application Layer, and Business Layer. The features of each layer are presented below:

1. *Objects layer*
   The Object layer represents the five sense organs of IoT and it is responsible for the acquisition and collection of data from the physical world. It is also known as "Device Layer," in fact it consists of the physical objects and sensor devices. This layer involves sensors and actuators to perform different operations, such as retrieving environmental information like temperature, wind speed, vibration, and weight. It requires plug-and-play mechanisms to configure the heterogeneous objects and transfer the data toward the next layers.
2. *Object Abstraction layer*
   This layer aims at transmitting and processing the information gathered from the Object layer through secure channels. Depending on the sensor devices, the transmission can be wired or wireless, while the technologies can be UMTS, ZigBee, Wi-Fi, or 3-4 G.

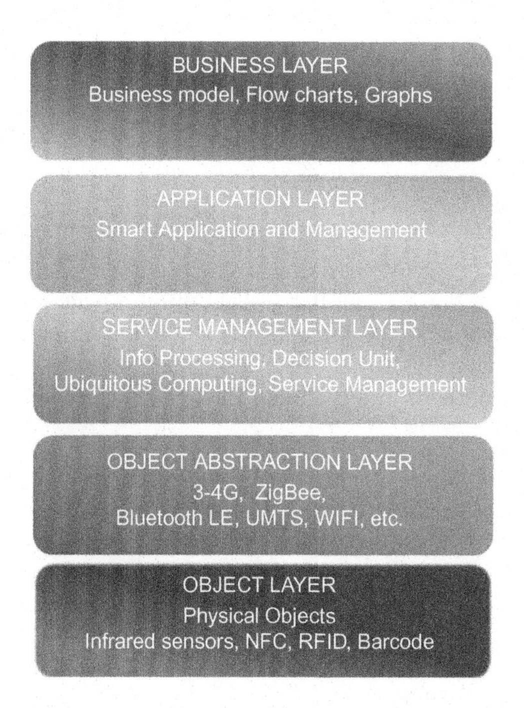


**FIG. 2**

IoT 5-layer architecture.

3. *Service Management layer*
   The Service Management layer processes data received from previous steps, makes decisions, and delivers the required services over the network. The devices involved in the IoT implement different services and, the Service Management layer enables the communication only with the devices that implement the same service. Also, this layer allows to the IoT developers to work with heterogeneous objects independently from the specific hardware platform (Khan et al., 2012; Atzori et al., 2010).
4. *Application layer*
   The Application layer implements the ability of the IoT to provide high-quality smart services to meet the users' requests. This level represents the convergence of IoT and industrial technologies to realize the intellectualization. Smart home, smart city, and automotives are only a small part of the market that this layer is able to cover. Data processing and services providing are the major purposes of the Application layer.
5. *Business layer*
   The Business layer can be seen as the coordinator of the IoT system activities including applications and services. It receives the data from the Application layer and produces a business model, flow charts, and graphs. This layer provides support to the decision-making process through Big Data analysis. It also evaluates the output from all the other four layers and compares

them with the expected output in order to enhance the service quality. The aim of Business layer is to identify future actions and business strategy. In this sense, the Business layer is fundamental for the real success of IoT technology.

## 3 IoT APPLICATION DOMAINS

The Internet-oriented vision of the IoT and the availability of a wide range of sensors and devices of different sizes and computational capability makes the IoT a prominent paradigm for enabling innovation in various application fields. The main application domains (Fig. 3) of the IoT solution can be grouped as follows: Infrastructure/Building, Environment, Industrial, Health, Transport, and Security.

1. *Infrastructure/Building domain*
   IoT applications are currently used in many infrastructure scenarios, such as smart homes and buildings, and smart cities. A smart home can be defined as a home-like environment that possesses ambient intelligence and automatic control able to meet the user needs in term of

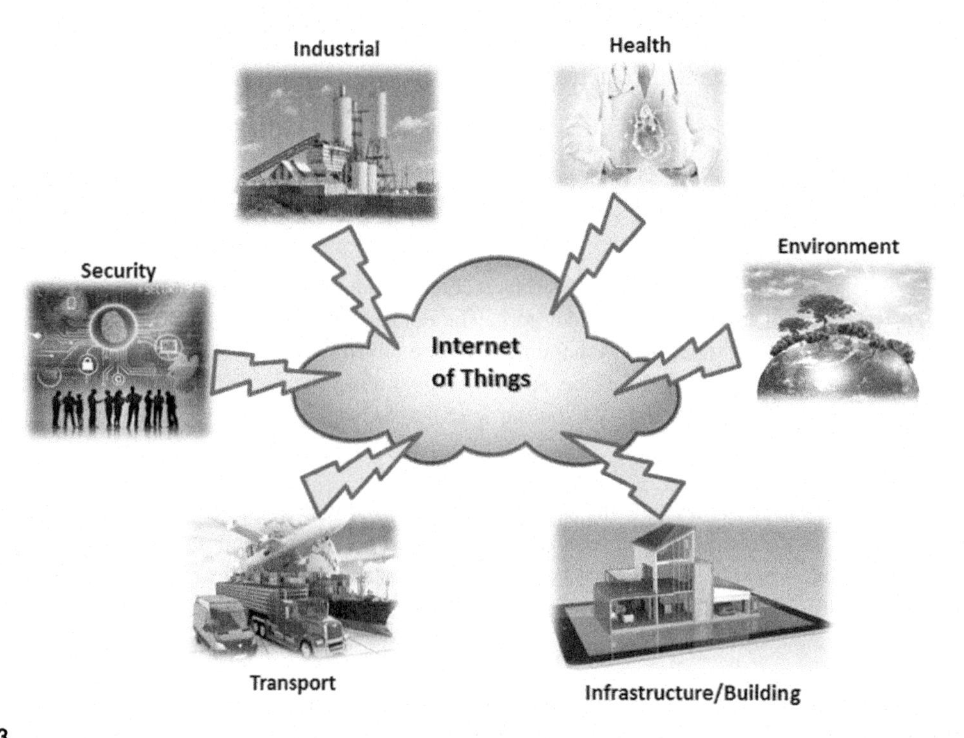


**FIG. 3**

IoT application domains.

comfort, physical integrity, and rational management of energy consumption (De Silva et al., 2012). In a smart home the high number of different subsystems and sensors used to monitor the resource consumption and to detect current user's needs, requires a high level of standardization to ensure interoperability (Miorandi et al., 2012). For instance in Wang et al. (2014), the physical integrity of people is provided through a fall detection system for elderly person monitoring, which is based on smart sensors worn on the body and operated through consumer home networks.
A smart city can be seen as an extension of a smart home where the role of "home" involves the physical city infrastructure (i.e., the water network, public lighting, and road network). Some services provided by a smart city range from the monitoring of highway traffic, the smart management of a parking (i.e., RFID or other technologies can allow the monitoring of available parking space) to the control of fine dust in the air.

2. *Environment domain*
The Environment domain is another significant field of application of IoT technology. These applications allow the monitoring of natural phenomena and processes (i.e., rainfall, river height, and ocean temperature). They use the process of miniaturization of modern devices and their self-management ability to reach critical areas for the human operators. The goal of these IoT application is to mitigate damage and, in general, reduce the level of risk for people and things in the case of a natural disaster. In Kitagami et al. (2016) a distributed cooperative IoT system for flood disaster prevention is proposed. This system provides both real-time flood alert and flood impact analysis. Another interesting situation in which the sensing ability of IoT can be used within an environment scenario is reported in Alphonsa and Ravi (2016). In this case an earthquake early warning system based on WSNs is presented. In particular a ZigBee transmitter placed over the earth surface detects the earth motions and transfers the alert signal to the gateway. The gateway is equipped with a ZigBee receiver and transfers the warning to a smartphone. The messages are shown to the user as location, time, and other parameters. Clearly, for saving humans, animals, and vegetation, an effective IoT platform needs real-time information processing ability and a large number of devices placed in environmentally-strategic points.

3. *Industrial domain*
Typically, the term industrial IoT refers to the use of IoT technology for the management of the manufacturing process. In this context, the IoT leverages machine-to-machine (M2M) communication and automation technologies implemented for years in the industry. Data acquired from sensors, for instance RFID tags placed on the items, enable enterprises to solve in fast and efficient way any issues during the supply and delivery chain process, saving time and money. Significant is the use of IoT solutions in the food industry where the production process requires strict working conditions, such as environments at temperature/humidity controlled and timed working processes. Another application of IoT solution in the industrial scope refers to the retail market. For instance, identification technologies can support the fighting of counterfeiting through the tracking of entire life-cycle of a products including information about its location and state during the transporting process.

4. *Health domain*
The Internet of Medical Things (IoMT) or healthcare IoT is one of the most attractive application fields. In the health domain the IoT plays a prominent role in several situations, such as clinical care and remote monitoring. In the first case, a hospitalized patient who suffers from a chronic disease can be monitored in real time through body sensors. These sensors wirelessly send

physiological information to the cloud where they are analyzed and stored. Hospital healthcare professionals can access this data for prompt action in cases of necessity or for further investigations. The adoption of the IoT-driven paradigm for clinical care improves the quality of care and reduces the overall cost for the hospital infrastructure. Remote monitoring allows people to use mobile medical devices from their home and to send health information to medical professionals. Sensors with wireless communication capabilities, such as blood pressure meters, glucose meters, and heart rate monitors, can send the collected data to a specialist using a specific software installed on the computer of the patient, on his smartphone, or through a telehealth system. Many platforms for remote health monitoring have been developed. In Prakash et al. (2016) a real-time remote monitoring human vital system is presented. This system makes use of IoT technologies, GSM connectivity, and an SMTP server for measuring pulse rate and body temperature of patient and sending emergency signals and messages.

5. *Transport domain*
   In the transport domain, the IoT finds large application in the so called Intelligent Transportation Systems (ITS). An EU directive defines the ITS as a system in which information and communication technologies are applied in the field of road transport, including infrastructure, vehicles and users, and in traffic management and mobility management, as well as for interfaces with other modes of transportation (EPC, 2010). In this context the integration of Cloud and IoT technologies is widely used in Vehicle to Vehicle (V2V) and Vehicle to Infrastructure (V2I) communications. V2V and V2I communications enable the wireless transmission of data between vehicles and roadside infrastructure. They make more effective the automotive embedded systems (i.e., cruise control, parking assistant, collision detection, and onboard cameras) both reporting alarms to the driver and taking physical control of the vehicle. From 2009 to 2012, Google has tested its self-driving technology on real vehicles for over 300,000 miles of testing in freeways. Later, the testing scenario has shifted on city streets. Finally, at the end of 2014, Google delivered its real prototype (Google, 2016).
6. *Security*
   The use of IoT-enabled technologies finds a great application for securing Critical Infrastructure, such as an energy production infrastructure, a stadium, a harbor, a bank, or another public ambient. Sensors monitoring can provide information about the presence of people and their behavior or detect dangerous substances in the air. RFID or other identification technologies can detect if a person is authorized to stay within a restricted area. Sensors can monitor the temperature of industrial equipment to check that it is working in safety condition both for the human operator and surrounding environment. Of course, the more information sources there are, the more effective the warning system will be. On the other hand, an excessively invasive use of such technologies may give rise to privacy infringements. For more details see Section 4.

## 4 SECURITY, SAFETY, AND PRIVACY ASPECTS

The study "Disruptive Civil Technologies: Six Technologies with Potential Impacts on US Interests out to 2025" carried out by The National Intelligence Council (2008), classified the Internet of Things among the top six "disruptive" technologies. Where "disruptive technology" means a technology with the potential to cause a significant—albeit temporary—degradation or improvement of one of the elements of US national power (political, military, economic, or social cohesion). Individuals, businesses, and governments were not prepared for a possible future in which Internet nodes reside in everyday things such as in the food packages, in furniture, and in paper documents. The IoT objects will be affected by the same security problems as the current systems connected in the network (as the underlying technologies are basically the same), but the risks associated will be far superior to those of current Internet network. There are two main reasons behind these issues:

1. They have a widespread distribution (which is likely to rise).
2. They control—and increasingly so—the physical world (while traditionally systems controlled only logical world).

In addition, threats to privacy are huge, as well as the possibility of social control and political manipulation. On November 2008, the Council conclusions on the Future Networks and Internet recognized that "the Internet of Things is poised to develop and give rise to important possibilities for developing new services but that it also represents risks in terms of the protection of individual privacy." An article published on Forbes in January 2014 (Steinberg, 2014) listed many devices connected to the Internet that can already "spy people in their homes," including televisions, kitchen appliances, cameras, and thermostats. Finally, an important aspect (perhaps the main aspect) of the IoT security is that in this domain the security problems have a direct impact on system safety. In all IoT objects, security is inseparable from safety, interferences (accidental or intentional) with the commands of a pacemaker, a car, or a nuclear reactor, and the short circuit of a smart plug are threats to human life. The x-by-wire systems in the automotive sector (computer-controlled devices in automobiles such as brakes, engine locks, horn, and the dashboard) have proven to be vulnerable to attacks that have access to the on-board network (e.g., if the network is overloaded or disrupted by a cyberattack the driven would be not able to steer the vehicle or use the brake) (Abramowitz and Stegun, 2015). Therefore many concerns have been raised about whether the IoT is developing too quickly, and without adequate consideration for the safety issues and regulatory changes that may be necessary. This perception of (in)security is pervasive: according to the Business Insider Intelligence Survey conducted in the last quarter of 2014, 39% of interviewees claimed that security is the most pressing problem in the IoT adoption (Insider, 2015). To address the growing concerns in terms of safety, on September 23, 2015 the Internet of Things Security Foundation (IoTSF) was launched. IoTSF has the mission to ensure the IoT security through the promotion of knowledge and "best practices." The founding members of the foundation are various ICT technology providers and telecommunications companies, including BT, Vodafone, and Imagination Technologies (IOTS, 2015b). IoTSF, considering that "The economic impact of the Internet of Things will be measured in trillions of dollars, the number of connected devices will be measured in billions, and the resultant benefits of a connected society are significant, disruptive, and transformational," has published a guide "Establishing Principles for Internet of Things Security" (IOTS, 2015a). The guide looks at questions that need to be considered when designing an IoT device, system, or network. A common theme is that investment into security at the design phase of IoT can save a lot of time, effort, and potential embarrassments. In this way, the developers can produce products with better security, and enhanced value, quality, and usability. These products will then be able to form part of a safe, secure, scalable, manageable, and transformative Internet of Things. Example of addressed questions are:

1. Does the data need to be private? The answer is yes if sensitive data is collected. "It is essential that this data is adequately protected at all times, and that the user is aware of what private data is being processed."
2. Does the data need to be trusted? Yes, "data may need to be protected from tampering and modification in transit. This may be a malicious attacker or simply poorly configured devices mishandling data."
3. Is the safe and/or timely arrival of data important? Yes, if a delayed of blocked data can have an impact on the provided service.
4. Is it necessary to restrict access to or control of the device? Yes, "prevention of unauthorized access or control is vital to secure devices. If an attacker gains control of the device he/she may be able to access sensitive data, or cause problems elsewhere in the network."
5. Is it necessary to update the software on the device? Yes, "If a device is running out-of-date software, it may contain unpatched security vulnerabilities. Such vulnerabilities may allow exploitation of the device and its data by attackers."

From a technical point of view the following security requirements must be addressed to allow a secure adoption of the IoT technologies:

1. Confidentiality—ensures that the content of the message being transferred is never disclosed to unauthorized entities. Sensor nodes may communicate highly sensitive data, such as key distribution, so it is extremely important to build a secure channel in a WSN.
2. Integrity—ensures that a message is never corrupted or modified during its "life" by unauthorized actor, this means also that an unauthorized modification must be detected if occurred.
3. Availability—ensures that the necessary functionalities or the services provided by the WSN are always carried out, even in the case of attacks/malfunctions.
4. Authentication—allows a receiver to verify that the data really is sent by the claimed sender.
5. Data Freshness—ensures that the data is recent and that no old messages have been replayed. This requirement is especially important when there are shared-key strategies employed in the design that need to be changed over time.
6. Self-Organization—allows to each WSN node to self-organizing and self-healing according to different situations, in this way the WSN network can adapt the topology and deployment strategy in case of need.
7. Time Synchronization—allows a common time reference. Most of the services in WSN require time synchronization and also security mechanism for WSN should be synchronized.
8. Secure Localization—allows the accurate and automatic location of each sensor in the network. This is useful for example to identify faults or to discover a sinkhole attack.
9. Auditing—ensures the possibility to investigate on facts happened on the WSN, this is possible only if each node provides auditing facilities (in the simplest case providing logs).
10. Secrecy—allows the secrecy of the network regarding nodes that leave it and also the secrecy of old messages regarding new nodes that join the network.
11. Nonrepudiation—implicates that a sensor cannot refute the reception of a message from the other involving party or the forward of a message to the other involving party in the communication.

## 5 ENABLING TECHNOLOGIES

The role of each IoT sensing node is to communicate information/command gathered and processed by local embedded processing features to the identified destinations. As described in Section 3, use cases are very heterogeneous, but they have one thing in common, namely that only few kilobytes of data for any node have to be carried (excluding high-bandwidth image processing or video data services). So, the communication techniques have evolved to overcome this new scenario. Many techniques have been implemented and they can be classified into two main families:

1. Wired This family uses a physical cable to connect all nodes of the network. In particular, two subfamilies can be identified: Existing wiring, that uses existing cables (like power, coaxial, and telephone wiring) to allow the communication (examples are: PowerLine and X10) and Structured wiring, that requires the installation of new cables (Ethernet).
2. Wireless This family identifies the protocols that do not require of a physical communication cable (example are: ZigBee, Thread, and Z-Wave).

In the follow a brief description of many of these protocols is provided.

- HomePNA (Phone line) (Existing wiring) (HomePNA, 2013)
  HomePNA is based on the frequency-division multiplexing (FDM), which uses different frequencies for voice and data on the same wire (phone line or coaxial cable) without interfering with each other. A standard phone line supports voice, high-speed DSL, and a landline phone. A HomePNA setup would include a HomePNA card or external adapter for each computer, an external adapter, cables, and software.
- X10 (Existing wiring)
  X10 is a protocol developed in 1975 as a one-way system where the communications were from controls to actuators. Subsequently it was updated, taking the name of x10-Pro, with the addition of the messages from the actuators to the controls. This protocol uses the conveyed waves for transmission and it uses the existing power line for the transmission. The transmitted packets are composed of 4 bits to indicate a house code, followed by one or more groups of 4 bits to designate the unit code, and to close always a 4 command bits. To make the process more "user friendly" the house code is expressed with a letter from "A" to "P" while the unit code by a number from "1" to "16." Each device is configured with one of 256 possible addresses and reacts to commands specifically sent to it, or even sent in broadcast. It is also allowed to target multiple units before giving the command.
- PowerLine (Existing wiring)
  PowerLine is a technology for the transmission of voice or data that uses the existing power supply network as transmission channel. It is also known as power-line carrier, power-line digital subscriber line (PDSL), mains communication, power-line telecommunications, or power-line networking (PLN). The power line is transformed into a data line via the superposition of a low-energy information signal to the power wave. Data is transmitted at least 3 kHz to ensure that the power wave (electricity is 50 or 60 Hz) does not interfere with the data signal (Semiconductor, 2011). The separation of the two types of current takes place with the filtering of the frequencies used. All components of the system are connected in parallel with each other and communicate through the electric network by means of suitable devices. The system is, however, limited by a low speed of transmission and by nonnegligible interference. Furthermore, problems for the

quality of the connection will still depend on the quality of the domestic electrical system. Improper wiring and circuit breakers in between the connected cables can negatively affect the performance and can cause connection interruptions.

- Ethernet (Structured Wiring)
  Ethernet was introduced in 1980 and standardized in 1983 as IEEE 802.3 (IEEE, 2013). It is a family of computer networking technologies that differs in the type of cable (from coaxial to optical fiber) and in maximum speed (3Mb/s to 100 Gbit/s). The original 10BASE5 Ethernet uses coaxial cable as a shared medium, while the newer Ethernet variants use twisted pair and fiber optic links. All computers on an Ethernet network are connected to the same transmission line, and the communication is done through a protocol called CSMA/CD (Carrier Sense Multiple Access with Collision Detection). CSMA/CD is a multiple access protocol with carrier surveillance (carrier Sense) and collision detection. A data stream is divided into shorter pieces called frames. Each frame contains source and destination addresses, and error-checking data so that the damaged frames can be detected, discarded, and retransmitted.
- ZigBee (Wireless) (Alliance, 2014)
  ZigBee is one of the most used two-way communication standard in the Wireless Personal Area Network (WPAN). It was introduced in 1998, standardized in 2003, and revised in 2006. It has been designed to be simpler and less expensive than other wireless personal area networks such as Bluetooth or Wi-Fi and it is based on the IEEE 802.15.4 standard. ZigBee devices are of three kinds:

  1. ZigBee Coordinator (ZC): It is the most capable device, the ZC forms the root of the network tree and might bridge to other networks. There is one ZigBee Coordinator in each network.
  2. ZigBee Router (ZR): A Router can act as an intermediate router, passing on data from other devices.
  3. ZigBee End Device (ZED): Contains only functionality to talk to the parent node (Coordinator or Router) and it cannot pass data from other devices. This limitation allows the node to be in a sleep mode for a significant amount of the time, thereby giving rise to long battery life.

  The supported network topologies are star, tree, and generic mesh networking.

- Z-Wave (Wireless) (Alliance, 2016)
  Z-Wave was initially developed in 2001 by Danish start-up Zen-Sys, with time it has become an international standard for the creation of interoperable and low-power wireless mesh networks. The protocol supports two-way communication between enabled devices, allowing products from different manufacturers to work together. Z-Wave uses a reduced data flow for design choice. This choice allows to obtain a low-latency communication with a data transmission rate up to 100 kbps. Z-Wave has been a proprietary protocol until September 2016 when the protocol has been released to the public (LWN, 2016).
- 6LoWPAN (Wireless) (IETF, 2007)
  6LoWPAN stands for IPv6 over Low-power Wireless Personal Area Networks. The concept behind 6LoWPAN is born from the idea that the Internet protocol could and should also be applied to smaller devices, and that the low-power devices with limited processing capacity should be able to participate in the Internet of Things. The 6LoWPAN defines encapsulation and compression

mechanisms that allow IPv6 packets to be sent and received on IEEE 802.15.4 networks (low-rate wireless personal area networks (LR-WPANs)).

- Thread (Wireless) (Thread-group, 2016)
  Thread is an IP-based IPv6 networking protocol (closed-documentation) aimed at the home automation environment. Launched in 2014 the protocol is based on different standards including IEEE802.15, IPv6 and 6LoWPAN, and offers a resilient IP-based solution for the IoT. Thread supports a mesh network using IEEE802.15.4 radio transceivers and is capable of handling up to 250 nodes with high levels of authentication and encryption.
- Bluetooth (Wireless) (Bluetooth-SIG, 2016b)
  Bluetooth is a short-range communications technology that has become very important in many markets of consumer products. It is the key technology for wearable products. In particular, a new version of the protocol, namely Bluetooth Low-Energy (BLE) (Bluetooth-SIG, 2016a) or Bluetooth Smart, has become one of the reference points for IoT applications. This version offers a range of transmission similar to the previous one but with a greatly reduced consumption of energy (key peculiarities in IoT environment).
- Wi-Fi (Wireless) (IEEE, 2016)
  WiFi is often an obvious choice for many developers, especially given the pervasiveness of WiFi within the home environment. The most commonly used WiFi standard is the 802.11n, which offers a throughput of hundreds of megabits per second, but that might be too exorbitant in terms of energy consumed in a context of the IoT.
- Cellular (GSM/3G/4G) (Wireless)
  This technology is the ideal solution for sensor-based low-bandwidth-data projects that will send very low amounts of data over long distances using the Internet. Cost and power consumption are too high for many applications, so the sensors need a constant power source or must be able to be recharged regularly.
- NFC (Wireless)
  Near Field Communication is a technology for two-way communications at shot distance (less than 4 cm) especially applicable for electronic badge and smartphones, allowing consumers to perform contactless payments, connect electronic devices, and access to specific areas.

Others important protocols are: Beacon, MQTT, WeMo, Sigfox, Neul, and LoRaWAN. A comparison between these technologies is showed in Table 1.

## 6 DEVICE CONSTRAINTS

Due to the low cost and its flexibility, Wireless Sensors Network (WSN) is become the most used sensors network in many domains such as Smart Home. More attention in traditional networks is posed on maximizing channel throughput, while in a sensor network the focus is on the extension of the system lifetime as well as the system robustness. Sensors in this network are characterized by limited energy and reduced physical size, consequently they have limited processing capability, very low storage capacity, and limited communication bandwidth. These characteristics pose also many limits in the functionalities of the sensors and must be taken in consideration also in the adoption of the

**Table 1 IoT Protocols Comparison**

| | Ethernet | HomePNA | PowerLine | Bluetooth LE | ZigBee | Z-Wave | Thread | WiFi | Cellular | NFC |
|---|---|---|---|---|---|---|---|---|---|---|
| *Type* | Structured Wiring | Existing wiring | | Wireless | | | | | | |
| *Freq.* | 600 MHz | 4–10 MHz | N/A | 2.4 GHz | 2.4 GHz | 900 MHz | 2.4 GHz | 2.4–5 GHz | 900–2100 Mhz | 13.56 MHz |
| *Max Range* | 100 m | 100 m–1 km | 300 m | 50–150 m | 10–100 m | 30 m | N/A | 1000 m | 200 km | 10 cm |
| *Max Data Rate* | 100 Gbs | 320 Mbps | +1 Gbps | 1 Mbps | 250 kbps | 100 kbps | N/A | 100 Gbps | 10 Mbps | 420 kbps |
| *Best use* | New builds, renovation | Stationary devices | | Mobile devices | | | | | | |
| *Cost* | High | Low | Low | Low | Low | Low | Low | Low | Low | Low |

security mechanisms (standard security mechanisms have to be adapted to these limitations). The main technical challenges that must be addressed are:

1. Unattended operation: In most cases, once deployed in remote regions, sensor networks have no continuous presence of a human, so sensors must identify its connectivity and distribution, accumulate data, and wait for the requests. In these cases, the sensor nodes themselves are responsible for reconfiguration in case of any changes. Furthermore, in unattended environments, the likelihood of a physical attack is very high and consequently the guarantee of the node security is not a simple task.
2. Unreliable communication: The communication in the sensor networks is based, typically, on connectionless protocols and is characterized by a broadcast forwarding of the messages. These features imply possible errors/drops/collisions on the communication packets and hence it may need retransmission.
3. Dynamic changes: Sensor network must be adaptable to changes. For example, a WSN should adapt its routing algorithm in case of new nodes or failures in the network.
4. Time synchronization: It is very important for basic communication and for security mechanism, furthermore, it provides the ability to detect movement, location, and proximity. There are many solutions for the classical networks, but the traditional synchronization techniques are not suitable for sensor networks because they do not consider the partitioning of the network and message delay (Youn, 2013).
5. Memory limitations: Memory in a sensor node is very limited, it typically includes a flash memory used to store the application code and a RAM used for storing application programs and data. The RAM size of current commercial/research sensor nodes is between 0.5 Kb of FemtoNode and 128Mb of RecoNode (Wikipedia, 2016). A dedicated data storage is usually available only via an external slot.
6. Energy limitation: The energy constraint is also the most stringent one. Sensors are tiny devices powered by a battery and hence any operation must consider the limited capacity of the battery. The communication is the most important, as well as the most energy consuming operation of a sensor. Laboratory tests have proved that the transmission of one bit in a WSN consumes about as much power as executing thousand of instructions (Hill et al., 2000; Akyildiz et al., 2002; Pottie and Kaiser, 2000). The energy limitation has a direct impact also on the sensor security, in fact strong security levels require cryptographic techniques that are computationally expensive and hence entails high power consumption.

## 7 ATTACKS

The WSN attacks can be classified in two categories, passive and active. A passive attack does not interrupt the communication, but the attacker listens and analyses the exchanged data. This kind of attack is not easy to detect and very simple to implement. The active attacks aim at spoiling and destroying the network functionality, removing or modifying the information(messages) transmitted over the network. The attackers can replay old messages, inject new ones or modify them, changing the standard network operations. A further classification is related to the Internet model, in this case one or more attacks can be assimilated (associated) at each layer. In Table 2 a list of layer attacks and some possible countermeasures are shown.

**Table 2 WSN Attacks and Countermeasures**

| Layer | Attacks | Countermeasures |
|---|---|---|
| *Physical Layer* | Jamming | Low transmission power |
| | Eavesdropping | Directional antennas |
| | Tampering | Channel surfing |
| | | Hidden nodes |
| *Link Layer* | Collision | Correcting codes |
| | Exhaustion | Reducing MAC admission control |
| | | Limiting transmission time slot |
| *Network Layer* | Selective forwarding | Multipath routing |
| | Sinkhole | Watchdog |
| | Sybil | Information flow monitoring |
| | Routing cycles | CPU usage monitoring |
| | Wormhole | Unique shared key |
| | Hello flood | Trusted certification |
| | Acknowledgment spoofing | Sinkhole countermeasures |
| | | Synchronized clock |
| | | 4 way handshaking |
| | | Identity verification protocol |
| | | Alternative routes |
| *Transport Layer* | Flooding | Client-Puzzles |
| | De-synchronization | Packet authentication |
| *Application Layer* | Data corruption | Application firewall |
| | Repudiation | Antivirus |
| | | Spyware |
| | | Accounting services |
| | | Packet filtering |

## 7.1 PHYSICAL LAYER ATTACKS

In a WSN the medium is the free space, the communication is broadcast and the signal could be easily intercepted or disrupted. Therefore, an attacker could physically destroy or manipulate the sensors of the WSN. Jamming, Eavesdropping and Tampering are the physical layer attacks described in this section.

1. Jamming
   Jamming is defined as the act of intentionally directing electromagnetic energy toward a communication system to disrupt or prevent the transmission (Adamy, 2004). The most used signals to perform the Jamming attack are the dom noise and pulse one. The Jamming attack consists in disturbing the radio channel, in fact the attacker sends useless information to corrupt or lost the message. This type of attack can be led from a remote location to the target network, and also the equipment useful for the attack purpose is easily available. A Jamming attack can be

viewed as a special case of Denial of Service (DoS) attack, and it can be temporary, intermittent, or permanent. Some jamming techniques (Spot Jamming, Sweep Jamming, Barrage Jamming, and Deceptive Jamming) and jamming types (Constant Jammer, Deceptive Jammer, Random Jammer, and Reactive Jammer) are described in Mpitziopoulos et al. (2009) and Uke et al. (2013). Possible solutions to defend the WSNs by the several jamming attacks involving the use of: low transmission power, directional antennas, frequency hopping spread spectrum, and channel surfing (Vadlamani et al., 2016).

2. Eavesdropping
   Eavesdropping attacks allow messages to be intercepted and read by the malicious operator, who can inject fake messages into the network. This kind of attack uses of the radio frequency (RF) channel and the broadcast nature of the wireless communications. The attacker can easily intercept the broadcast signals through an RF receiver opportunely tuned on the frequency of the transmitted channel. Access control, reduction in sensed data details, distributed processing, access restriction, and strong encryption techniques (Mohammadi and Jadidoleslamy, 2011) are some countermeasures to the Eavesdropping.
3. Tampering
   Tampering is the physical access to the sensors performed by the attacker. In this way he/she can obtain information like encryption/decryption keys or other sensitive data. The nodes could be hidden in strategical positions to avoid the simple access by an attacker. Therefore the sensors could be located in an environment exposed to adverse conditions and to prevent physical damaging, the sensors could be made with resistant materials.

## 7.2 LINK LAYER ATTACKS

The link layer protocols allow to manage the connection of the neighbor nodes over the WSN, the attacks can interrupt, damage, or disrupt the communication among them. Below are listed some link layer attacks against the WSNs.

1. Collisions
   The attacks related to this level aim at interfering in the communication channel and replicating/altering data frames. Collision attacks are performed through an interfering signal. The attacker listens to the transmitting frequency of the WSN and forwards its own signal, it causes the collision, and the receiver obtains an incorrect message. The collision of a single byte can create a CRC (Cyclic Redundancy Check) error and cripple the message received. The collision attack can be identified when the receiver reads the incorrect messages. Jamming countermeasures and error correcting codes are some possible defensive methods. Communication overheads and additional processing are the disadvantages using the error correcting code (Fatema and Brad, 2014).
2. Exhaustion
   Similar to the collisions, this attack exploits the checksum fault of the forwarded messages exhausting the resources. The attacker consumes the energy of the victim nodes forcing them to retransmit unnecessary data. The first defensive technique reduces the MAC admission control rate (Bartariya and Rastogi, 2016) and another one, limits the time slot in which each node can access the channel.

## 7.3 NETWORK LAYER ATTACKS

The goal of these attacks is to control the traffic over the network (i.e., creating a nonoptimal path, introducing a significant delay), to generate a routing loops or clog the network. The attacker could interpose itself with the destination node preventing the source node to reach it. The main network layer attacks are:

1. Sinkhole
   The Sinkhole attack tries to manage all of the traffic over the network. The attacker provides a very attractive node compared to the other nodes of the network (i.e., a node with a higher power transmission). In this way, the surrounding nodes will forward the data through the harmful ones. The attacker offers a false sink to the nodes to prevent the delivery of the messages to the base station, causing the partial or total damage of the WSN. Typically, a sinkhole attack is launched by a laptop or a PDA, exploiting the nonauthentication of links and their strong power radio transmitter. These features make them a very attractive target over the network. A technique to detect the Sinkhole attack leverages on the network flow of information to find the suspected nodes and refine the interested area. Another strategy consists in the monitoring of the nodes, observing their CPU level usage. Comparing this parameter with a predefined threshold is possible to identify the malicious nodes. Further countermeasures against the Sinkhole attack are presented in Soni et al. (2013).
2. Routing Cycles
   In the routing cycles attack, the attacker tries to make an endless path between source and network nodes. Each message contains the number of the hop necessary to reach the base station, so this type of attack is easily detectable. Also, this kind of attack needs more than one player to have success. The Routing Cycles attack presents the same properties of the Sinkhole attack, therefore can be adopted by similar countermeasures.
3. Selective Forwarding
   The Selective Forwarding attack allows the attacker to disperse a certain amount of messages. It aims at altering and/or losing a part of the messages. The malicious nodes forward the messages they did not want, and block all those who should not be propagated. The attacker chooses what and how many messages are to be sent, limiting the suspicion about its malicious identity. The Black-Hole attack is a simple form of the Selective Forwarding, in this case the malicious nodes will forward no packets, consequently the adjacent nodes could consider them as dead. The multipath routing can be used to protect the network by this type of attack. Watchdogs (Baburajan and Prajapati, 2014), in addition, can be implemented to monitor and control the traffic between the neighbor nodes over the network.
4. Sybil
   In a Sybil attack the malicious node presents multiple false identities and appears to the other nodes as a set of nodes. The victims send the messages to the malicious node, assuming that the forwarding is performed to different nodes. The attacker would like to fill the memory of the adjacent nodes through the forwarding of useless data, while the malicious one sends incorrect information (i.e., position of nodes, signal strengths). The goal of this attack is the partial or total degradation of the service. One possible countermeasure consists in to provide a unique shared key (Karlof and Wagner, 2003) between the node and the base station. The Sybil attack can be fought

with a set of trusted certification authorities, which provide an unique node-Ids to each node (Saha et al., 2010).

5. Wormhole
   The Wormhole is launched by multiple attackers positioned in different strategic points of the network; typically the best places are the ends of the WSNs. The malicious user tunnels the messages in another place, over the network, from where these messages were recorded. The tunnel established between two collude attackers is the wormhole. This attack is a low-latency link between two nodes that can be exploited to launch other kinds of attacks. The wormhole is most dangerous for the WSNs routing protocol. A first solution to detect this attack using synchronized clock (Patel and Aggarwal, 2013), which allows to establish if the message forwarded has took more time to be received. Another solution is based on 4 way handshaking message mechanism.
6. Hello Flood
   The "HELLO" packet, in the routing protocol, allows the discover of nodes over the network, so a Hello Flood attack consists in forwarding of a large amount of this specific message in order to flood the network and thus avoid the exchange of other types of messages. A solution to this attack is the authentication of the neighbor nodes through an identity verification protocol, and checking if the communication for each node is bidirectional (Sharma and Ghose, 2010).
7. Acknowledgement Spoofing
   The malicious operator, in the Acknowledgement Spoofing attack, hides the real status of one or more nodes to the neighbor ones. The attacker spoofs the real condition of the node, convincing the forwarder that a weak link is strong, while a disabled node is enabled. The messages sent through the disabled or dead links will be lost. Authentication, link layer encryption, and global shared key techniques (Singh and Verma, 2011) are the main countermeasures used to defense the WSNs against this kind of attack. A prevention strategy relies on the use of alternative routes to forward the messages.

## 7.4 TRANSPORT LAYER ATTACKS

The Transport layer should ensure the efficiency of the services, such as setting of the end-to-end connection, the flow and traffic control and the delivery reliability of the forwarded messages. The most common attacks for this layer are the SYN Flood and the De-synchronization.

1. Flooding
   In a TCP SYN flood attack the malicious node forwards several connection requests, without completing this operation. The SYN flood is a Denial of Service attack. The goal is the exhaustion of energy and memory of the nodes through the flooding of them. The attacker sends multiple SYN requests to start a TCP connection with a host, but it never will confirm the request. The victim will wait for the ACK packet response, in this way no connection will be provided with others legitimate nodes causing the Denial of Service. The Flooding attack effects can be mitigated by limiting/reducing the connection numbers, due to the resolution of a puzzle. This technique is known as Client Puzzle (Aura et al., 2000).

2. De-synchronization
   An attacker forwards some fake sequence numbers or control flags for de-synchronizing the endpoints and producing the data retransmission. The authentication of each forwarded packet can be the solution to mitigate this attack (Zhao, 2012).

### 7.5 APPLICATION LAYER ATTACKS

Application layer contains the user data. This layer works with some protocols, such as HTTP, FTP and others, which represent an attractive target for the attackers due to their vulnerability. These attacks aim at exhausting the energy of the node wasting the bandwidth. Data Corruption and Repudiation are the main attacks against this layer.

1. Data Corruption
   Malicious codes, such as viruses, spy-wares, worms, and Trojan Horses, are possible attacks to this layer. The malicious code can corrupt the data collected by the sensors, in this case the base station receives the altered data and performs wrong actions. Other objectives of this kind of attack ranging from slowing-down and damaging of the services to confidential information theft.
2. Repudiation
   This type of attack presents a partial or full denial participation of a specific node to the communication (Mohd et al., 2014).

Application layer firewalls, packet filtering, accounting services, antivirus, and spy-ware detection software are the countermeasures used to detect and defeat these attacks.

## 8 CONCLUSION

In the last decade, the IoT technology took on a prominent role in the modern society. Bughin et al. (2015) estimated the potential economic impact of IoT technologies to be 2.7–6.2 trillion of dollars per year by 2025. From one hand, the adoption and the wide spread of IoT bring with it several benefits increasing the quality of provided services and, therefore, a better quality of life. On the other hand, risks for privacy and security emerged. This work has presented a survey of the IoT, its key elements, main application domains, and enabling technologies. Among the introduced technologies, we have focused our attention on the Wireless Sensor Networks showing their features, device constraints, and limits. Finally, we have provided an overview on WSN attacks and possible countermeasures. In the near future, the IoT and wearable technologies will start to play a major impact globally and will lead a revolution for the daily life of people. The progress of the IoT applications and technologies will change the current perception of the world giving rise to a "smart world."

## ACKNOWLEDGMENT

This research has been partially founded by University of Naples "Parthenope" in the context of cooperation with "Istituto Superiore delle Comunicazioni e delle Tecnologie dell'Informazione" (ISCOM).

# ACRONYMS

**6LoWPAN** IPv6 over Low Power Wireless Personal Area Networks
**ACK** acknowledgement
**ARM** Advanced RISC Machine
**BLE** Bluetooth Low Energy
**CPU** Central Processing Unit
**CRC** Cyclic Redundancy Check
**CSMA/CD** Carrier Sense Multiple Access with Collision Detection
**DoS** Denial of Service
**DSL** Digital Subscriber Line
**EPC** Electronic Product Code
**EU** European Union
**FDM** Frequency Division Multiplexing
**FTP** File Transfer Protocol
**GSM** Global System for Mobile communications
**HTTP** Hypertext Transfer Protocol
**ICT** Information and Communication Technologies
**IoMT** Internet of Medical Things
**IoT** Internet of Things
**IoTSF** Internet of Things Security Foundation
**ITS** Intelligent Transportation System
**ITU-T** ITU Telecommunication Standardization Sector
**LoRaWAN** Long Range Wide Area Network
**M2M** Machine to Machine
**MAC** Media Access Control
**MIT** Massachusetts Institute of Technology
**MQTT** MQ Telemetry Transport
**NFC** Near Field Communication
**PDA** Personal Digital Assistant
**PDSL** Power line Digital Subscriber Line
**PLN** Power Line Networking
**QoS** Quality of Service
**RAM** Random Access Memory
**RFID** Radio Frequency IDentification
**RTOS** Real Time Operating System
**SMTP** Simple Mail Transfer Protocol
**SOA** Service Oriented Architecture
**TCP** Transmission Control Protocol
**UMTS** Universal Mobile Telecommunications System
**UPC** Universal Product Code
**URI** Uniform Resource Identifier
**URL** Uniform Resource Locator
**UUID** Universally Unique IDentifier

**V2I** Vehicle-to-Infrastructure
**V2V** Vehicle-to-vehicle
**WPAN** Wireless Personal Area Network
**WSN** Wireless sensor networks
**ZC** ZigBee Coordinator
**ZED** ZigBee End Device
**ZR** ZigBee Router

## GLOSSARY

**Big Data** paradigm for enabling the collection, storage, management, analysis, and visualization, potentially under real-time constraints, of extensive datasets with heterogeneous characteristics.

**Critical Infrastructure** systems and assets, whether physical or virtual, vital to national security, governance, public health and safety, economy and public confidence. Example of Critical Infrastructure are facilities for: electricity and gas production, water supply, financial services, public health, and transportation.

**Denial of Service** cyber attack in which an attacker tries to make a device or a network resource unavailable for its users. The attack is perpetrated flooding the target machine with a massive amount of requests that aim at saturating the target's resources.

**Internet of Things** network of physical objects that contains embedded technology to communicate and sense or interact with their internal states or the external environment.

**Interoperability** the interoperability refers to the ability of a system to interact with other systems (or some parts of them) without special efforts by users.

**Near Field Communication** set of communication protocols that enable two devices, typically smartphones, bidirectional communication by bringing them within four centimeters of each other.

**RFID** the Radio Frequency IDentification is a technology that uses the electromagnetic fields to identify and track tags attached to objects, animals or people.

**Safety** measure of the absence of catastrophic influences on the environment, in particular on human life.

**Scalability** the scalability, in computer science, is the feature of a system able to rebalance and/or redistribute the available resources to satisfy an increase of the workload.

**Security** security is the absence of unauthorized access to, or handling of, system state.

**Smart city** city in which traditional networks and services reaches more efficiency through the use of digital and telecommunication technologies. A smart city aims at providing better public services for the citizens, better use of resources and less impact on the environment.

**Smart home** home that incorporates advanced automation systems to provide the inhabitants with sophisticated monitoring and control over the building's functions.

**Wireless Sensor Networks** wireless network of spatially distributed autonomous sensors for monitoring and recording physical/environmental conditions, organizing the collected data at a central location.

**Z-Wave** wireless standard primarily targeted for home automation. It allows to create low power wireless mesh networks with low latency communications and throughput up to 100 kbps.

**ZigBee** wireless technology based on the IEEE 802.15.4 specification for low-power area networks. It is often associated with machine-to-machine communication and Internet of Things and operates in unlicensed radio frequency bands, including 2.4 GHz, 900 MHz, and 868 MHz.

## REFERENCES

Abramowitz, M., Stegun, I.A., 2015. Cyber-Physical Attacks, A Growing Invisible Threat, first ed. Butterworth-Heinemann, London.

Adamy, D., 2004. EW 102: A Second Course in Electronic Warfare. Artech House, Norwood, MA.

Akyildiz, I., Su, W., Sankarasubramaniam, Y., Cayirci, E., 2002. Wireless sensor networks: a survey. Comput. Netw. 38 (4), 393–422.

Alliance, Z., 2014. ZigBee Specifications. http://www.zigbee.org/download/standards-zigbee-specification/. (Online; Accessed 05 October 2016).

Alliance, Z.W., 2016. About Z-Wave. http://z-wavealliance.org/about_z-wave_technology. (Online; Accessed 05 October 2016).

Alphonsa, A., Ravi, G., 2016. Earthquake early warning system by IoT using wireless sensor networks. In: International Conference on Wireless Communications, Signal Processing and Networking (WiSPNET). IEEE, New York, pp. 1201–1205.

Anithaa, S., Arunaa, S., Dheepthika, M., Kalaivani, S., Nagammai, M., Aasha, M., Sivakumari, S., 2016. The Internet of Things—a survey. World Sci. News 41, 150.

Atzori, L., Iera, A., Morabito, G., 2010. The Internet of Things: a survey. Comput. Netw. 54 (15), 2787–2805.

Aura, T., Nikander, P., Leiwo, J., 2000. Dos-resistant authentication with client puzzles. In: International Workshop on Security Protocols. Springer-Verlag, Berlin, Heidelberg, pp. 170–177.

Baburajan, J., Prajapati, J., 2014. A review paper on watchdog mechanism in wireless sensor network to eliminate false malicious node detection. Int. J. Res. Eng. Technol. 3 (1), 381–384.

Bartariya, S., Rastogi, A., 2016. Security in wireless sensor networks: attacks and solutions, International Journal of Advanced Research in Computer and Communication Engineering 5 (3).

Bluetooth-SIG, 2016a. Bluetooth Low Energy. https://www.bluetooth.com/what-is-bluetooth-technology/bluetooth-technology-basics/low-energy. (Online; Accessed 05 October 2016).

Bluetooth-SIG, 2016b. What is Bluetooth Technology. https://www.bluetooth.com/what-is-bluetooth-technology/bluetooth. (Online; Accessed 05 October 2016),

Bughin, J., Chui, M., Manyika, J., 2015. An executive's guide to the Internet of Things. McKinsey Quarterly, McKinsey&Company.

Chaqfeh, M.A., Mohamed, N., 2012. Challenges in middleware solutions for the Internet of Things. In: 2012 International Conference on Collaboration Technologies and Systems (CTS). IEEE, New York, pp. 21–26.

De Silva, L.C., Morikawa, C., Petra, I.M., 2012. State of the art of smart homes. Eng. Appl. Artif. Intel. 25 (7), 1313–1321.

EPC, 2010. Directive 2010/40/EU of the European Parliament and of the Council of 7 July 2010.

Fatema, N., Brad, R., 2014. Attacks and Counterattacks on Wireless Sensor Networks. arXiv preprint arXiv:1401.4443

Gigli, M., Koo, S., 2011. Internet of Things: services and applications categorization. Adv. Internet Things 1 (02), 27.

Google, 2016. Google Self-Driving Car Project.

Hill, J., Szewczyk, R., Woo, A., Hollar, S., Culler, D.E., Pister, K.S.J., 2000. System architecture directions for networked sensors. In: Architectural Support for Programming Languages and Operating Systems, pp. 93–104. citeseer.ist.psu.edu/382595.html.

HomePNA, 2013. Home Phoneline Networking Alliance. http://www.homepna.org/. (Online; Accessed 05 October 2016).

IEEE, 2013. IEEE 802.3 'Standard for Ethernet' Marks 30 Years of Innovation and Global Market Growth. http://standards.ieee.org/news/2013/802.3_30anniv.html. (Online; Accessed 05 October 2016).

IEEE, 2016. 802.11. http://www.ieee802.org/11/. (Online; Accessed 05 October 2016).

IETF, 2007. 6LoPAN. https://datatracker.ietf.org/wg/6lowpan/documents/. (Online; Accessed 05 October 2016).

Insider, B., 2015. We Asked Executives About the Internet of Things and Their Answers Reveal That Security Remains a Huge Concern. http://uk.businessinsider.com/internet-of-things-survey-and-statistics-2015-1. (Online; Accessed 05 October 2016).

IOTS, 2015a. Establishing Principles for Internet of Things Security. https://iotsecurityfoundation.org/establishing-principles-for-internet-of-things-security/. (Online; Accessed 05 October 2016).

IOTS, 2015b. IoT Security Foundation. ttps://iotsecurityfoundation.org/. (Online; Accessed 05 October 2016).

ITU-T, 2012. Recommendation y.2060: Overview of the Internet of Things.

Karlof, C., Wagner, D., 2003. Secure routing in wireless sensor networks: attacks and countermeasures. Ad Hoc Netw. 1 (2), 293–315.

Khan, R., Khan, S.U., Zaheer, R., Khan, S., 2012. Future internet: the Internet of Things architecture, possible applications and key challenges. In: 2012 10th International Conference on Frontiers of Information Technology (FIT). IEEE, New York, pp. 257–260.

Kitagami, S., Thanh, V.T., Bac, D.H., Urano, Y., Miyanishi, Y., Shiratori, N., 2016. Proposal of a distributed cooperative IoT system for flood disaster prevention and its field trial evaluation. Int. J. Internet Things 5 (1), 9–16.

LWN, 2016. Z-Wave Protocol Specification Now Public. https://lwn.net/Articles/699241. (Online; Accessed 05 October 2016).

Madakam, S., Ramaswamy, R., Tripathi, S., 2015. Internet of Things (IoT): a literature review. J. Comput. Commun. 3 (05), 164.

Miorandi, D., Sicari, S., De Pellegrini, F., Chlamtac, I., 2012. Internet of Things: vision, applications and research challenges. Ad Hoc Netw. 10 (7), 1497–1516.

Mohammadi, S., Jadidoleslamy, H., 2011. A comparison of physical attacks on wireless sensor networks. Int. J. Peer Peer Netw. 2 (2), 24–42.

Mohd, N., Annapurna, S., Bhadauria, H., 2014. Taxonomy on security attacks on self configurable networks. World Appl. Sci. J. 31 (3), 390–398.

Mpitziopoulos, A., Gavalas, D., Konstantopoulos, C., Pantziou, G., 2009. A survey on jamming attacks and countermeasures in WSNS. IEEE Commun. Surv. Tut. 11 (4), 42–56.

Patel, M.M., Aggarwal, A., 2013. Security attacks in wireless sensor networks: a survey. In: 2013 International Conference on Intelligent Systems and Signal Processing (ISSP). IEEE, New York, pp. 329–333.

Pottie, G.J., Kaiser, W.J., 2000. Wireless integrated network sensors. Commun. ACM. 43 (5), 51–58.

Prakash, R., Girish, S.V., Ganesh, A.B., 2016. Real-time remote monitoring of human vital signs using Internet of Things (IoT) and GSM connectivity. In: Proceedings of the International Conference on Soft Computing Systems. Springer, New Delhi, pp. 47–56.

Saha, H.N., Bhattacharyya, D., Banerjee, P., 2010. Semi-centralized multi-authenticated RSSI based solution to Sybil attack. Int. J. Comput. Sci. Emerg. Technol. I (4), 338–341.

Semiconductor, C., 2011. What is Power Line Communication? http://www.eetimes.com/document.asp?doc_id=1279014&. (Online; Accessed 05 October 2016).

Sharma, K., Ghose, M., 2010. Wireless sensor networks: an overview on its security threats. . In: IJCA, Special Issue on "Mobile Ad-hoc Networks" MANETs, pp. 42–45.

Singh, S., Verma, H.K., 2011. Security for wireless sensor network. Int. J. Comput. Sci. Eng. 3 (6), 2393–2399.

Soni, V., Modi, P., Chaudhri, V., 2013. Detecting sinkhole attack in wireless sensor network. Int. J. Appl. Innov. Eng. Manag. 2 (2), 29–32.

Steinberg, N., 2014. These Devices may be Spying on You (Even in Your Own Home). http://www.forbes.com/sites/josephsteinberg/2014/01/27/these-devices-may-be-spying-on-you-even-in-your-own-home. (Online; Accessed 05 October 2016).

The National Intelligence Council, 2008. Disruptive Civil Technologies: Six Technologies with Potential Impacts on US Interests out to 2025. https://fas.org/irp/nic/disruptive.pdf. (Online; Accessed 05 October 2016).

Thread-group, 2016. What is Thread. https://www.threadgroup.org/what-Is-thread. (Online; Accessed 05 October 2016).

Uke, S., Mahajan, A., Thool, R., 2013. UML modeling of physical and data link layer security attacks in WSN. Int. J. Comput. Appl. 70 (11).

Vadlamani, S., Eksioglu, B., Medal, H., Nandi, A., 2016. Jamming attacks on wireless networks: a taxonomic survey. Int. J. Prod. Econ. 172, 76–94.

Wang, J., Zhang, Z., Li, B., Lee, S., Sherratt, R.S., 2014. An enhanced fall detection system for elderly person monitoring using consumer home networks. IEEE Trans. Consum. Electron. 60 (1), 23–29.

Wikipedia, 2016. List of Wireless Sensor Nodes. https://en.wikipedia.org/wiki/List_of_wireless_sensor_nodes. (Online; Accessed 05 October 2016).

Wu, M., Lu, T.J., Ling, F.Y., Sun, J., Du, H.Y., 2010. Research on the architecture of Internet of Things. In: 2010 3rd International Conference on Advanced Computer Theory and Engineering (ICACTE), vol. 5, p. V5-484.

Xiaojiang, X., Jianli, W., Mingdong, L., 2010. Services and key technologies of the internet of things. ZTE Commun. 2, 011.

Youn, S., 2013. A comparison of clock synchronization in wireless sensor networks. Int. J. Distrib. Sens. Netw. 9. http://dblp.uni-trier.de/db/journals/ijdsn/ijdsn2013.html#Youn13.

Zhao, X., 2012. The security problem in wireless sensor networks. In: 2012 IEEE 2nd International Conference on Cloud Computing and Intelligence Systems, vol. 3. IEEE, New York, pp. 1079–1082.

CHAPTER

# SMART ACCESS CONTROL MODELS IN SENSOR NETWORK

# 5

**Christian Esposito***, **Jian Shen**[†], **Chang Choi**[‡]

*University of Salerno, Salerno, Italy** *Nanjing University of Information Science & Technology, Nanjing, China*[†]
*Chosun University, Gwangju, South Korea*[‡]

## 1 INTRODUCTION

Several current research efforts are paving the way for an evolution of our society and cities by letting ICT pervasively support every daily human activities and optimize the processes that govern a city. Such smart city applications, as introduced in Nam and Pardo (2011), are strongly information-intensive since the right decisions can be made only if a suitable volume of information is collected and properly processed. Therefore their main supporting technology is the one provided by the sensor networks that have been deployed in our objects and cities in order to measure the key factors determining the correct behavior of a given process, such as air pollution, garbage collection, or road traffic management. To this aim, in the literature the vision of the Internet of Things, described in Al-Fuqaha et al. (2015), is enforced as a key element of a smart city. The current initiatives of smart cities do not assume to deploy newly designed sensor networks to support their applications, but to use the already existing ones by augmenting them with novel functionalities and/or adding more portions of sensor networks, so as to augment their functional and spatial extension. Therefore the current issue is related to how to integrate multiple sensor networks in a seamless manner among themselves.

A crucial aspect to be investigated for the effective application of those upcoming innovations in our daily activities is security and privacy, as seen in Martinez-Balleste et al. (2013). In fact, the sensor networks used to realize a smart city or the Internet of Things may gather sensitive data on citizens, such as their movements and habits, but also collect information that may be valuable to a malicious adversary in order to properly plan cyber or terrorist attacks. Within this aspect, access control plays a crucial role in protecting information systems from misuses; therefore, there is the need for an efficient access control in order to make sure that sensory data are accessible only to authorized users. However, when integrating multiple legacy systems, each one managed by a given organization and characterized by a given access control model, it is unfeasible to impose a single access control strategy but it is crucial to integrate the existing models and to let them interoperate. In fact, it is possible that a given sensor network may be used by multiple organizations in order to feed their relative applications with proper sensory data. In addition, it is also possible that a given application may

**Security and Resilience in Intelligent Data-Centric Systems and Communication Networks. https://doi.org/10.1016/B978-0-12-811373-8.00005-7**

Copyright © 2018 Elsevier Inc. All rights reserved.

need the data of multiple networks, deployed in the different parts of a city by different organizations. Therefore there is a need to design access control in an interoperable manner among heterogeneous enterprises.

The presented work addresses such issues by equipping a given sensor network with the following innovative contributions: introducing a flexible ontology-based access control, matching different ontologies in an automatic manner, and integrating citizen consent within the applied access control model. This approach represents an evolution of our previous works on flexible access control models presented in Esposito et al. (2016), applied to the context of sensor networks. The rest of the paper is structured as follows. Section 2 introduces the key aspects of access control models within the context of sensor networks, and Section 3 highlights the many issues in the current literature on this topic. Section 4 describes the proposed semantic approach to describe an access control model and how to match diverse ontologies. The following Section 5 describes the prototype of the presented approach so as to prove its usability within a service-oriented sensor networks. We conclude this paper with Section 6 where some final remarks and a plan for future work on the addressed topics are given.

## 2 BACKGROUND AND RELATED WORK

As mentioned, sensor networks are being used in several application scenarios, which are critical and require a high degree of security by all of its constituents. A secure sensor network is the one able to keep its data private and confidential, to preserve its resources and information from malicious uses and alterations, and to defend itself from possible attacks by adversaries. Security is a complex concept made of a combination of several different properties with some sort of interleaving among themselves. Among those properties we can find that the demanding ones are Authenticity and Authorization. The first one consists in the validation of the identity of users requesting certain services and/or accessing certain resources, such as data stored at certain nodes of the network. The authentication is required so as to avoid unauthorized users invoking the functionalities provided by sensor network and/or accessing data hosted by sensors or other nodes in the network. The second one consists in granting the rights to access certain data or even to use certain functionalities to given authenticated users depending on several factors, such as their role within a given organization or properly formalized security policies.

Two main techniques are adopted to offer these properties: user authentication and access control. The user authentication indicates the process of confirming the truth about a stated identity, by means of proper information provided by the user. An identity is the representation of an entity in a particular context, as indicated in Torres et al. (2013), made of a unique identifier and/or a set of attributes related to the user. The identity needs to be managed with a proper service to return a properly formatted identify claim, to manage their life-cycle from creation to revocation, to protect the identities from adversary interested in stealing them, and to verify that an identify claim is truthful. Therefore ICT infrastructures are typically equipped with special subsystems in charge of implementing identity management, which are typically defined in the literature as Identity Manager (IdM). Identify management is tightly related to access control so that in certain academic papers and industrial practice they are used interchangeably. Access control disciplines which resources of the sensor network a given entity is authorized to access, so as to implement authorization. During the last few years, several different access control models have been proposed, and most of them have been applied within the context of the sensor networks, as reviewed in Maw et al. (2014). Fig. 1 describes a possible

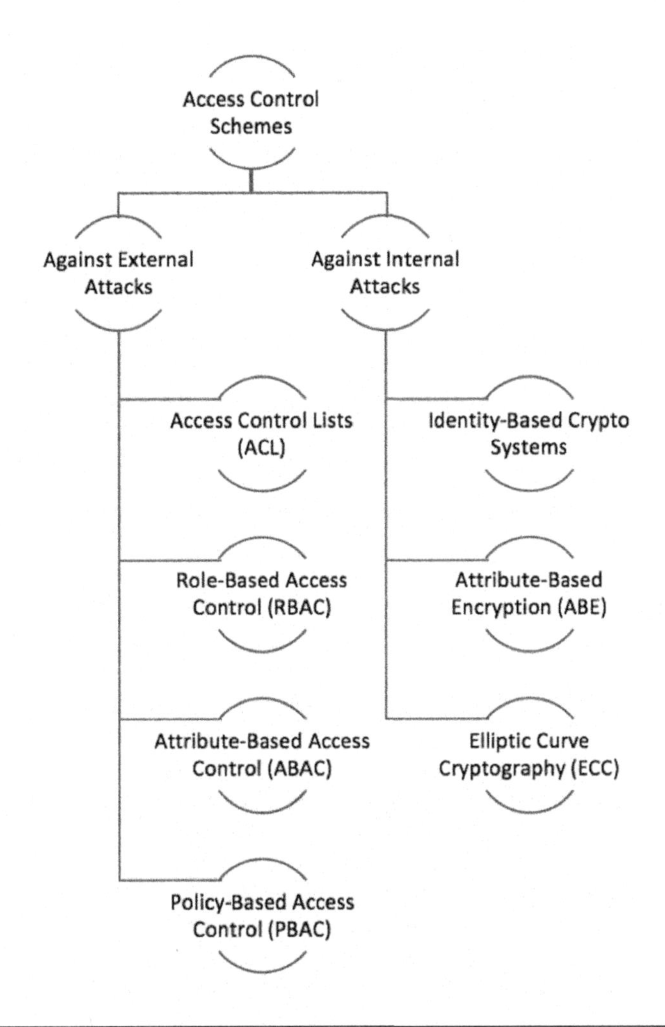


**FIG. 1**

Taxonomy of the access control schemes.

classification of the main access control models, based on the provenance of the possible attacks that may compromise the sensor network and may be treated by the adopted access control model. An eventual attack may be caused by an adversary taking advantage of a vulnerability of the network, such a Denial of Service (DoS) attack aiming to overload a certain node and compromise its availability, as in Wood and Stankovic (2002). Those kinds of attacks are named as external, so as to differentiate them from another class of attacks that are conducted by compromised (as illustrated in Zhang et al., 2008) or camouflaged (as described in Wang and Bhargava, 2004) nodes of the network that deviated from their correct behavior, or adversaries that are able to eavesdropping the radio channel used for communications among the nodes to steal data or to inject false data, as in Dai et al. (2013). This second class of attacks is called internal, according to Ahmed et al. (2012). Dealing with these two classes of attacks requires complementary solutions: a first class of solutions aims at realizing access control by verifying the security claims of the requesting entities, while the second one is based on

**Table 1 Main Access Control Schemes: Pros and Cons**

| Scheme | Pros | Cons |
|---|---|---|
| Access Control Lists (ACL) | Simple Implementation | Scalability and Expressivity Limitations |
| Role-Based Access Control (RBAC) | Grouping Rights into Categories | Coarse-grained Scheme |
| Attribute-Based Access Control (ABAC) | Fine-grained Scheme | Issues with Heterogeneous Organizations and Complexity |
| Policy-Based Access Control (PBAC) | Standardized Access Rules | Complicated Model |
| Identity-Based Crypto Systems | Similar to ACL | Affected by Key Escrow Problem |
| Attribute-Based Encryption (ABE) | Fine-grained Scheme like ABAC | Too complex for Sensors |
| Elliptic Curve Cryptography (ECC) | Effective for Sensors | Tricky to Implement Securely |

cryptographic primitives. The main techniques of those two classes are briefly compared in Table 1, by highlighting the pros and cons of each technique.

The most basic means for an access control method is the one known as *Access Control Lists* (ACL), presented in National Computer Security Center (1987), sometimes known in the literature as *Identity-Based Access Control* (IBAC), consisting of a list of permissions attached to a resource of the system to be protected. Within the context of a sensor network (Benenson et al., 2005), each object, or datum, within the network is associated with an ACL that indicates which entities are able to access it and which operations are allowed. When an entity wants to access a given object, it must send its identity to the node holding the object of interest or to the base station of the network. At the reception of the request the ACL of the requested object is accessed and queried, if the ACL indicates that the request can be satisfied then the object is accessed and the requested operation performed. ACLs provide a simple solution to access control, but they present several drawbacks, especially when a large number of users and permissions need to be managed. A more advanced method than the ACL has been designed so as to overcome these limitations, where the access to a resource is based on the roles that individuals play within the organization in control of the resource. Such a solution is called *Role-Based Access Control* (RBAC), described in Ferraiolo et al. (2007), and substitutes the resource-focused approach of the ACL with the approach of grouping individuals sharing the same permissions into the same category of people fulfilling a particular role. The RBAC has been extensively used in sensor networks, as overviewed in Panja et al. (2008), since it is more efficient than defining the ACL in every node for every possible entity requesting resources to the sensor network. In these access control models a set of roles is defined and distributed among the nodes of the network, when requesting a resource an entity must be claimed as belonging to a certain role, for which the resource is related. If the claimed role is verified and the requested operation is associated to this role, then the request is grated. The RBAC model is more scalable than ACLs since it is not necessary to indicate all the permissions of the individuals to access each resource, but only the ones related to a role, associating individuals to one, or even more, roles. Regardless of its advantages, RBAC also suffers from several limitations, such as a coarse-grained access control and an unsuitability for cross-domain accesses. In order to achieve a more fine-grained access control than RBAC, *Attribute-Based Access Control* (ABAC), described in Priebe et al. (2004), has been introduced. In such a model, grants to access certain resources are decided based on the attributes possessed by the requester, the context and/or the resource itself. The benefit of using an ABAC is the absence of the need to know the requester in advance (as required by

ACL): as long as the provided attributes match with the requested ones, access is permitted. An ABAC can meet some issues in a large environment where disparate attributes and access control mechanisms exist among the organizational units. However, it has not met great attention within the context of sensor networks due to its high complexity in managing the attributes, performing access control and exchange claimed attributes. In order to have a more harmonized and lightweight access control across the enterprise and to achieve a more uniform access control model than ABAC, *Policy-Based Access Control* (PBAC), briefly presented in Zhi et al. (2009), has been proposed, where the attribute-based approach of an ABAC is standardized by defining access rules in terms of policies written in a formal and precise language (such as the eXtensible Access Control Markup Language (XACML) standard, described in Moses, 2005), which allows policy negotiation among the distinct units of an enterprise or even different federated enterprises, as in Bistarelli et al. (2010). PBAC has been recently applied to the context of sensor networks, as in Zhu et al. (2009) or Manifavas et al. (2014), with the intent of a flexible adaptive access control by supporting dynamic loading, enabling and disabling of policies without shutting down nodes.

The second class of access control means are based on certain encryption schemes and the use of proper cryptographic primitives. Encryption consists of the process of transforming the content of messages or data objects, using a proper mathematical encoding algorithm, so as to turn them into artifacts unintelligible without previously applying a mathematical decoding algorithm to recover the original data. The encoding process is carried out by using an algorithm and providing an encryption key as input. Such a key indicates how data can be encoded by performing a series of transformations and substitutions. The adopted encoding algorithm has to be strong enough to avoid an adversary figuring out the used encryption key by discovering the algorithm and some encoded data. The opposite process is carried out by running another algorithm (essentially the encoding algorithm run in reverse), giving as input a decryption key, typically kept secret, which specifies the opposite operations to be performed. Encryption is typically used in order to provide privacy and confidentiality within a communication system, but recently it has also been used as a means to realize access control. The basic idea is to allow only authorized entities the ability to decrypt what it has been encoded before, so that the possession of the decryption key or privileges is equivalent to a grant in the traditional access control schemes. The first of those solutions are the so-called *Identity-Based Crypto Systems*, introduced in Shamir (1985). In such a method, a key is not the typical random string of classic encryption schemes, but is a string related to the identity of the one performing the encoding or the dual decoding operations. Such a string is easily computable from the user identity by means of the bilinear pairings presented in Zhang et al. (2004), so that a data source can use such functions in order to obtain an encryption key to encode data that only a certain identity is able to obtain. Such a solution is the realization of ACLs by means of encryption. Such an approach was given a practical implementation with *Pairing-Based Cryptography* (PBC) in Barreto et al. (2002), where any valid string that uniquely identifies a user can be his/her public key in the sense of asymmetric encryption keys. A key manager is present in order to assign private keys to data sources and consumers, which are kept separate. The user's private key is computed from his/her known public identity and the secret information is obtained from the key manager, while such a manager is not strictly necessary to obtain the public key of a user. Due to the involvement of a manager in generating the private key, Identity-Based Crypto Systems are affected by the key escrow problem, consisting in the need for all the users to trust fully the key manager and the possibility of having a key disclosure to adversaries. This problem makes such systems less secure, stimulating several research efforts to resolve it. The PCB has not found successful application in sensor networks due to its security problems but also scalability limitations, which as the ones we have considered when presenting the ACLs. If the user identity is not only expressed by a given string, but by a set of attributes, we can extend identity-based cryptography by making the decryption of a cipher-text possible only if at least $k$ of the attributes of the user key matches the attributes of the cipher-text. Such a solution is known in literature as *Attribute-Based Encryption* (ABE), presented in Goyal et al. (2006), and provides a semantically rich tool for implementing access control over encrypted data. Currently, there is a rich literature on ABE, and the available solutions can be grouped into two classes: Key-Policy ABE (KP-ABE), overviewed in Goyal et al. (2006), where the encrypted data are annotated with attributes and the user keys are associated with access policies, and Cipher-text-Policy ABE (CP-ABE), surveyed in Bethencourt et al. (2007), where the user keys are associated with a set of attributes and the encrypted data are annotated with an access policy over the attributes. KP-ABE has the problem that the data owner cannot decide who is allowed to decrypt its data. Such a problem is not present in CP-ABE; therefore, most of the existing ABE solutions are based on CP-ABE. Such schemes are typically not sufficiently expressive in the formalization of the access policies and computationally expensive. As a practical example, the CP-ABE scheme in Bethencourt et al. (2007) supports only policies with logical conjunction, and presents the size of the cipher-texts and secret keys linearly increasing with the total number of attributes. ABE is an efficient and simple way to implement ABAC and has been extensively being used in sensor networks, such as in Tan et al. (2011). The last cryptographic access control solution is the one based on *Elliptic Curve Cryptography* (ECC), described in Koblitz et al. (2000), which is an encryption scheme based on the algebraic structure of elliptic curves over finite fields. ECC is starting to have a considerable application to sensor networks, such as in Chatterjee and Das (2015) or Wang et al. (2006), since it reduces the required computational intensity with respect to the other traditional schemes, so as to be suitable for sensors with limited computation capability and energy budget. ECC is a recent encryption scheme, and the relative standards aren't state-of-the-art and there are still has some patent problems. Moreover, it is complicated and tricky to implement securely, particularly the standard curves, as argued in Bos et al. (2014).

In this work we have focused on the first kind of access control means, rather than the ones based on cryptographic schemes and primitives. However, our work is complementary to the research efforts and attempts to provide security guarantees toward internal attacks by using encryption.

## 3 PROBLEM STATEMENT

Traditionally, an ICT infrastructure adopts a single and precise access control model, which has been agreed and specified by the organization owning it. Sensor networks do not represent an exception to this traditional practice, and such a situation may be valid for the traditional sensor networks that are realized and deployed by a given organization in order to carry out a very specific application. However, it may not be suitable for the upcoming uses of sensor networks, as depicted in Fig. 2, where a given network may be used by multiple organizations in order to support and be part of disparate applications, or several legacy sensor networks may be federated and integrated by means of brokering nodes conveying data from one network to the others, in order to cope with novel uses than the ones for which they have been designed. A concrete example of those upcoming uses of sensor networks is provided by the novel conception of smart cities, illustrated in Jin et al. (2014), where sensor networks may play a key role in collecting multiple kinds of information about pollution measures, citizen habits,

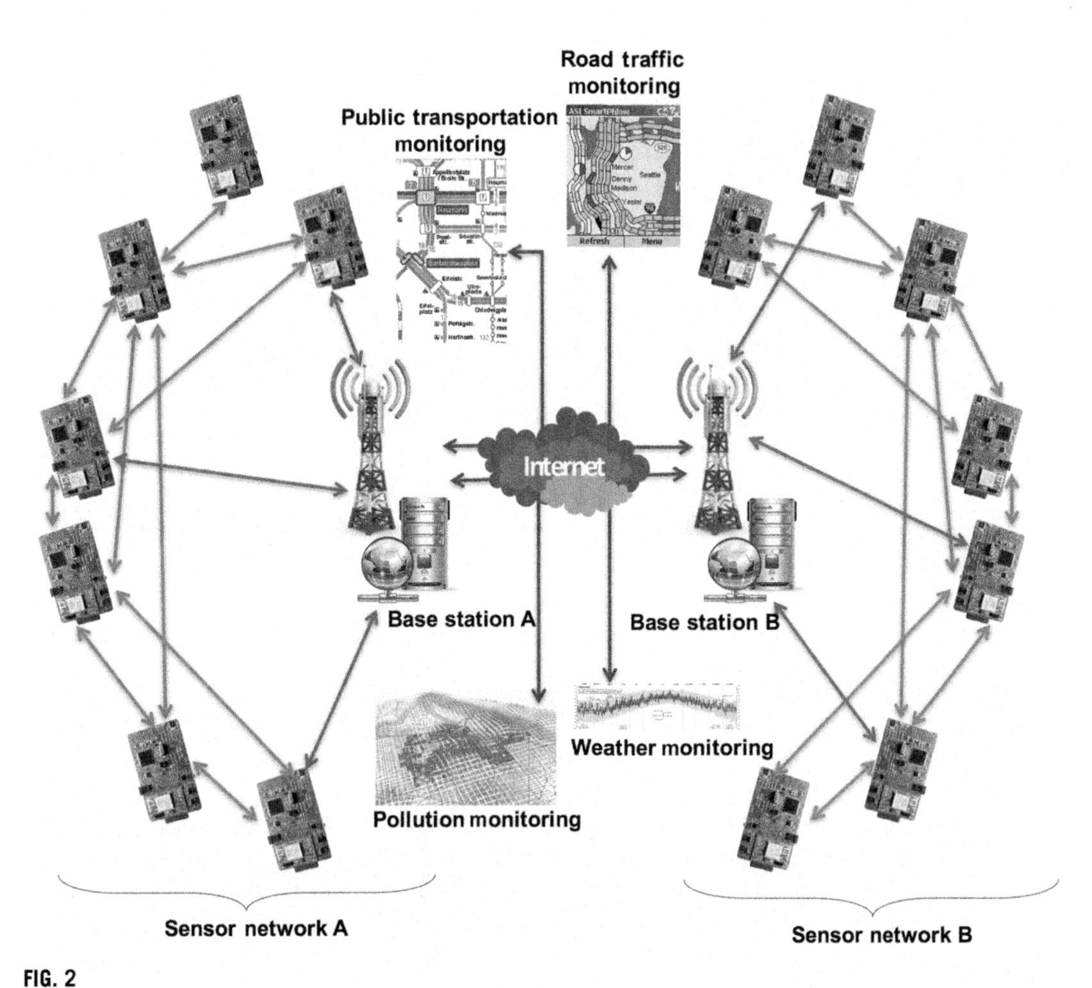


**FIG. 2**

Integrated sensor networks supporting some applications of the "smart city" vision.

traffic state, and so on. A smart city cannot be created by scratch, but it is more reasonable to have existing legacy sensor networks being integrated and novel ones being deployed where needed, and seamlessly federated to the existing ones. Moreover, in this vision, the sensory data are required to feed multiple kinds of applications spanning traffic and garbage management, mobility and road traffic improvement or citizen enhancement. In such a novel context, security and access control are required mainly for privacy reasons (Martinez-Balleste et al., 2013) and to lower the success of an attack, but it is not possible to impose a unique access model within the overall integrated sensor networks. This is due to the fact that there is no central management of those networks, but the management is distributed among several authorities which are in charge for parts of it, where each of such parts is characterized by a given access control model. Within a typical urban-wide sensor network made of the integration of multiple subnetworks, there may be large segments of legacy systems that ended up having implemented some sort of RBAC, and most of prototyping initiatives related to the "smart city" concept and public city infrastructures have already agreed on a set of the basic role-based access control rules, as in Steuer et al. (2016), Kawada et al. (2013), and Chifor et al. (2016). However, there is no agreement on the most suitable and effective access model within the context of "smart city" prototypes and supporting technologies, so that we can find also other models being applied, such as the PBAC one in Kos et al. (2012), Apolinarski et al. (2014), and Preuveneers and Joosen (2016). In these cases, it is not reasonable or profitable to impose a different model than the one already in place, such as the PBAC where the RBAC has been applied, since it will consist of rethinking the internal access control rules of these networks and municipal authorities. On the contrary, it is strongly desirable to have a flexible authorization solution that can welcome any given access control model with which a particular entity is confident. Furthermore, in large scale ecosystems, such as the one supporting the vision of "smart city," it is not possible to have a single access control model but multiple ones must coexist and be used effectively.

## 4 APPROACH

The problem of having multiple access control models that must coexists in the ecosystems of independent "islands" of sensor networks, composing the overall "smart city" cannot be resolved by imposing a single standard model. In fact, it is preferred to have such islands able to orchestrate their security policies so that a given organization can allow or deny access to the resources requested by users remotely with minimal changes to its own access control scheme and policies. The interoperability of the access control models can be achieved by resolving the differences in the semantics of the terms used to express the access control policies of heterogeneous organizations. Due to the scale and complexity of the public and municipal organizations integrated by our envisioned large-scale integrated sensor networks, it is not feasible to have a manual resolution of the above-mentioned issues done by one or more experts, but an automatic tool is necessary. To this aim, a formal description of each access control model is required, so that a computer program can resolve the differences and automatically match the different models. In computer science, the means to precisely describe (classes of) entities of a domain are the ontology, so we propose to use them in order to describe a given access control model by specifying subjects, their attributes and contexts. The use of ontologies in the current literature is not new, as present in Khan et al. (2011), Chen (2008), and He et al. (2011), but in our work they are used in a different manner. The first element of heterogeneity among the access control models is related to their class of affinity, e.g., RBAC or PBAC, and existing ontology-based access control approaches are couples with a specific king of model (e.g., Choi et al., 2014, He et al., 2011, Chen, 2008 follow a RBAC model, while Khan et al., 2011 considers a PBAC model). In our work, we do not want to impose a given class of models but leaving the freedom of specifying any possible access control model. This is a first level of interoperability that our solution can provide. Moreover, each model can use a given term to indicate the elements of the models, for a concrete example, a subject can be indicated as citizen or user of a given municipal service. An ontological representation of these terms is able to relate terms that are syntactical different but share the same semantic. The decision of granting or denying a request is traditionally implemented as policies on the attributes held by the requesting subject. If the access control model is formalized as an ontology, the access control decision logic needed for granting or denying a request from the users can be realized by means of permission rules formalized as queries expressed in the SPARQL language, formalized in Prud'hommeaux and Seaborne (2008). SPARQL is a query language able to retrieve and manipulate data stored in the Resource Description Framework (RDF) format of an ontology.

The ontological representation of an access control model for a given organization can be structured in two distinct parts: the first one is related to a description of the context of usage of the system of interest, and the second one is related to the set of security policies and restrictions agreed by an organization. Specifically, a *Domain Ontology* contains the specification of the set of users that are interested to request services and data to a sensor network or the entities monitored by the network. Such an ontology does not only specify such entities, but also their attributes, and provides also a hierarchical structure of these concepts and possible relations among them. On the other hand, a *Control Ontology* is a declarative description of the access control rules adopted by the municipal authority running the integrated sensor networks, by having the concepts expressed in this rules, their attributes and relations. Such a semantically-rich policy representation has the advantage of having a description of the access control policies at a very high level of abstraction and promoting a common understanding of security policies among heterogeneous entities that do not share the same information model for their adopted access control strategies. This has the benefit of making the interpretation of the policies across different organizations much easier. Fig. 3 contains a concrete example of a Domain Ontology for a given local municipal authority, where the main entities within an typical organization, such as staff, citizens or managers, and their relations are represented. In such an ontology, we have the smart city composed of a series of organizations, all derived from the generic concept of "municipal organization." Such organizations can be (1) a given front office where citizens can go to in order to receive information or documents, (2) a proper public infrastructure, such as for public transportation, or road management or garbage collection, and (3) an entity providing a kind of public services, such as real-time notification on the traffic conditions along the city roads, or the congestion of certain metro lines, or the latency of given bus lines. Each organization encompasses a proper administration office and several technical staff, derived from the generic concept of "technical staff," which is made of workers at the public infrastructures, or at the public services, or at the organization front desk. The citizen is the central player in this ontology, by using the offices, services and infrastructures of the organization and interacting with the organization staff. In addition, the organization owns a set of sensors, that are useful for its mission and its offices, services and infrastructures in order to better organize the work, or to have data about their perceived quality. Such sensors may collect information related to the citizens, or not.

Such an ontology has to be coupled with a proper Control Ontology for modeling the knowledge required for realizing the access control strategy applied by the given local municipal authority. In the figure, we have two distinct control ontologies, so as to illustrate that the two ontologies are independent and interchangeable. In particular, in Fig. 3A we have that the control ontology is specific for the RBAC case applied to a generic sensor network for the smart city scenario. In particular, each member of the personnel employed by the organization is associated with a specific role within such an ontology, and such roles can be structured in a hierarchy, with roles containing other roles. Based on the associated roles, it is possible to have access to specific topics that group the sensory data generated by the sensor network. The figure shows only the primary entities of the ontology, but each of them are also characterized by a series of properties that are not shown in order to simplify the figure. For example, the citizen and the technical staff are characterized by the following attributes: name, surname, unique identifier, address and so on. If we consider the PBAC access model, we have our ontology modified

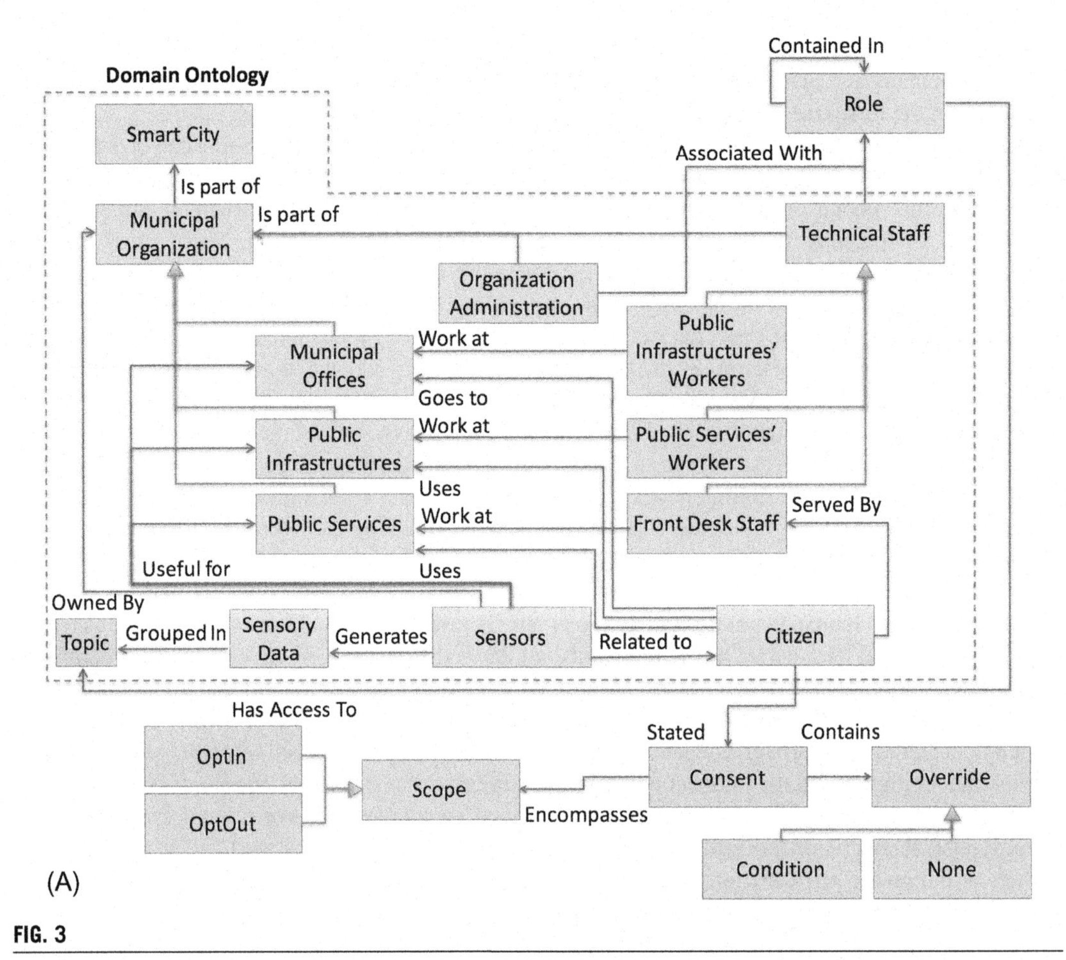


**FIG. 3**

See legend on opposite page.

as shown in Fig. 3B. In such an ontology, we have that the organization has issued a series of policies involving citizens and staffs, and based on specific contextual information related on where, when and how the sensory data have been acquired. Such policies are applied to the sensory data in order to determine if an access can be granted or denied, as traditionally done in the PBAC. These two examples are just explicative of the possible use of our semantic approach, but any possible user is free to model its domain and access control model as it pleases.

As mentioned previously, such ontologies can be queried so as to verify the satisfaction of access rules, expressed as SPARQL predicates. Specifically, the SPARQL is a graph-matching query language, and a predicate expressed in such a language consists of several constituents: (1) a prefix (indicating the namespace for the used terms); (2) a dataset definition clause (expressing where the data to be processed reside); (3) a result clause (specifying the output to return to the user, such as how to construct new triples to be returned); (4) pattern matching (such as optional, union, nesting, filtering); and (5) solution

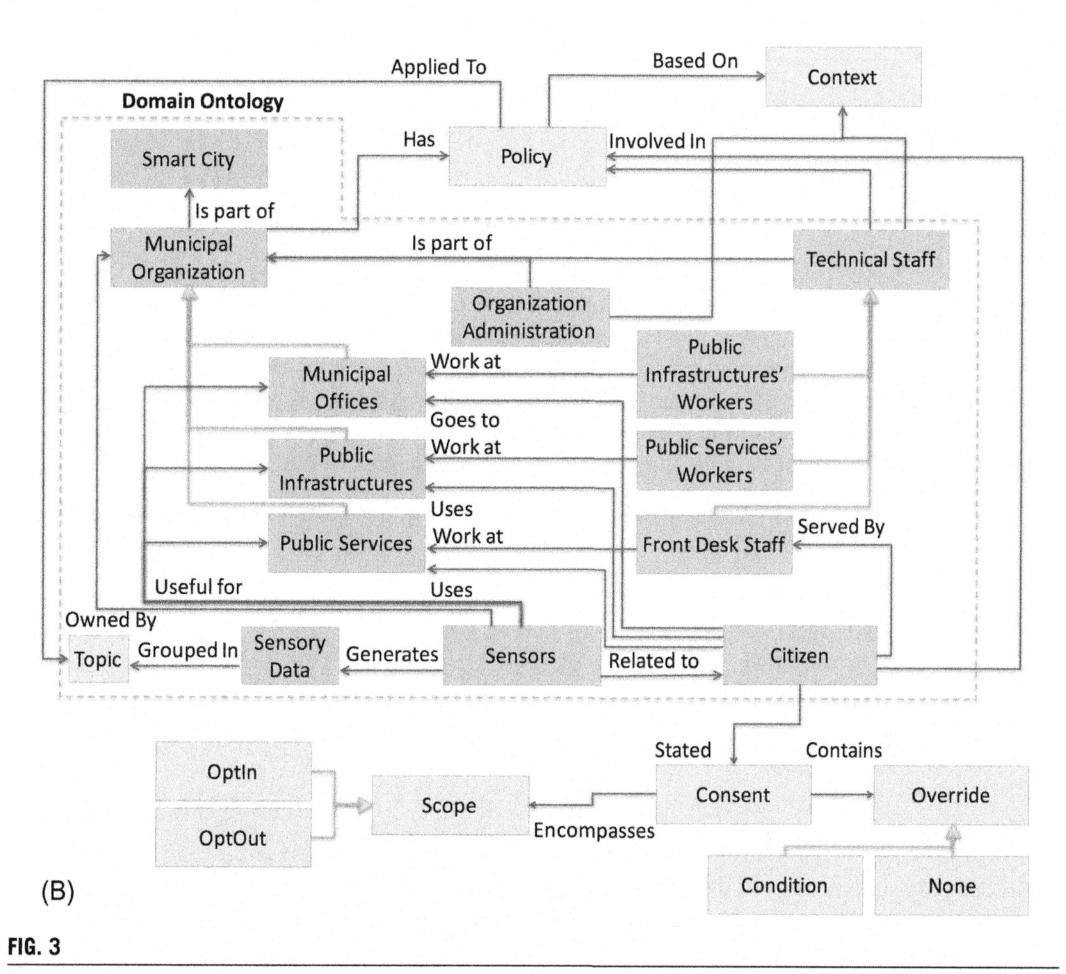


(B)

**FIG. 3**

Example of a Domain Ontology for healthcare augmented respectively with RBAC and PBAC Control Ontologies.

modifiers (such as, projection, distinct, order, limit, offset). There are three main query forms: (1) the SELECT form returns variable bindings, (2) the CONSTRUCT form returns an RDF graph specified by a proper template, and (3) the ASK form return a boolean value indicating the existence of a solution for a graph pattern. In our work we have used only ASK forms as a means to express access rules. The boolean return of the ASK queries are intended as a permission to access the requested resource or not. Such reasoning features allow deducing new information from the existing knowledge represented by the ontologies and to detect possible conflicts between policies before being used in the system. For a concrete example, a public infrastructure's worker can assume the role of bus driver, which allows him/her to obtain all the events containing the results of road traffic monitoring. Such a RBAC rule can be formulated as a SPARQL predicate:

```
ASK WHERE
{ ?pub_inf_worker HasName ?name;
        AssociatedWith ?role.
  ?role HasAccessTo ?sensory_data.
  ?sensory_data GroupedIn ?topic;
        HasId ?id.
}
```

where *?id* and *?name* are respectively the identifier of the requested sensory data and the name of the worker requesting it. On the other hand, when the PBAC model is applied, a request can be granted or denied based on proper policies, which can be decided by the given municipal organization or local authority and applied to the users of our sensor network. Such policies can be defined in terms of the current context and attributes of the user. For concrete examples, let us consider the following items, where a policy can be one of the following ones, properly represented by a SPARQL predicate:

- First policy: Sensory data with the results of real-time road traffic monitoring can be retrieved by a bus driver located within their respective bus route (here the context is the driver location);

```
ASK WHERE
{ ?bus_driver HasName ?name;
          AssociatedWith ?role;
          CharacterizedBy ?context.
  ?context Equal "Bus Route".
  ?role HasAccessTo ?topic.
  ?sensory_data GroupedIn ?topic;
          HasId ?id.
}
```

- Second policy: A member of the administrative staff of a municipal infrastructure can gain access to the sensory data generated in the given portion only within the accounting period in order to infer all the operations performed (such as the number of served passengers within the public transportation or the amount of collected garbage) and calculate the costs incurred (here the context is if the current date falls within preestablished accounting periods).

```
ASK WHERE
{ ?admin HasName ?name
          AssociatedWith ?role;
          CharacterizedBy ?context.
  ?context Equal ''Accounting Period".
  ?role HasAccessTo ?topic.
  ?sensory_data GroupedIn ?topic;
          HasId ?id .
}
```

- Third policy: A member of the front desk staff can obtain access to the sensory data related only to the citizens currently managed at his/her respective office (here the context is not explicit in an instance of the relative entity of the ontology, but on the existence of the relationship "Goes to" between the citizen in the requested sensory data and the office where the requesting officer is employed).

```
ASK WHERE
{ ?officer HasName ?name
        IsPartOf ?office.
  ?sensory_data RelatedTo ?citizen;
        HasId ?id.
  ?citizen GoesTo ?office.
}
```

We do not limit the adopted ontology to only describe the access control rules, but to also model the citizen consent to share the related sensory data through the dissemination of proper messages (i.e., the lower part of the ontologies shown in Fig. 3). Specifically, we have considered the semantic modeling of citizen consent in Khan et al. (2011), based on the study described in Coiera and Clarke (2004) to express specific conditions for controlling accesses to the electronic sensory information of a citizen. Specifically, a citizen allows the sharing of his/her sensitive information if he/she has enabled the Opt-in attribute. Such a consent is absolute or subject to proper overriding condition based on the context and/or the data to be shared. With Opt-out, the citizen refuses to share his/her data, and also this attribute can be subject to proper overriding Conditions. Let us consider the RBAC ontology shown in Fig. 3A, and the citizen consent in the formulation of the following role-based access control rule: a managing officers/workers can to obtain all the sensory data containing the results of road traffic monitoring, if the relative citizen have stated an Opt-in consent. Such a rule can be expressed as a SPARQL predicate as follows:

```
ASK WHERE
{ ?officer HasName ?name;
                AssociatedWith ?role.
  ?role HasAccessTo ?topic.
  ?sensory_data GroupedIn ?topic;
          HasId ?id;
          RelatedTo ?citizen.
  ?citizen Stated ?consent.
  ?consent Encompasses ?x.
  ?x type OptIn.
}
```

Fig. 4 shows the usage of an ontology for access control and its possible use to mediate heterogeneous security strategies by depicts the interactions of two access controllers in different municipal organizations. Specifically, an administrator has to insert the access control policies within the subsystem of its network, by instantiating the Domain and Control Ontologies and loading the set of parameterized SPARQL predicates representing the security policies. Furthermore, it must populate the ontologies by instantiating RDF objects with the details of the authorized users and their attributes. From this moment, the controller is able to take decision on granting or denying incoming requests by verifying if the active SPARQL predicates populated with data extracted from the receive requests are fully or partially satisfied. If an infrastructure encompasses multiple organizations, it means that there

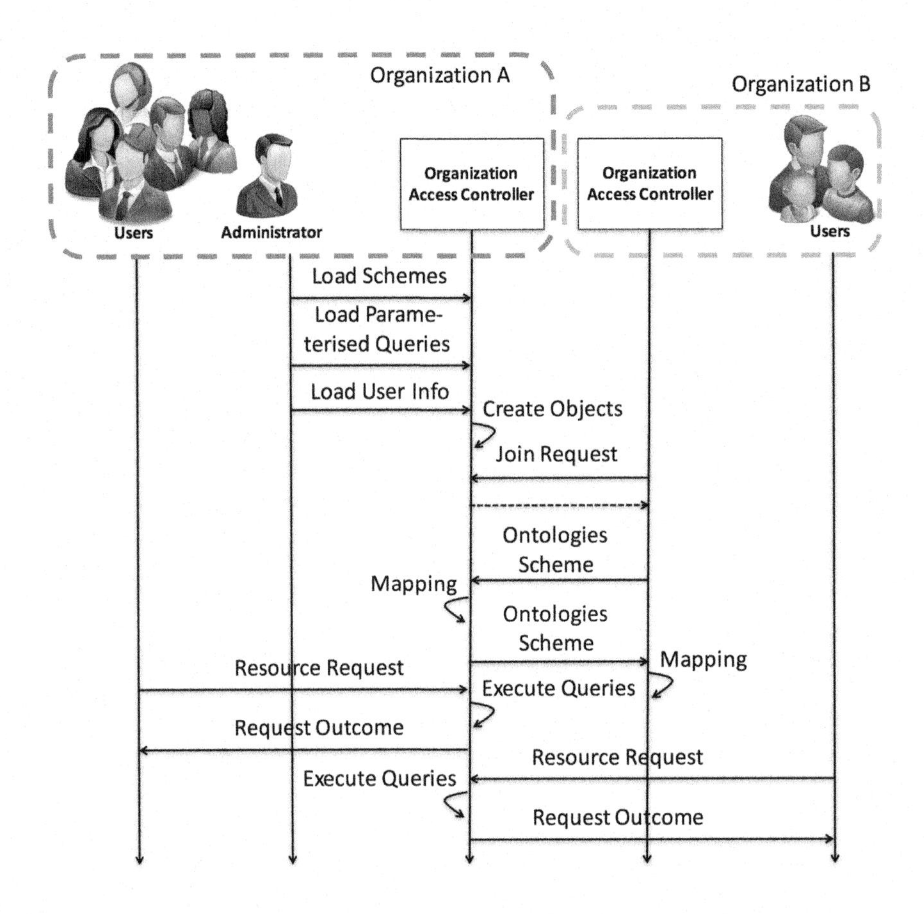


**FIG. 4**

Interactions to setup and execute access control.

are multiple access controllers that must coexists. Those controllers must discover themselves within the network, and this is possible with widely known protocols and peer sampling services, as in Mian et al. (2009). Then, they must exchange their information on the loaded ontologies, and automatically map the received schemes with their own ones. Matching diverse ontologies is still an open issue in the current literature and a survey on this topic is available in Rahm and Bernstein (2001). In our work, we have adopted a simple approach based on the semantic similarity of the terms composing two diverse ontologies. Specifically, let us consider two ontologies, namely $S$ and $T$, and two terms, namely $\nu$ and $\mu$, one per each ontology and holding a given semantic. The similarity among those two terms can be quantified according to four different operators: equivalence ($=$), if the two terms hold the same semantic; less general ($\sqsubseteq$), if the first term has a more specific meaning than the second one; more general ($\sqsupseteq$), if the first term has a broader semantic of the second one; and disjointness ($\perp$), if the two terms have opposing or different meaning. During the ontology matching, one term, $\nu$, of the first ontology is compared with one term, $\mu$, of the second one, and the two terms are mapped if the following holds:

$$\forall \nu \in S, \mu \in T : \nu \approx \mu \quad \textit{iff} \quad \nu = \mu \vee \nu \sqsubseteq \mu. \tag{1}$$

In other words, the two terms are mapped if they are equivalent or if $\nu$ is less general than $\mu$ (since all the values of $\nu$ will belong to the domain of $\mu$). If $\nu$ is more general than $\mu$, they cannot be mapped (since some values of $\nu$ might not belong to the domain of $\mu$). After such a mapping is applied, then the requests from an organization can be transformed by using the mapped terms of the receiving organization and verified on this organization's ontology. To this aim, there is no difference if such a request is received by a user belonging to the same organization of the controller or by a remote one, whose access control has joined the one that has received the request. In both cases, from the claim inserted by the user in the header of the request, the controller extract some values to populate the parameters of its SPARQL queries; in addition, if the user belongs to a different organization, the controller can retrieve the user information so as to create proper instances of the classes in its ontologies. If at least one of the queries returns a result, then the request can be granted. Otherwise, the request is refused.

## 5 PROTOTYPE

Fig. 6 shows our tentative solution for access control interoperability within a sensor network. Our starting point is that a sensor network is composed of sensors and base stations, running proper applications that are exposed as Web Services (WS). This is not our choice in order to simplify the development of our solution, but is also the supporting assumption for the Internet of Things (IoT) concept. Specifically, each sensor may host applications for measuring specific environment aspects (such as temperature, atmospheric pressure, luminosity, and air pollution), storing and processing sensory data and providing WS APIs to obtain the sensory data by users and base stations. Concrete examples of this kind of SOA applications for sensors are described in Ludovici et al. (2013) and Perera et al. (2014). Also the base station hosts a WS application, as described in Kyusakov et al. (2013), that collects data from the sensors by accessing to their offered interface numbered as 2 in the figure. The base station is able to store such data and to perform complex analytics on them, and to let users access both the collected sensory data and the obtained derived information by providing an offered interface numbered as 4 in the figure.

The starting point for the realization of our solution is not to design a novel tool needing a radical change to the existing software for an SOA-based sensor networks, but to allow the soft and "painless" adoption of our solution by simply plugging it as needed. By considering the paradigm of the Web services, the concept of handlers represented the best option for out scope, since it allows a simple insertion of additional logic to an existing Web service by avoiding any changes to its internal code. Specifically, WS-related standards and any relative development platforms allows developers to inserts handlers to a given Web service, which act as message interceptors. Traditionally, the handlers can be of two types. The first ones are specific to a given communication protocols and have the responsibility of implementing specific operations that process the context of the protocol-specific headers of the intercepted messages. The second ones, named as logical, are protocol-agnostic and are not interested to the information contains in any protocol-specific parts (like headers) of a message. On the contrary, logical handlers uses as their input only the payload of the message. Fig. 5 schematically shows the use of handlers (both SOAP and logical ones) to intercept incoming and outgoing SOAP messages and implement additional operations in order to add horizontal functionalities to an existing web service

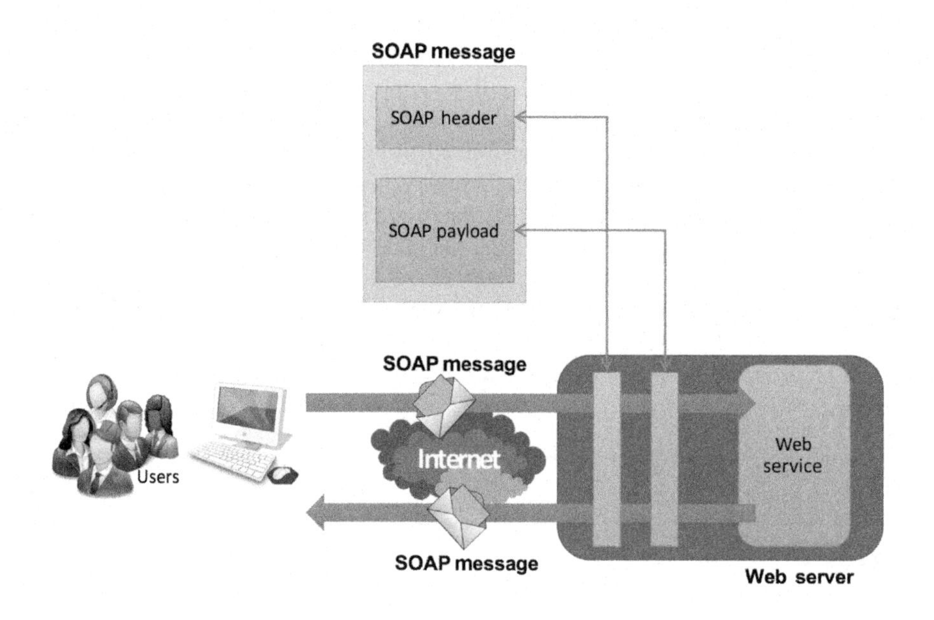


**FIG. 5**

Architecture overview of the SOAP and logical handlers.

without requiring any changes. This is a way to realize the design pattern called as Decorator, described in Gamma et al. (1995), within the context of web services.

Fig. 6 shows that our solution is composed of two handlers where the first one is of the SOAP kind, since they take their input data from the content of the message's header, while the second is a logical one. Specifically, the first one, named as Access Control Handler (ACH) in orange in the figure, is responsible for implementing the logic for authenticating users and controlling the accesses to the functionalities provided by the applied web server, so that it operates only on the incoming messages. Such a logic consists of the following actions:

1. The ACH checks if the header of the incoming messages contains an identify claim, which indicates the identify and attributed of the requesting entity.
2. If an identify claim is present, the ACH contacts a proper external service to verify its validity and authenticity.
3. If the first test is passed, the ACH must decide to allow or deny the request by contacting an external service implementing the access control model. The ACL passes the identify claim and a description of the request to the service and receives a decision. Based on the received response and the result of the first test, the SOAP message is passed to the web service or returned to the requesting entity.

The second is names as Audit Handler (AH), in blue in the figure, and implements the logic of auditing the decisions taken and to log the received requests. Since the handler can be applied to services hosted on sensors, the audit logs are not stored locally, but passed to a proper external service in charge of managing them. As evident from the description of ACH and AH, out solution does not encompass

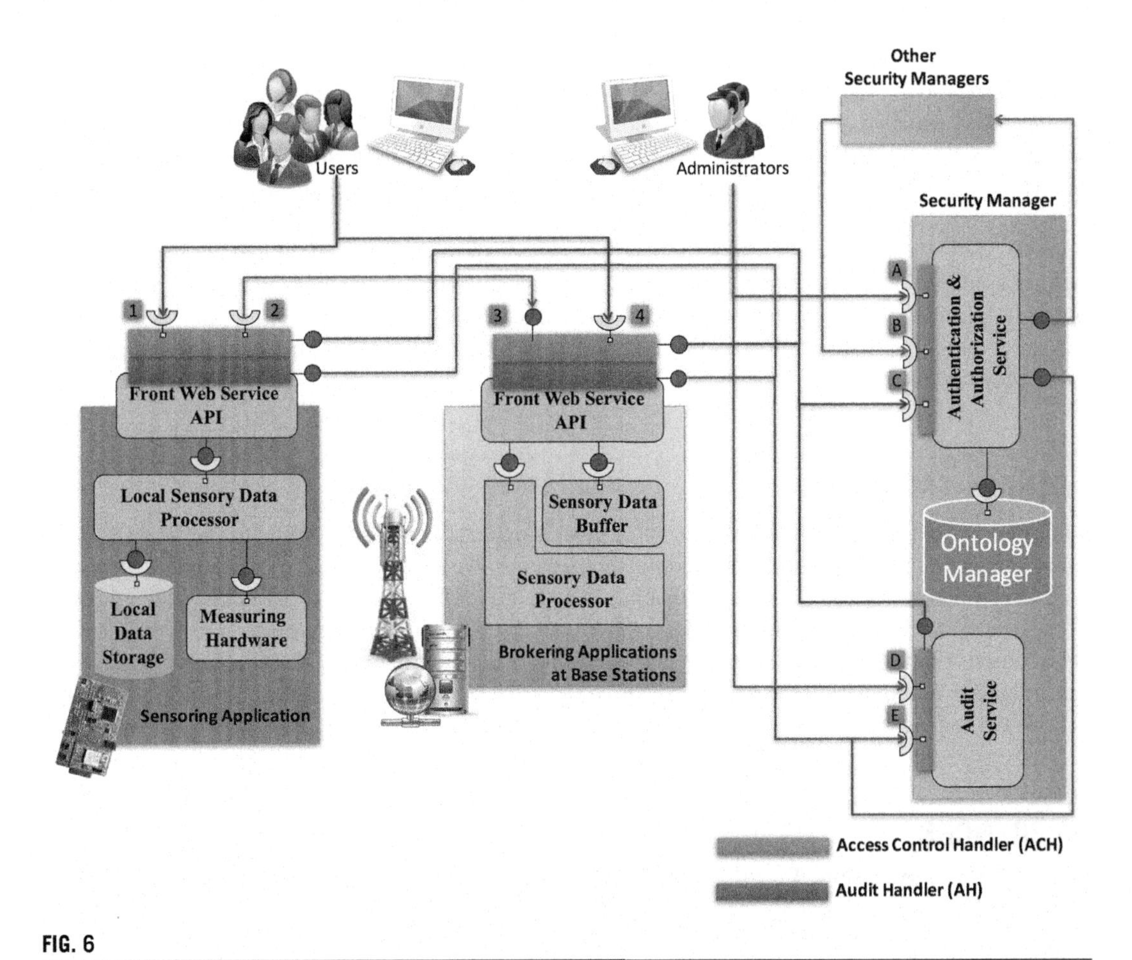


**FIG. 6**

Architecture overview of the proposed security solution (all the arrows indicate flows of SOAP messages).

only the two handlers, but also two external web services. The first one is called Authentication and Authorization Service (AAS), for managing identities and taking decisions on access requests. The ACH is also responsible to manage the ontological representations of access control models by properly creating an ontology, inferring data from its by means of SPARQL predicates and populating the ontology with proper object indicating the user identities as inserted by administrators. As ontology manager, we have used Apache Jena,[1] which is a Java library for manipulating the RDF models, and query RDF data using ARQ, a SPARQL 1.1 compliant engine. The second one acts as a repository of audit logs, and is named Audit Service (AS). AS does not only manage the audit logs coming from the AH but also the decisions taken by AAS, so that it is possible to realize postmortem analysis to assess the validity of our approach. These two services compose the so-called trusted third party used

[1] https://jena.apache.org/.

to secure the interactions among sensors, base stations and the users. In our implementation, there are two key issues to be resolved, related to how structuring the messages exchanged between the ACH and the AAS, and how to define the audit logs exchanged between the AH and the AS and stored locally by AS. In the first case, we have used a widely accepted formalisms within the Web service community and names *Security Assertion Markup Language* (SAML), which is the reference standard for exchanging authentication and authorization information for web services. The second issue is not simple to resolve as the first one, since there is no standard specification of audit logs for web services. We have used the Distributed Audit Service (XDAS)[2] specification as the architectural reference for the design of AS and the effective storage of such logs, and its set of generic events of relevance, in order to obtain a common portable audit record format and a set of well-defined interfaces for the components of a generic audit service.

## 6 FINAL REMARKS

In this paper, we have presented the problem of interoperable access controls models that must coexist within the context of urban-wide federated sensor networks. We have proposed an ontological approach to deal with this problem by having a matching of the different ontologies that describe the diverse access control models to be orchestrated. We have also described a prototype we have implemented of the proposed solution and how it has been applied to a generic sensor network. Our future work is to deal with other possible issues to provide security and privacy guarantees in sensor networks, such as a scalable key management, certificate-less signatures, and/or interoperable identity management.

## REFERENCES

Ahmed, M., Huang, X., Sharma, D., 2012. A taxonomy of internal attacks in wireless sensor network. World Acad. Sci. Eng. Technol. 62, 427–430.

Al-Fuqaha, A., Guizani, M., Mohammadi, M., Aledhari, M., Ayyash, M., 2015. Internet of things: a survey on enabling technologies, protocols, and applications. IEEE Commun. Surv. Tut. 17 (4), 2347–2376.

Apolinarski, W., Iqbal, U., Parreira, J.X., 2014. The GAMBAS middleware and SDK for smart city applications. In: Proceedings of the 2014 IEEE International Conference on Pervasive Computing and Communication Workshops (PERCOM Workshops), pp. 117–122.

Barreto, P., Kim, H., Lynn, B., Scott, M., 2002. Efficient algorithms for pairing-based cryptosystems. In: Proceedings of the 22nd Annual International Cryptology Conference on Advances in Cryptology, Lecture Notes in Computer Science, vol. 2442, pp. 354–369.

Benenson, Z., Gartner, F.C., Kesdogan, D., 2005. An algorithmic framework for robust access control in wireless sensor networks. In: Proceedings of the Second European Workshop on Wireless Sensor Networks, pp. 158–165.

Bethencourt, J., Sahai, A., Waters, B., 2007. Ciphertext-policy attribute-based encryption. In: Proceedings of the IEEE Symposium on Security and Privacy, pp. 321–334.

Bistarelli, S., Martinelli, F., Santini, F., 2010. A formal framework for trust policy negotiation in autonomic systems: abduction with soft constraints. In: Proceedings of the 7th International Conference on Autonomic and Trusted Computing, pp. 268–282.

[2]http://openxdas.sourceforge.net/.

Bos, J., Halderman, J., Heninger, N., Moore, J., Naehrig, M., Wustrow, E., 2014. Elliptic curve cryptography in practice. In: Proceedings of the 18th International Conference on Financial Cryptography and Data Security (FC 2014), pp. 157–175.

Chatterjee, S., Das, A., 2015. An effective ECC-based user access control scheme with attribute-based encryption for wireless sensor networks. Secur. Commun. Netw. 8 (9), 1752–1771.

Chen, T.Y., 2008. Knowledge sharing in virtual enterprises via an ontology-based access control approach. Comput. Ind. 59 (5), 502–519.

Chifor, B.C., Bica, I., Patriciu, V.V., 2016. Sensing service architecture for smart cities using social network platforms. Soft Comput. 1–10.

Choi, C., Choi, J., Kim, P., 2014. Ontology-based access control model for security policy reasoning in cloud computing. J. Supercomput. 67 (3), 711–722.

Coiera, E., Clarke, R., 2004. e-consent: the design and implementation of consumer consent mechanisms in an electronic environment. J. Am. Med. Inform. Assoc. 11 (2), 129–140.

Dai, H.N., Wang, Q., Li, D., Wong, R.W., 2013. On eavesdropping attacks in wireless sensor networks with directional antennas. Int. J. Distrib. Sens. Netw. 9 (8).

Esposito, C., Castiglione, A., Palmieri, F., 2016. Interoperable access control by means of a semantic approach. In: Proceedings of the 30th International Conference on Advanced Information Networking and Applications Workshops (WAINA), pp. 280–285.

Ferraiolo, D., Kuhn, D., Chandramouli, R., 2007. Role-Based Access Control, second ed. Artech House, Norwood, MA. Artech Print on Demand.

Gamma, E., Helm, R., Johnson, R., Vlissides, J., 1995. Design Patterns: Elements of Reusable Object-Oriented Software. Addison Wesley, Reading, MA.

Goyal, V., Pandey, O., Sahai, A., Waters, B., 2006. Attribute-based encryption for fine-grained access control of encrypted data. In: Proceedings of the 13th ACM Conference on Computer and Communications Security, pp. 89–98.

He, Z., Wu, L., Li, H., Lai, H., Hong, Z., 2011. Semantics-based access control approach for web service. J. Comput. 6 (6), 1152–1161.

Jin, J., Gubbi, J., Marusic, S., Palaniswami, M., 2014. An information framework for creating a smart city through internet of things. IEEE Internet Things J. 1 (2), 112–121.

Kawada, Y., Yano, K., Mizuno, Y., Terada, H., 2013. Data model and data access control method on service platform for smart public infrastructure. In: Proceedings of the International Conference on e-Business (ICE-B), pp. 1–9.

Khan, A., Chen, H., McKillop, I., 2011. A semantic approach to secure electronic patient information exchange in distributed environments. In: Proceedings of the Annual Conference of the Northeast Decision Sciences Institute (NEDSI).

Koblitz, N., Menezes, A., Vanstone, S., 2000. The state of elliptic curve cryptography. In: Towards a Quarter-Century of Public Key Cryptography, pp. 103–123.

Kos, A., et al., 2012. Open and scalable IoT platform and its applications for real time access line monitoring and alarm correlation. In: Proceedings of the 12th International Conference on Next Generation Wired/Wireless Advanced Networking (NEW2AN 12), pp. 27–38.

Kyusakov, R., Eliasson, J., Delsing, J., van Deventer, J., Gustafsson, J., 2013. Integration of wireless sensor and actuator nodes with it infrastructure using service-oriented architecture. IEEE Trans. Ind. Inform. 9 (1), 43–51.

Ludovici, A., Moreno, P., Calveras, A., 2013. Tinycoap: a novel constrained application protocol (COAP) implementation for embedding restful web services in wireless sensor networks based on TINYOS. J. Sens. Actuator Netw. 2, 288–315.

Manifavas, C., Fysarakis, K., Rantos, K., Kagiambakis, K., Papaefstathiou, I., 2014. Policy-based access control for body sensor networks. In: Proceedings of the 8th IFIP WG 11.2 International Workshop on Information Security Theory and Practice: Securing the Internet of Things (WISTP 14), pp. 150–159.

Martinez-Balleste, A., Perez-Martinez, P.A., Solanas, A., 2013. The pursuit of citizens' privacy: a privacy-aware smart city is possible. IEEE Commun. Mag. 51 (6), 136–141.

Maw, H., Xiao, H., Christianson, B., Malcolm, J., 2014. A survey of access control models in wireless sensor networks. J. Sens. Actuator Netw. 3, 150–180.

Mian, A.N., Baldoni, R., Beraldi, R., 2009. A survey of service discovery protocols in multihop mobile ad hoc networks. IEEE Pervasive Comput. 8 (1), 66–74.

Moses, T., 2005. Extensible Access Control Markup Language (XACML)—OASIS Standard. http://docs.oasis-open.org/xacml/2.0/access_control-xacml-2.0-core-spec-os.pdf. (Accessed July 2013).

Nam, T., Pardo, T.A., 2011. Smart city as urban innovation: focusing on management, policy, and context. In: Proceedings of the 5th International Conference on Theory and Practice of Electronic Governance, pp. 185–194.

National Computer Security Center, 1987. A Guide to Understanding Discretionary Access Control in Trusted Systems. NCSC-TG-003, Version 1.

Panja, B., Madria, S., Bhargava, B., 2008. A role-based access in a hierarchical sensor network architecture to provide multilevel security. Comput. Commun. 31 (4), 793–806.

Perera, C., Zaslavsky, A., Liu, C.H., Compton, M., Christen, P., Georgakopoulos, D., 2014. Sensor search techniques for sensing as a service architecture for the internet of things. IEEE Sens. J. 14 (2), 406–420.

Preuveneers, D., Joosen, W., 2016. Security and privacy controls for streaming data in extended intelligent environments. J. Ambient Intell. Smart Environ. 8 (4), 467–483.

Priebe, T., Fernandez, E., Mehlau, J., Pernul, G., 2004. A pattern system for access control. In: Proceedings of the 18th Annual IFIP WG 11.3 Working Conference on Data and Application Security.

Prud'hommeaux, E., Seaborne, A., 2008. Sparql Query Language for rdf. W3C, www.w3.org/TR/rdf-sparql-query/. (Accessed July 2013).

Rahm, E., Bernstein, P., 2001. A survey of approaches to automatic schema matching. VLDB J. 10 (4), 334–350.

Shamir, A., 1985. Identity-based crypto systems and signature schemes. In: Advances in Cryptology, Lecture Notes in Computer Science, vol. 196, 47–53.

Steuer, S., Benabbas, A., Kasrin, N., Nicklas, D., 2016. Challenges and design goals for an architecture of a privacy-preserving smart city lab. Datenbank-Spektrum 16 (2), 147–156.

Tan, Y.L., Goi, B.M., Komiya, R., Tan, S.Y., 2011. A study of attribute-based encryption for body sensor networks. In: Proceedings of the International Conference on Informatics Engineering and Information Science (ICIEIS 11), pp. 238–247.

Torres, J., Nogueira, M., Pujolle, G., 2013. A survey on identity management for the future network. IEEE Commun. Surv. Tut. 15 (2), 787–802.

Wang, W., Bhargava, B., 2004. Visualization of wormholes in sensor networks. In: Proceedings of the 3rd ACM Workshop on Wireless Security (WiSe 04), pp. 51–60.

Wang, H., Sheng, B., Li, Q., 2006. Elliptic curve cryptography-based access control in sensor networks. Int. J. Secur. Netw. 1 (3–4), 127–137.

Wood, A., Stankovic, J., 2002. Denial of service in sensor networks. Computer 35, 54–62.

Zhang, F., Safavi-Naini, R., Susilo, W., 2004. An efficient signature scheme from bilinear pairings and its applications. In: Public Key Cryptography—PKC 04, Lecture Notes in Computer Science, vol. 2947, pp. 277–290.

Zhang, Q., Yu, T., Ning, P., 2008. A framework for identifying compromised nodes in wireless sensor networks. ACM Trans. Inform. Syst. Secur. 11.

Zhi, L., Jing, W., Xiao-su, C., Lian-Xing, J., 2009. Research on policy-based access control model. In: Proceedings of the International Conference on Networks Security, Wireless Communications and Trusted Computing, vol. 2, pp. 164–167.

Zhu, Y., Keoh, S.L., Sloman, M., Lupu, E.C., 2009. A lightweight policy system for body sensor networks. IEEE Trans. Netw. Serv. Manage. 6 (3), 137–148.

CHAPTER

# SMART SENSOR AND BIG DATA SECURITY AND RESILIENCE

**Salvatore Venticinque*, Alba Amato***

*University of Campania "Luigi Vanvitelli", Aversa, Italy**

## 1 INTRODUCTION

Digitization is becoming an integral part of everyday life. People contribute to the greatest part of data generation, using application such as social media sites, digital pictures and videos, commercial transactions, advertising applications, and games. Currently, according to IDC,[1] an estimated 80 percent of existing data is unstructured. On the other hand Al Nuaimi et al. (2015) clearly discusses how smart phones, computers, environmental sensors, cameras, and GPS (Geo-graphical Positioning Systems) have contributed to accelerate data generation in the past few years.

Data sources are around us everywhere and are connected to the Internet. Such a technological revolution in data collection has been called the Internet of Things (IoT). Gartner (2015) defines the IoT as "the network of physical objects that contain embedded technology to communicate and sense or interact with their internal states or the external environment." The growth of the IoT represents one source of increased data, in many application domains, with all properties characterizing Big Data. According to Gartner (2015), "high-volume, high-velocity and high-variety information assets that demand cost-effective, are the most innovative forms of information that can be processed for enhanced insight and decision making."

In Burrus (2015) the authors claim that the value created by intersection of the IoT with Big Data is at the junction of collecting and leveraging data. The real value of such data is gained by new knowledge, that can be acquired by performing data analytics and used to improve the well-being of citizens and businesses alike. For example, sensor applications in multiple fields such as smart power grids, smart buildings, and smart industrial process control, feeding intelligent applications, can significantly contribute to more efficient use of resources and thus provide a reduction of greenhouse gas emissions and other sources of pollution. At the same time water, energy, and other utilities face intensifying pressure to improve customer service, strengthen resilience, and deliver security of supply.

Al Nuaimi et al. (2015) states that effective analysis and utilization of Big Data is a key factor for success in many business and service domains, including the smart city domain. Big Data analytics represents one of the recent technologies that has a huge potential to extract value and enhance services.

[1] www.idc.com/.

**Security and Resilience in Intelligent Data-Centric Systems and Communication Networks. https://doi.org/10.1016/B978-0-12-811373-8.00006-9**

Copyright © 2018 Elsevier Inc. All rights reserved.

On the other hand, to enable the flow of data from source to application architectural challenges need to be addressed and some relevant requirements such as security and resilience need to be met. In particular, the increasing complexity of Information and Communication Technologies (ICT), hyperconnectivity, as well as the generation of significant amounts of data, will also mean increased vulnerability, both to malicious attacks and unintentional incidents.

Such issues are critical because the safety and security of citizens depends on functionality, continuity of services, and the integrity and availability of data. Security can affect safety, and the two terms differ just in the triggering events of a disaster.

For this reason ICT architectures need to be designed, from inception, with security and resilience in mind. The system must be able to withstand and recover from deliberate attacks, accidents, or naturally occurring threats or incidents. It must be able to adapt to changing conditions, and withstand and recover rapidly from disruptions.

In this paper, we present an overview about smart sensors and Big Data security and resilience in different application domains, providing a critical analysis, and summarizing design guidelines and recommendations.

The next section introduces a reference model for the IoT architectures and provides details for each layer. Section 3 focuses on issues related to (big) data driven management of such kind of systems in terms of security and resilience. In Section 4 the application context is presented deepening the most relevant smart applications. Section 5 discusses security risks and provides recommendations to implement resilient systems which exploit smart sensors and big data to deliver smart applications.

## 2 THE IoT SYSTEM ARCHITECTURE

In the introduced scenario a new generation of systems, which are conceived to deliver ubiquitous services with increased awareness, arise. These kind of systems are characterized by fully distributed networks embracing the Internet of Things (IoT) and feeding applications with huge amount of data.

Security prevents system breaches or compromises of privacy and safety. Effective prevention requires the definition and enforcement of security policies at many layers. Actions range from the physical hardening of sensors to firewall installation and configuration. Continuous monitoring and assessment of the system security are necessary because working conditions can dynamically change, new vulnerabilities may be discovered, and degradation and obsolescence of techniques and technology may affect the security level.

System resilience prevents the spread of problems, allowing the system to still continuing to work, even if with a degradation of service levels. When a fault or a breach occurs it is necessary to limit the damage and recover operating condition at normal or acceptable level. It needs to protect information based value across all the layers providing *security* and *resilience*.

Security and resilience must be addressed both in terms of system design and deployment in the physical world, and in terms of data driven management.

A data driven management of the kind of system that is required by the flow of data from sensors to applications according to the value loop shown in Fig. 1. Sensors are the main source of information and can introduce faults or security breaches, which eventually propagate to applications along the direction of the value loop. However, at each step of the loop new threats and failures can occur, affecting the dependability of the system, and eventually delivering malicious or unsafe actions to the environment

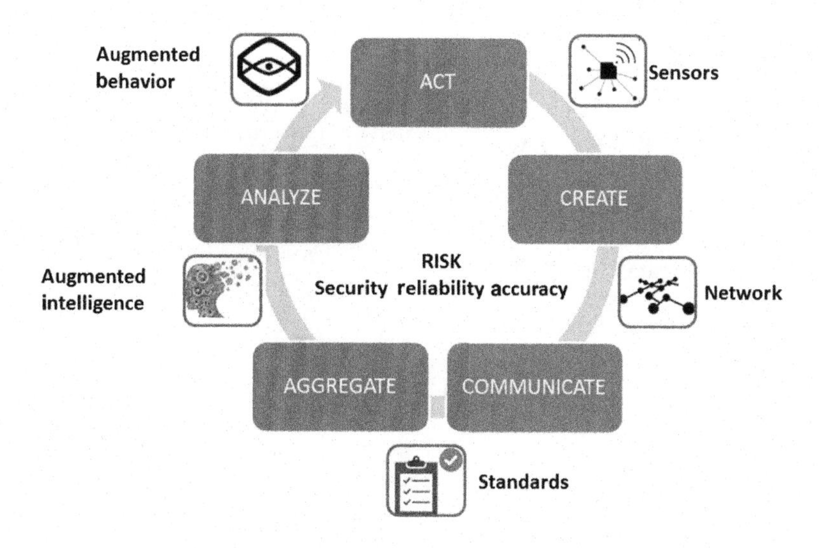


**FIG. 1**

The value loop.

by actuators at the end of the loop. For this reason, security and resilience must be addressed in each phase to provide specific protection of information-based value.

Fig. 2 shows a typical architecture where techniques and technologies are exploited to manage the phases of the value loop. Information and computation are spread across different layers and can move from cloud to edge analytics. The layered organization is exploited to dominate the complexity and allow for isolating different kinds of issues and risks.

In the following subsections, details are provided about how it needs to protect information based value across all the layers providing *security* and *resilience*.

## 2.1 SENSOR NETWORK

At a lower level we have a number of heterogeneous and ubiquitous smart sensors with low-power processing capability to enable edge-analytics. At this level information is not consciously produced by humans, but devices create information about individuals' behavior, analyze it, and eventually take actions on their behalf, autonomically or controlled by local and remote applications implemented at higher levels. Information value is created here in new ways, from data produced at a faster velocity, with greater heterogeneity and above all in the form that is always structured. Smart sensors implement the building blocks of the Internet of Things (IoT). They are not merely equipped with sensors and processing capability, but they are also connected to share information with other sensors or with higher levels. Wireless sensors and actuators communicate the information through wireless links "enabling interaction between people or computers and the surrounding environment" (OECD, 2010).

As data are created and transmitted, this represents a new opportunity for that information to be compromised (Saif et al., 2015). More sensitive data available across a broad network means the risks are higher and that data breaches could pose significant dangers to individuals and enterprises alike.

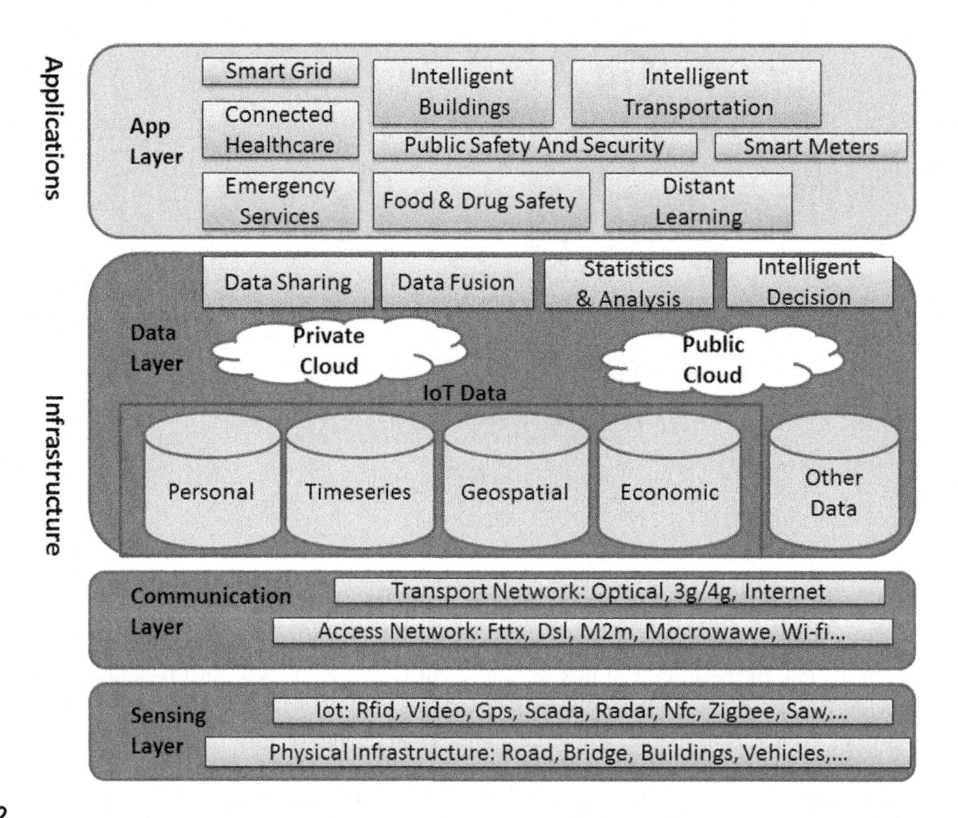


**FIG. 2**

IoT system architecture.

Data security risks could very likely go beyond privacy leaks to, potentially, the hacking of important public systems. According to the World Economic Forum, "Hacking the location data on a car is merely an invasion of privacy, whereas hacking the control system of a car would be a threat to a life."[2]

Specifically, the ability of the IoT to create and communicate data, in the first two stages of the value loop, makes the IoT valuable, but it introduce new risks. In fact these stages, nodes and networks, become vulnerable to security breaches. Some examples are introduced in Saif et al. (2015). Sensors are susceptible to counterfeiting (fake products embedded with malware or malicious code); data exfiltration (extracting sensitive data from a device via hacking); identity spoofing (an unauthorized source gaining access to a device using the correct credentials); and malicious modification of components (replacement of components with parts modified to generate incorrect results or allow unauthorized access).

On the other hand communication networks can be hacked, allowing data to be intercepted or their flow disrupted through denial-of-service attacks. In fact, many of the systems already in place use old sensors (water meters and gas sensors) with minimal security protocols because they were not designed

[2]https://www.weforum.org/.

to be connected to a more generally accessible network. Using these old, nonstandard technologies, at this layer of the IoT may introduce additional vulnerabilities and affects system reliability.

## 2.2 INTEGRATION

The data gathered by the different nodes of the network is usually sent to a sink which either uses the data locally, or which is connected to other networks (e.g., the Internet) through a gateway (Verdone et al., 2008).

The main issue addressed at this layer is the heterogeneity. It means the heterogeneity of devices and communication technologies, protocols, and data.

New metros and buses collects automatic passenger counting, bicycle counters collect real-time information on the number of bicycles that pass them, environmental sensor networks collect thermal and air quality data, the Bluetooth and Wi-Fi detectors provide information about crowds in a city, third party GPS-based vehicle probe data allows for traffic monitoring.

Saif et al. (2015) identifies as a relevant issue the lack of a single, generally accepted standard governing the functioning of IoT-enabled devices, which is therefore frequently a barrier to the interoperability required to realize the IoT deployments that many envision, with necessary level of security and resilience.

The integration layer uses mechanisms to collect and align these kind of data according to an uniform model, which allows for storing and processing them, individually or together, in order to extract valuable information.

Usually developers choose ad hoc solutions to create the interoperability that a given IoT solution needs. Unfortunately, without investing the time and money required to harden and test these solutions at the same level as formally developed standards, they are potentially more vulnerable to attacks.

The integration layer usually works as a gateway to route information to a higher layer where data are permanently stored and made available to applications and services.

## 2.3 BACKBONE NETWORKING

Information provided by smart sensors and smart meters needs to be transmitted via a communication backbone (OECD, 2010). This backbone is characterized by a high-speed and two-way flow of information. Different communication applications and technologies form the communication backbone. These can be classified into communication services groups as described in EPRI (2006).

Developers have to choose among several and diverse technologies in the area of communication network technologies. Usually, several network technologies are deployed. Different wide-area networks (WAN) and local-area networks (LAN) can be integrated to provide a means for such a necessary two-way information flow: from sensors to applications and from applications to actuators.

Multiple technologies are available, which provide both broadband and narrow-band solutions, resulting in a highly fragmented market. The choice of network technologies will depend on factors such as performance, reliability, cost, security, and the network infrastructure that is already available. It is likely that developers will rely on several network technologies when they build such infrastructure as they have to cope with differences in geography and population densities as well as availability and competition of different network technologies in their services areas. Some of these will require broadband, others will not. It will depend on amount of data and real-time requirements of applications.

## 2.4 BIG DATA STORAGES AND SERVICES

At this layer, data are collected from different sources and they are stored and made available to a higher level by the most suited technologies, according to the kind of data and the application which will consume it. Data is generated from many different sources in many different formats. There are a lot of new data formats many of which are unstructured. Not only data produced by sensors are collected stored here, but also input from humans, who produce unstructured data, search terms, posts, semantic tags, social information, are collected here.

Big Data is defined as data with high volume, velocity, and variety. The sampling frequency from perception devices can make the data size very large. Data velocity reflects the required speed for collecting and processing the data. Hence, Big Data management and processing techniques (hardware, software, algorithms, and AI) can be borrowed and applied in the domain of the IoT.

This data needs to be managed and classified into a structured format using some form of advanced database systems (big tables, columnar databases, graph databases, time-series, document storages, object storages, and key-value stores). In fact, according to Al Nuaimi et al. (2015) the current methodologies or data mining software tools cannot handle the large size and complexity. Technical and technological choices must be driven by the requirements to be met: availability, reliability, performance, integrity, or coherence.

## 2.5 SMART APPLICATIONS AND SERVICES

At these layers, applications and services exploit data to extract values and improve their own awareness and effectiveness. A Big Data analytic is identified as the main enabler to extract values that can be directly exploited by applications. The most relevant application domain for the IoT and Big Data are addressed in Section 4.

# 3 (BIG) DATA DRIVEN MANAGEMENT AND RISKS OF THE VALUE LOOP

The main goal pursued by applications is to exploit *value*, which is the best possible advantage Big Data can offer a business, based on good Big Data collection, management, and analysis. However, all companies use it to collect as much information as possible, including information which is currently not useful, believing that it could be exploited in the future.

For this reason, these times have been defined the Era of Big (Wasted) Data. Danowitz et al. (2012) estimates that ICT collect actually 23% Useful Data, 3% of which are Tagged only 0.5% are Analyzed.

Some features have been identified as characteristics of Big Data. *Volume* refers to the size of data that has been created from all the sources. *Velocity* refers to the speed at which data is generated, stored, analyzed, and processed. An emphasis has recently been put on supporting real-time Big Data analysis. *Variety* refers to the different types of data being generated. It is common now that most data is unstructured and cannot be easily categorized or tabulated. Other properties contribute to increase the complexity related to data management such as *Variability*. Variability refers to how the structure and meaning of data constantly changes especially when dealing with data generated from natural language analysis, for example.

The applications of Big Data can be classified into two types, off-line Big Data applications and real-time Big Data applications. In Mohamed and Al-Jaroodi (2014) the authors discuss how real-time Big Data applications are different because they rely on instantaneous input and fast analysis to arrive at a decision or action within a short and very specific timeline. In many cases, if a decision cannot be made within that timeline, it becomes useless. As a result, it is important to make all data necessary for such a decision available in a timely fashion and that the analysis is done in a fast and reliable way. As a result, real-time Big Data applications usually need higher technological requirements.

Al Nuaimi et al. (2015) present some controversies, discussed in the scientific literature, related to the definition, use, and benefits of Big Data for the IoT application domains and in the current innovation horizon. These deal with available Big Data tools, real-time analytics, accuracy, representation, cost, and accessibility.

However, many risks actually exist due to the complexity of Big Data, above all related to security and resilience. Some of them are specifically related to the Big Data properties as it is shown in Fig. 3.

Sharing data and information among different stakeholders is another challenge. Each government and city agency or department typically has its own warehouse or silo of confidential or public information. Some data may be governed by certain privacy conditions that make them hard to share across different entities. The challenge is to make sure not to cross the fine line between collecting and using Big Data and ensuring citizens' rights of privacy.

Security and privacy are major challenges in the IoT applications when using Big Data. In basic terms, this mean that databases may include confidential information related to the government and

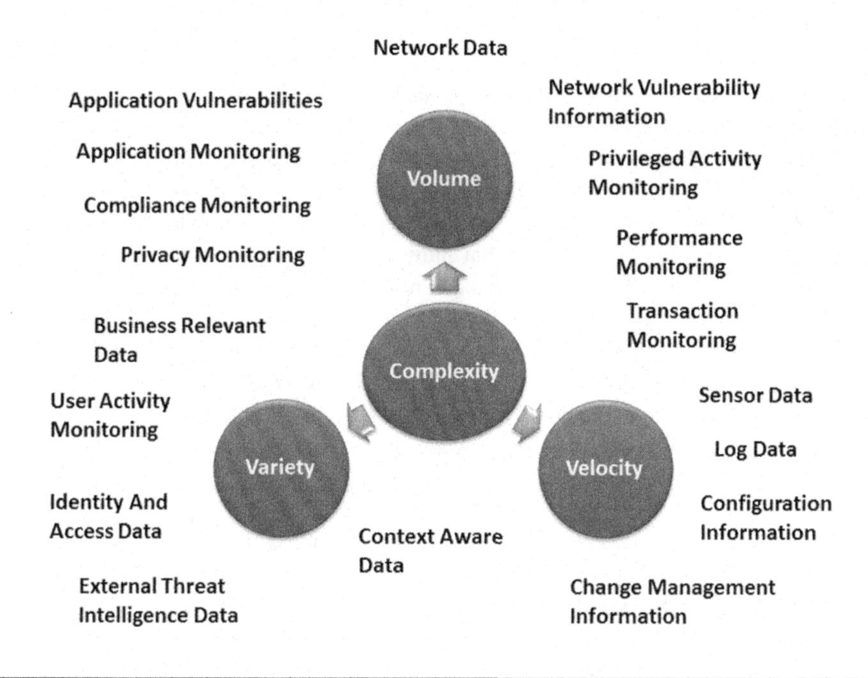


**FIG. 3**

Risks in big data management.

people, so they need high levels of security policies and mechanisms to protect this data against unauthorized use and malicious attacks. In addition, smart applications integrated together across agencies also require high security because the data will move over various types of networks, some of which may be open or unsecure as discussed in Khan et al. (2014a). Kim et al. (2014) demonstrates that most Big Data technologies today, including Cassandra and Hadoop, suffer from a lack of sufficient security, and make such an issue more complex.

In addition to the need to secure data as it travels and as it is being used by the different components of applications, there is also the need to clearly identify and protect privacy rights of organizations and individuals this data represents. Although specific entities can claim ownership of most Big Data, a lot of it includes personal and private information about individuals. Health and medical records, financial and bank records, retail history, and much more all provide intimate views of the people they represent. Many view access to this type of data as a violation of a person's legal rights for privacy. Making sure that stringent privacy policies are put in place and properly enforced represents a major challenge for Big Data smart city applications developers and users.

Security and privacy issues of Big Data are addressed also in Elmaghraby and Losavio (2014). A mathematical model is presented depicting the interaction between people, the IoT, and servers, which are vulnerable to information security threats. Using the proposed model they conclude that the benefits do and will far outweigh the risks when the rights and liberties in a democratic society are observed and protected. In particular, in the Smart Health sector there are several attempts of overcoming security and privacy limitations. Tarouco et al. (2012) discuss today's issues, including benefits and difficulties, as well as approaches to circumvent the problems of employing and integrating the Internet of Things devices in healthcare systems. It also discusses the challenge of data leakage and describes an innovative technological solution; namely the Secure Medical Workspace (SMW) that is a System that enables researchers to use clinical data securely for research.

According to Bertot and Choi (2013) there is, however, also a range of policy challenges to address regarding Big Data, including access and dissemination, digital asset management, archiving and preservation, privacy, and security. The requirement that organizations disclose their decisional criteria highlights an important fault line between law and technology. Authors of Tene and Polonetsky (2013) claim that fairness and due process mandate that individuals are informed of the basis for decisions affecting their lives, particularly those made by machines operating under opaque criteria. In Catteddu (2015) the authors discuss how weaknesses and threats are mainly linked to the lack of governance and control over IT operations and the potential lack of compliance with laws and regulations. National laws and regulations in the Member States of the European Union currently impose some restrictions on the movement of data outside the national territory; moreover, a problem exists in the determination of the applicable body of law (governing laws) when data is being stored and processed outside the European Union or by a non-EU service provider.

Data uncertainty and trustworthiness at lower levels also affect security and resilience of data, than affecting applications. For example, sensor data collected through a third party without a centralized control could have been produced by sensors that are faulty, wrongly calibrated, or beyond their lifetime. The challenge may also extend to the outputs of analyzing existing data (given the possibility of errors) and reporting the results for use by others, who may not be aware of such issues. Therefore continuously updating data gathering and usage policies, sharing and discussing them among all stakeholders of an applications, ensuring that the users understand and apply the policies correctly is vital and challenging at the same time, as stated in Bertot and Choi (2013).

# 4 APPLICATION DOMAINS

The Libelium[3] website presents a top 50 ranking IoT sensor applications. It identifies some relevant domains and related applications providing references to focus on them. We address some of them here discussing security requirements and resilience.

## 4.1 SMART CITIES

ICT is the prime enabler for smart cities, transforming application specific data into useful information and knowledge that can help in city planning and decision-making. Khan et al. (2014b) claims that concerning the ICT perspective, the possibility of the realization of smart cities is being enabled by smarter hardware and software, such as IoTs, and capacity to manage and process large scale data using cloud computing without compromising data security and citizens privacy. Smart parking, monitoring of vibrations and material conditions in buildings, monitoring bridges and historical monuments, urban noise, electromagnetic field levels, traffic congestion, art and goods preservation are just a few examples of these applications. The main objective is usually the optimization of public services and utilization of resources. Collected information are used by intelligent applications to optimize driving and walking routes, intelligent and weather adaptive lighting in street lights, optimize the trash collection routes, intelligent highways with warning messages and diversions according to climate conditions and unexpected events like accidents or traffic jams.

The Smart City is therefore characterized by being a *network of networks*, a network of infrastructure based on highly heterogeneous technologies. For this reason the Smart City is prone to the risks of service disruption due to natural or man-made causes, with the additional risk of a cascading effect. It is clear that a Smart City requires the definition of a policy based on strategies and models of prevention, protection, and mitigation of the effects of security or catastrophic events. There is, therefore, a significant convergence between the issues of resilience and continuity of service of the Smart City and those of Critical Infrastructures, with the important difference that the Smart Cities always are densely populated areas, with all the inherent problems in terms of management of supply, evacuation, and public health.[4]

Moreover like any other ICT system, the smart city technological and communication environment, the network infrastructure, and the IoT, will present vulnerabilities to cyberattacks. The higher complexity and heterogeneity of these environments could, in fact, determine an even higher exposure, and the need for more sophisticated protection strategies. Symantec (2016) claims that the sum of hyper connectivity, hypercomplexity, and hyperinformation volumes is equal to hypervulnerability. ICT infrastructure supporting smart cities applications must be conceived with high security against cyberattacks in mind, in order to guarantee service availability and continuity, data management and protection, as well as network resilience in the case of severe incidents.

The compromise between security level and resources in city governance will need to identify the most critical areas to protect, the types of threat they could be subject to, categories of attackers and likely motivations (financial, criminal, or political). At the Big Data layer information management, protection systems, and backup and recovery systems for mission-critical administration data should

[3]http://www.libelium.com/.
[4]http://www.piattaformaserit.it.

be implemented. Citizens' privacy and identities must be protected across domains, including local tax, health-care, education, and utilities.

## 4.2 SMART GRIDS

Smart Grids represent the most relevant application in the wider Smart Metering context. It has emerged to provide an intelligent power infrastructure. Smart grid technology promises to make the world power systems more secure, reliable, efficient, flexible, and sustainable. It achieves these goals through the integration of information and communication networks.

Smart devices and smart metering include sensors and sensor networks. Sensors are used at multiple places along the grid, such as at transformers and substations or at customers' homes. Shargal and Houseman (2009) explains how they play an outstanding role in the area of remote monitoring and they enable demand-side management and thus new business processes such as real-time pricing. Spread over the grid, sensors and sensor networks monitor the functioning and the health of grid devices, monitor temperature, provide outage detection, and detect power quality disturbances. Control centers can thus immediately receive accurate information about the actual condition of the grid. Consequently, as shown in OECD (2010), maintenance staff can maintain the grid just-in-time in the case of disruptions rather than rely on interval-based inspections.

Wireless sensor networks (WSNs) can be used as compared to traditional communication technologies because of its low-cost, rapid deployment, flexibility, and aggregated intelligence via parallel processing. However, better security solutions are needed to prevent the network from any malicious behaviors, sniffing, or attackers.

The integration of WSNs, actuators, smart meters, and other components of the power grid together with information and communication technology (ICT), is referred to as the Internet of Energy (IoE). IoT technology integrated within the smart power grid comes with a cost of storing and processing large volume of data every minute. This data includes end users load demand, power lines faults, network's components status, scheduling energy consumption, forecast conditions, advanced metering records, outage management records, and enterprise assets. Hence, utility companies must have the software and hardware capabilities to store, manage, and process the collected data efficiently. In Witt (2015) the author explains how the high-volume data gathered in a smart grid is similar in size and characteristics to the concept of Big Data.

IoE uses the bidirectional flow of energy and information within the smart grid to gain deep insights on power usage and predicts future actions to increase energy efficiency and low overall cost. Jaradat et al. (2015) cites reports, according to which, 800 millions smart meters are expected to be installed globally by 2020. In order to achieve fine-grain monitoring and scheduling, information from the power grid needs to be collected within short intervals. New regulation, at least in Italy, will ask for installed smart meters to provide a new sample every 15 minutes, this leads to about 77 billions of readings globally during one day. Such huge amount of data will obviously require the adoption of Big Data technologies.

The resilience of the ICT infrastructure is so relevant to make available data from which information is extracted in order to detect faults, isolate the fault, and then resolve faults. Also the security of the ICT infrastructure becomes a requirements to avoid failures caused by cyberattacks or by combined attacks to the power grid and to the ICT infrastructures. Jaradat et al. (2015) explains how in this situation the reliability of the power system will depend more and more on online monitoring of power lines and infrastructures. Performing energy restoration in smart grid must take into account the location criticality of blackouts. For example, it is critical to guarantee high reliability for the health and industrial systems. The restoration problem becomes a very complex problem when taking into the consideration the large number of combinations of switching operations which exponentially increases with the increase in system's components.

## 4.3 SMART BUILDINGS

Smart buildings are a field closely linked to smart grids. Smart buildings rely on a set of technologies that enhance energy-efficiency and user comfort as well as the monitoring and safety of the buildings. IoT technologies are used in building management systems which monitor heating, lighting, and ventilation, software packages which automatically switch off devices such as computers and monitors when offices are empty and security and access systems.

The smart home provides a particularly resonant example of the risks involved when multiple brands, devices, and stakeholders aggregate and analyze multiple data sets and are knit together to form an ecosystem. For example, the garage door opener provides access to not just the garage, but also the primary home. In some configurations, opening the garage door deactivates the home alarm. This, however, means that the entire alarm system is deactivated if only the garage door opener is compromised.

Major concerns regarding smart home technology are those of security and data privacy, and these issues must be addressed also from vendors to promote their products. One question is whether different demographic age groups will introduce additional vulnerabilities by sharing information about their location, habits, and even home temperature settings with, at least, application providers. Even if they are feeling more and more comfortable with this way of information sharing, they must be aware about correct behaviors and related risks.

## 4.4 DISASTER MANAGEMENT, EMERGENCY, AND RECOVERY

Disaster management is another application context where the huge amount of devices implementing the IoT can be exploited. Typical applications are early warning in forest fire detection or in earthquake detection. But also liquid detection in data centers is another example. Smart sensors and their data are exploited also for early detection of disasters and for the enforcement of emergency strategies; in warehouses and sensitive building grounds, to prevent break downs and corrosion; for access control to restricted areas and detection of people in nonauthorized areas; for distributed measurement of radiation levels in nuclear power stations surroundings to generate leakage alerts; for detection of gas levels and leakages in industrial environments, surroundings of chemical factories, and inside mines.

Usually, in these contexts, a distributed infrastructure, such as the IoT architecture, is preferred to centralized ones, in order to avoid single points of failure. In an emergency situation, "smart" distributed infrastructure can be a strength if they allow a better management of crises, also through the use of the population as an additional "information source" (e.g., citizen as a distributed sensor, timely communication of context). Just because the connection of the citizens in the area through mobile networks is a source of information, its resilience is still relevant. Its availability must be guaranteed, even in emergency conditions, as not to lose a channel of acquisition and communication of information essential for the management of the emergency itself.

In fact, such a distributed system of sensors and networks also represents a point of weakness if a poor resilience creates additional problems such as lack of information, energy, transport capacity (e.g., deficiencies in electrical distribution which prevent operation of transport networks thus complicating efficient evacuation of the population in crisis scenarios). Moreover preserving users' privacy (it fosters resilience, but affect privacy and security) becomes an additional requirements that must be addressed before, during, and after the emergency.

## 4.5 SMART TRANSPORTATION AND LOGISTICS

Sensors and sensor networks play a vital role in the increase of transport efficiency. For example, sensor technology contributes to better tracking of goods and vehicles, which might result in lower level of inventories and thus energy savings from less inventory infrastructure as well as a reduced need for transportation (Atkinson and Castro, 2008). Data collected from sensors and advance analytics are used for road traffic monitoring systems and to provide information for traffic lights which are then controlled. These sensors are further able to detect whether, for example, public buses are approaching so that the green phase of traffic lights can be extended, allowing buses to keep their schedules (Veloso et al., 2009). They also transmit information to update public transport panels, for motorway tolling purposes where they detect vehicle tags and retrieve the required information, for speed control and others. Monitoring of vibrations, strokes, container openings or cold chain maintenance for insurance purposes allow for improving the quality of shipment conditions. RFID and other kind of sensor networks allow also for searching individual items in big surfaces like warehouses or harbors. Warning emission on containers storing inflammable goods closed to others containing explosive material and control of routes followed for delicate goods like medical drugs, jewels, or dangerous merchandises are other examples.

## 4.6 OTHER APPLICATION DOMAINS

Environmental degradation and global warming are among the major global challenges facing us. These challenges include improving the efficient use of energy as well as climate change. Various examples illustrate the role of the IoT as a provider of solutions to a smart environment domain.

The review of the studies assessing the impact of sensor technology in reducing greenhouse gas emissions reveals that the technology has a high potential to contribute to a reduction of emissions across various fields of application. Relevant applications are the monitoring of air pollution, snow level monitoring, potable water monitoring, chemical leakage detection in rivers, and pollution levels in the sea and river floods.

In the Retail domain the Supply Chain Control allows for supply chain and product tracking for traceability purposes. Contact-less payment in location or activity duration for public transport, gyms, and theme parks or RFID-based theft prevention and access control systems are other examples. In this case the security of the RFID tag, according to Ijaz et al. (2016), it is still considered risky to give away sensitive information through unauthorized access, which could create problems with data confidentiality, and privacy. The problems to data integrity may also occur due to information leakage. The security of tags and of readers is still an issue that limits the spread of this kind of services.

Intelligent Shopping Applications exploit information provided by personal devices to deliver recommendation to customers according to their habits, preferences, location. In the e-health domain, examples of application are fall detection, control of conditions inside freezers storing vaccines, medicines and organic elements, ports-men care, patients surveillance inside hospitals and in elderly housing, and the measurement of UV sun rays to warn people not to be exposed in certain hours. In this context the potential of wearables devices and smart home technology could be exploited.

In fact, wearable technology has already begun collecting data on the habits and behavior of individuals and the transmission of data to third parties. Examples are smart-watches, or wristbands that track the wearers' steps each day as well as sleep quality to motivate activity and physical health. However, in this case, such sensors were conceived for playing sports and having free-time with a smarter approach, and they do not meet security requirements or provide the necessary resilience, which are necessary in safety critical applications.

Industrial control systems are changing nowadays, moving to fully-distributed networks and embracing the Internet of Things (IoT). Khan et al. (2013) explains how handling interconnected communication infrastructures, to access contextual information in smart applications and physical spaces, to support good decision-making processes, requires attention to various aspects of connectivity, security, and privacy. Using cloud computing for SCADA systems raises security and network issues caused by the fact that the storage in the cloud is shared among several users. This makes it more vulnerable to attacks. For this reason new solutions are investigated, moving from cloud to edge analytics, and moving beyond traditional ideas of resiliency. Control systems can be defined as IoT networks, which are remotely repairable, able to mitigate failures to ensure uptime and safety. Control is done in the cloud at the center of the network.

Sensors and especially sensor networks are used in multiple ways in industrial applications. They enable real-time data sharing on industrial processes, on the *health state* of equipment and the control of operating resources to increase industrial efficiency, productivity, and reduce energy usage and emissions. An example of applications are machine auto-diagnosis and assets control, indoor air quality, temperature monitoring, ozone presence, indoor location, smart agriculture, and smart animal farming.

# 5 DISCUSSION AND ANALYSIS

The lesson learned tells us that the promise of the IoT today concerns more possibilities to create more value, because of more data, but it also introduces new liabilities in terms of security and system reliability.

## 5.1 SECURITY CONCERNS AND RECOMMENDATIONS

The complex nature of IoT ecosystems may lead enterprises, or even users, to assume that all the players involved can share responsibility for security. However, it is a risk to assume that partners, much less customers, should or will take responsibility for maintaining data privacy and protecting against breaches. Enterprises should consider behaving as if the responsibility for security were theirs alone.

Low-cost sensors and increasing flexibility make it easy to collect more data than is currently needed. Data availability by smart sensors technologies leverage also the collection and aggregation of any kind information of wide scope from heterogeneous devices to foster an unforeseen future utility. This common practice can be observed when, during installations, applications for mobile devices ask users for a set of permissions, to use on-board peripherals and data, which is larger than the one really necessary, breaching the user's privacy without benefits for him.

Moreover, the Big Data technologies allow more and more processing capabilities of data, also in real-time that, combined with such a wider scope of information, allows for unauthorized inferences causing a false perception of one's own privacy. Such a perception is also affected by the huge number and variety of companies that collect, share, or sell data, or just delegate their management to third parties, making it difficult for the users and the companies themselves to understand how and if data or applications have been breached. Safeguarding user's data is a critical issue and the transparency of a peoples' lives is increasing, as more systems are holding information. At the data layer, because of huge volumes, small thefts of information could be unnoticeable until the total amount become evident enough, however, a small quantity of data loss can affect the security and the reliability of the system.

Al Nuaimi et al. (2015) claims that security and privacy policies and procedures must be conceived as an integral part of the design and implementation of both hardware/software architecture and data, and along the data value loop.

In particular, in order to ensure that all technology and applications components include and maintain the right levels of security and privacy mechanisms *risks*, *controls*, and *metrics* must be defined.

At sensors layer the following guidelines have been identified to address security concerns:

- Sensor and sensor network security must be hardened, to avoid malicious modification, counterfeiting, a man-in-the-middle attack, or sniffing. Cryptography and authentication are security mechanisms to be applied. Usage of storing authentication mechanisms also prevents from accidental disclosing of credentials and from attaching unauthorized devices to the infrastructure.
- Human awareness about the management of his own data and device must be improved. Installing sensors, actuators, networks, and application must support standard authentication and authorization policies. This reduces the risk of eavesdropping, phishing, and password-based attacks
- Standards must be developed to allow for the connection of only certified devices to build IoT infrastructures. This reduces the introduction of risks, such as threats and vulnerabilities.
- Players must agree on benchmarks to increase interoperability even among systems used by different industries.
- Limits on what can be collected in the first place (sensor or device) must be set to mitigate many risks. A good practice will be the collection of only those data that will generate enough value to justify the risk. It prevents disclosure or inference of private information.
- Physical security must be granted.
- Security must be embedded within data-utilizing encryption.
- Infrastructure must be protected securing endpoints, messaging, and web environments.
- Fog computing or edge computing should be exploited in a distributed manner rather than using the centralized cloud computing model. It means more computing at the edge and localized decision-making.
- It needs to work to define and use standards for interoperability.
- It is preferable to use purpose-built devices or add-ons, rather than pre-IoT solutions.
- It must develop advanced mobile computing capabilities that can integrate new IoT information tracking and capture requirements.

At the data layer the main security control is data protection, but the establishment of data governance can help to mitigate some risks arising from aggregation.

- A fine definition of data ownership must allow for monitoring, where a stakeholder within the ecosystem owns each piece of information and enforces authorization.
- The length of the data's lifecycle must be established to ensure that data cannot be retained beyond a suitable time-frame or used for nonprescribed purposes.
- Data storages must be isolated and not shared among stakeholders.
- Physical security must be granted by well-defined data documentation and codebooks to ensure informed use of the datasets as required in Bertot and Choi (2013).
- The beneficial uses of data against individuals' privacy concerns must be leveraged addressing some of the fundamental concepts of privacy laws. This includes defining "personally identifiable information," and the role of individual control as stated in Tene and Polonetsky (2013).
- Security services must be managed or outsourced to security providers. The ICT leadership can, in that way, focus on their functional duties of the system.
- Threat intelligence must be exploited in order to understand the major trends in terms of potential attackers, through analyzing trends on malware, security threats, and vulnerabilities.

The introduced protection mechanisms must be complemented with continuous and fine vigilance facilities. Remaining vigilant to new or unexpected challenges is crucial to maintaining security. According to Bertot and Choi (2013), vigilance in this case means:

- observing changes of infrastructures.
- evaluating introduction of new technologies for protection.
- evaluating new technologies available to attackers.
- monitoring the discovery of new vulnerabilities.
- identifying and profile new data sources and data sinks.
- analyzing all gathered data and, eventually correlating data against a normal condition or behavior.
- periodically evaluating security metrics, such as active/inactive user accounts, accounts with(without) password expiration, mean password expiration, and account expiration dates.
- reviewing and recalibrating information and data policies as necessary by focusing on privacy, data reuse, data accuracy, data access, archiving, and preservation.

## 5.2 RESILIENCE CONCERNS AND GUIDELINES

Although a smart IoT application provides many positive advantages for its users, it also poses several threats to their safety and wellbeing because of security breach or unpredictable accidents.

The IoT has meaningful potential to increase the implementation and efficacy of disaster-safety, resilience, and mitigation efforts. Emergency preparedness, loss estimation, and decision support software tools are improved by remote sensing technologies. On one hand, IoT has the potential to create an enormous amount of (Big) Data, and that data can revolutionize how we understand risk. On the other hand, we observed that resilience of application and services rely on the resilience of ICT platforms, which allows for their development and delivery.

The real danger of a such a complex system is the so called *data-chain domino effect*. Any component can disrupt the entire system by a self-propagating disaster in case of failure of some elements.

No one can guarantee that a failure will never occur at some point. And in the face of almost certain failure, a system's resilience defines how quickly a realized risk can be addressed and normal operations restored. It requires that consequences of failures are not catastrophic and can be recovered, at least partially, moving the system to even degraded, but safe working conditions.

Even the need for resilience of the IoT deployments is relevant, both in terms of system design and deployment in the physical world, and in term of data driven design and management.

Regarding the system design and deployment some recommendations are summarized:

- External devices and databases may be under the control of third-party organizations, however, it needs to guarantee that they implement themselves resilient strategies, because the system is only as strong as its weakest link.
- A loosely coupled architecture must be designed and developed in order to contain threats to small areas without compromising the full system.
- Stronger security-event-monitoring controls at the hub to effectively shut down the affected smart components in a fail-safe manner must be implemented.
- Reliable communication must be granted installing physically redundant communication systems and using multiple transmission channels (fiber and wireless). As the system grows it needs to plan for further redundant communications.
- Identify critical components and ensure 24x7 availability of the critical infrastructure.
- Use appropriate integration software and platforms. Consider adopting the latest integrated IoT platforms.

About the information-centric approaches we identify some relevant recommendations.

- Data storages and data services must be replicated to improve reliability, availability, data loss prevention, archiving, and disaster recovery.
- Key stakeholders must be identified and organized. Governance, Risk and Compliance (GRC) must be defined to ensure service continuity, even exploiting solutions and methodologies on cyber security.
- It must be aligned with national coordination on cyber-incidents and natural disasters.
- Users' awareness to use ICT solutions safely must be improved. This will also finance the quality of collected data and the performance of the applications. An important aspect is their knowledge and practice of good safety and well-established procedures to address and recover from failures.
- Governing entities of smart cities must establish guiding principles of openness, transparency, participation, and collaboration to keep the exchange and flow of Big Data under control.
- Simulation systems must be developed to help predict and view possible changes and forecast potential problems. According to Al Nuaimi et al. (2015), this will help avoid or at least reduce some of the risks involved and in many cases also help reduce implementation and testing costs.
- The correlation between different layers of the infrastructure must be investigated.
- An information management strategy must be developed.
- Advanced data aggregation and processing capabilities must be provided. It needs to select most suited advanced Big Data tools to aggregate large data sets, according to data properties, and data type (structured and unstructured).
- Data must be processed in multiple ways.
- Built platform must be flexible to exploit the continuous evolution of data standards, which render prior implementations obsolete and make them failure prone.

## 6 CONCLUSION

This paper provided a critical survey about security and resilience in those application domains and systems where the information-based value must be protected across different architectural layers and during the data management process. In particular, it has been discussed how the Internet of Things has accelerated both the installation of ubiquitous devices, which behave as distributed smart sensors and actuators, and accordingly the generation of Big Data. The heterogeneity of devices and the huge amount of collected data suggested to exploit the extracted knowledge to improve the well-being of citizens by improving effectiveness and resilience of services themselves, which become smarter and smarter.

There are several recent initiatives examining the interrelation of Big Data and community-driven resilience. The United Nations' Global Pulse is a focus on the intersection of Big Data, sustainable development, and humanitarian action. Assisting in future disasters through supporting research leveraging big data and data analytics is another goal defined in Henderson et al. (2015). Processing disaster data and improving the resilience and responsiveness of computer systems and networks in disasters to facilitate real-time data analytics is proposed as a mean for such an achievement. The exploitation of Big Data and IoT by communities and governments is also considered to end extreme poverty and boost shared productivity.

However, this contribution aimed at explaining that connection of smart sensors and collection of Big Data is not enough to provide value-added smart services. In fact, security and safety risks for users and applications are improved by increasing awareness and knowledge, but are affected also by security and resilience of each layer of the ICT infrastructure and of (Big) Data management strategy. This paper pointed that all tangible benefits provided by IoT and Big Data must be compared with introduced risks and vulnerabilities due, above all, to the complexity of both data and technologies.

Here the authors would like to conclude that security and resilience cannot be neglected for optimizing the benefit obtained from the intersection of the IoT and Big Data. Issues and guidelines summarized in Section 5 must be addressed providing the necessary security level and resilience, and evaluating the trade-off between cost or complexity and the expected value promised by big data analytics.

## ACRONYMS

**BMS** building management system
**CERT** computer emergency response teams
**CRE** customer relations executive
**GRC** Governance, Risk and Compliance
**GPS** Geo-graphical Positioning Systems
**ICT** information and communication technology
**IoE** Internet of Energy
**IoT** Internet of Things
**LAN** local area networks
**RFID** Radio-Frequency IDentification
**SCADA** supervisory control and data acquisition
**WAN** wide area networks
**WSN** wireless sensor network

## GLOSSARY

**Big Data** high-volume, high-velocity, and/or high-variety information assets that demand cost-effective, innovative forms of information processing that enable enhanced insight, decision making, and process automation.

**Dependability** the ability to deliver service that can justifiably be trusted. The alternate definition is the ability to avoid service failures that are more frequent and more severe than is acceptable. It is a measure of a system's availability, reliability, and its maintainability, and maintenance support performance, and, in some cases, other characteristics such as durability, safety, and security.

**IoT** the network of physical objects that contain embedded technology to communicate and sense or interact with their internal states or the external environment.

**Resilience** Operational resilience is a set of techniques that allow people, processes, and informational systems to adapt to changing patterns.

**Security** measures taken to protect a system from unauthorized access to, and accidental or willful interference of, regular operations.

**Smart Device** an electronic device, generally connected to other devices or networks via different wireless protocols such as Bluetooth, NFC, Wi-Fi, 3G, etc., that can operate to some extent interactively and autonomously.

## REFERENCES


Al Nuaimi, E., Al Neyadi, H., Mohamed, N., Al-Jaroodi, J., 2015. Applications of big data to smart cities. J. Internet Serv. Appl. 6 (1), 25. doi:10.1186/s13174-015-0041-5.

Atkinson, R.D., Castro, D.D., 2008. Digital Quality of Life: Understanding the Personal and Social Benefits of the Information Technology Revolution. Information, Technology and Innovation Foundation, Washington, DC. http://library.bsl.org.au/showitem.php?handle=1/1278.

Bertot, J.C., Choi, H., 2013. Big data and e-government: issues, policies, and recommendations. In: Proceedings of the 14th Annual International Conference on Digital Government Research. ACM, New York, NY, USA, pp. 1–10. doi:10.1145/2479724.2479730.

Burrus, D., 2015. The internet of things is far bigger than anyone realizes. Wired. http://www.wired.com/2014/11/the-internet-of-things-bigger/. (Online; Accessed 29 September 2016).

Catteddu, D., 2015. Security & resilience in governmental clouds, making an informed decision, European. ENISA. http://www.enisa.europa.eu/activities/riskmanagement/emerging-and-future-risk/deliverables/security-and resilience-in-governmental-clouds. (Online; Accessed 29 September 2016)

Danowitz, A., Kelley, K., Mao, J., Stevenson, J.P., Horowitz, M., 2012. CPU DB: recording microprocessor history. Queue 10 (4), 10:10–10:27. doi:10.1145/2181796.2181798.

Elmaghraby, A.S., Losavio, M.M., 2014. Cyber security challenges in smart cities: safety, security and privacy. J. Adv. Res. 5 (4), 491–497. Cyber Security.

EPRI, 2006. 2006 Annual Report. Together... Shaping the Future of Electricity. Tech. rep. EPRI. http://mydocs.epri.com/docs/CorporateDocuments/Mission&History/AnnualReport2006.pdf. (Online; Accessed 29 September 2016).

Gartner, 2015. It Glossary. Big Data. Gartner, Inc. http://www.gartner.com/it-glossary/big-data. (Online; Accessed 29 September 2016).

Henderson, L.C., Audrey, K., Rierson, J., 2015. Understanding the intersection of resilience, big data, and the internet of things in the changing insurance marketplace. Federal Alliance for Safe Homes (FLASH). http://flash.org/intersectionofresilience.pdf.

Ijaz, S., Ali Shah, M., Khan, A., Ahmed, M., 2016. Smart cities: a survey on security concerns. Int. J. Adv. Comput. Sci. Appl. 7 (2). doi:10.14569/IJACSA.2016.070277.

Jaradat, M., Jarrah, M., Bousselham, A., Jararweh, Y., Al-Ayyoub, M., 2015. The internet of energy: smart sensor networks and big data management for smart grid. Procedia Comput. Sci. 56, 592–597. doi:10.1016/j.procs.2015.07.250.

Khan, Z., Anjum, A., Kiani, S.L., 2013. Cloud based big data analytics for smart future cities. In: Proceedings of the 2013 IEEE/ACM 6th International Conference on Utility and Cloud Computing, UCC '13. IEEE Computer Society, Washington, DC, USA, pp. 381–386. doi:10.1109/UCC.2013.77.

Khan, M., Uddin, M.F., Gupta, N., 2014a. Seven v's of big data understanding big data to extract value. In: American Society for Engineering Education (ASEE Zone 1), Conference of the IEEE, pp. 1–5.

Khan, Z., Pervez, Z., Ghafoor, A., 2014b. Towards Cloud Based Smart Cities Data Security and Privacy Management. IEEE, New York, pp. 806–811. doi:10.1109/UCC.2014.131.

Kim, G.H., Trimi, S., Chung, J.H., 2014. Big-data applications in the government sector. In: Commun. ACM 57 (3), 78–85.

Mohamed, N., Al-Jaroodi, J., 2014. Real-time big data analytics: applications and challenges. In: 2014 International Conference on High Performance Computing Simulation (HPCS), pp. 305–310.

OECD, 2010. Smart sensor networks for green growth. In: OECD Information Technology Outlook 2010. doi: 10.1787/it_outlook-2010-8-en. /content/chapter/it_outlook-2010-8-en.

Saif, I., Peasley, S., Perinkolam, A., 2015. Safeguarding the Internet of Things Being Secure, Vigilant, and Resilient in the Connected Age. Deloitte University Press. https://dupress.deloitte.com/content/dam/dup-us-en/articles/internet-of-things-data-security-and-privacy/DUP1158_DR17_SafeguardingtheInternetofThings.pdf. (Online; Accessed 29 September 2016).

Shargal, M., Houseman, D., 2009. Why Your Smart. Grid Must Start with Communications. Smart Grid News. http://www.smartgridnews.com. (Online; Accessed 29 September 2016).

Symantec, 2016. 2016 Internet Security Threat Report. Tech. rep. Symantec. https://www.symantec.com/. (Online; Accessed 29 September 2016).

Tarouco, L.M.R., Bertholdo, L.M., Granville, L.Z., Arbiza, L.M.R., Carbone, F.J., Marotta, M.A., de Santanna, J.J.C., 2012. Internet of things in healthcare: interoperability and security issues. In: ICC. IEEE, New York, pp. 6121–6125.

Tene, O., Polonetsky, J., 2013. Big data for all: privacy and user control in the age of analytics. Nw. J. Tech. Intell. Prop. 11 (5).

Veloso, M., Bento, C., Pereira, F.C., 2009. Multi-sensor data fusion on intelligent transport systems. In: Multi-Sensor Data Fusion; Intelligent Transport Systems, ITS-CM-09-02.

Verdone, R., Dardari, D., Mazzini, G., Conti, A., 2008. Preface. In: Wireless Sensor and Actuator Networks, first ed. Academic Press, Oxford. doi:10.1016/B978-0-12-372539-4.00011-7. http://www.sciencedirect.com/science/article/pii/B9780123725394000117.

Witt, S., 2015. Data Management and Analytics for Utilities. Smart Grid Update. http://www.smartgridupdate.com. (Online; Accessed 29 September 2016).

CHAPTER

# LOAD BALANCING ALGORITHMS AND PROTOCOLS TO ENHANCE QUALITY OF SERVICE AND PERFORMANCE IN DATA OF WSN

7

**Arif Sari***, **Ersin Caglar***

*Girne American University, Kyrenia, Cyprus**

## 1 INTRODUCTION

Due to the recent innovation of technology, radio communication, Internet connection, and micro electrical mechanical systems (MEMS), monitoring of the environment becomes simple, and is important (Min et al., 2001). With these innovations, it is possible to create a small tiny device, called a sensor node, to monitor, compute, communicate, and sense the environment in a possible range (Haenggi, 2005).

Environmental monitoring consists of measuring and analyzing environmental parameters like oxygen, temperature, humidity, and height (Ho et al., 2005). Environmental monitoring plays a vital role to understanding human beings and creating a better society. So, environmental monitoring is done by a sensor node, because sensor nodes can easily sense a wide range of factors that are impossible to for human beings to easily detect (Sharma and Jain, 2015). The MIT Technology Review and Global Future mentioned that sensor node technology is one of the top ten developing technologies (Werff, 2003).

Sensors are tiny devices that have the ability to gathering information from the environment, such as heat, oxygen, and light (Amrinder and Sunil, 2013) Each sensor has low battery power, and limited storage and processing power (Kun et al., 2006). Sensors can be anywhere in the environment because of their size. When the sensors are used in the environment, their job is to gather information and send this collected information to a more powerful node, called sink (Almomani et al., 2011).

In Fig. 1, the WSN is a collection of hundreds or thousands of these sensor nodes (Yu et al., 2006). In other words, the WSN is a special type of Ad-Hoc network with energy, storage, and processing for environmental monitoring and communicating with other nodes that typically have heavy data transmission loads, moving on the sink via transceiver (Jabbar et al., 2011). The characteristics of WSNs are (Xiangyu and Chao, 2006);

- Not any infrastructure
- Self-organized network
- Network topology is dynamic

Security and Resilience in Intelligent Data-Centric Systems and Communication Networks. https://doi.org/10.1016/B978-0-12-811373-8.00007-0
Copyright © 2018 Elsevier Inc. All rights reserved.

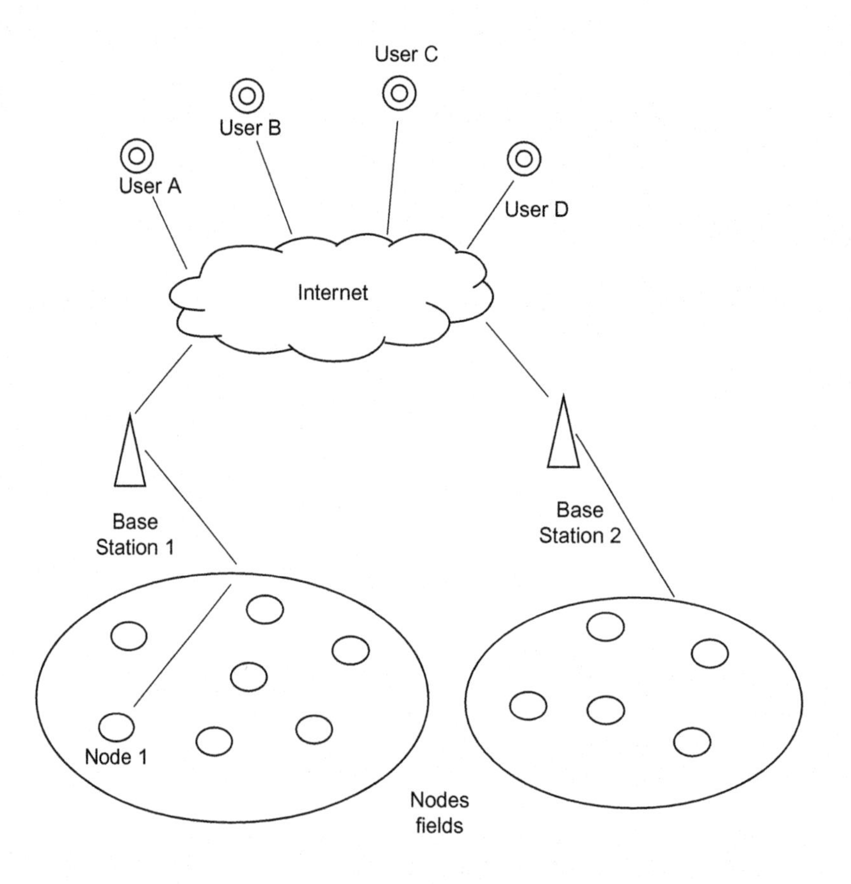


**FIG. 1**

Accessing WSNs through Internet.

- Multihop or single-hop communication
- A large number of sensor nodes

The application areas of WSNs are divided into two groups that are monitored and tracked. As shown in Fig. 2, military security detection, industrial factory monitoring, and health patient monitoring are examples of monitoring applications. Business human tracking and military enemy tracking are examples of tracking applications (Sharma and Ghose, 2010; Li and Gong, 2008).

Generally, the WSN has two types: structured and unstructured. In structured types of WSNs, all sensors organize as a structure or are preplanned. But unstructured types consists of an intensive collection of sensors and that are not organized, either preplanned in an ad-hoc manner. The big difference between structured and unstructured WSNs is security. In unstructured types of WSNs, it is hard to detected failure nodes or any other security problem because the sensors number in the hundreds or thousands, many of which are organized randomly (Yick et al., 2008). A WSN has five layers. In Fig. 3, the physical layer is the responsibility of signal detection and the selection of frequencies (Ganesan et al., 2003). The Data Link layer, or in other words, the MAC layer, is responsible for finding a reliable communication between sensors and sink. The Network layer is responsible for routing and network topologies. The Transport layer is responsible for transferring reliable and trusted data. And

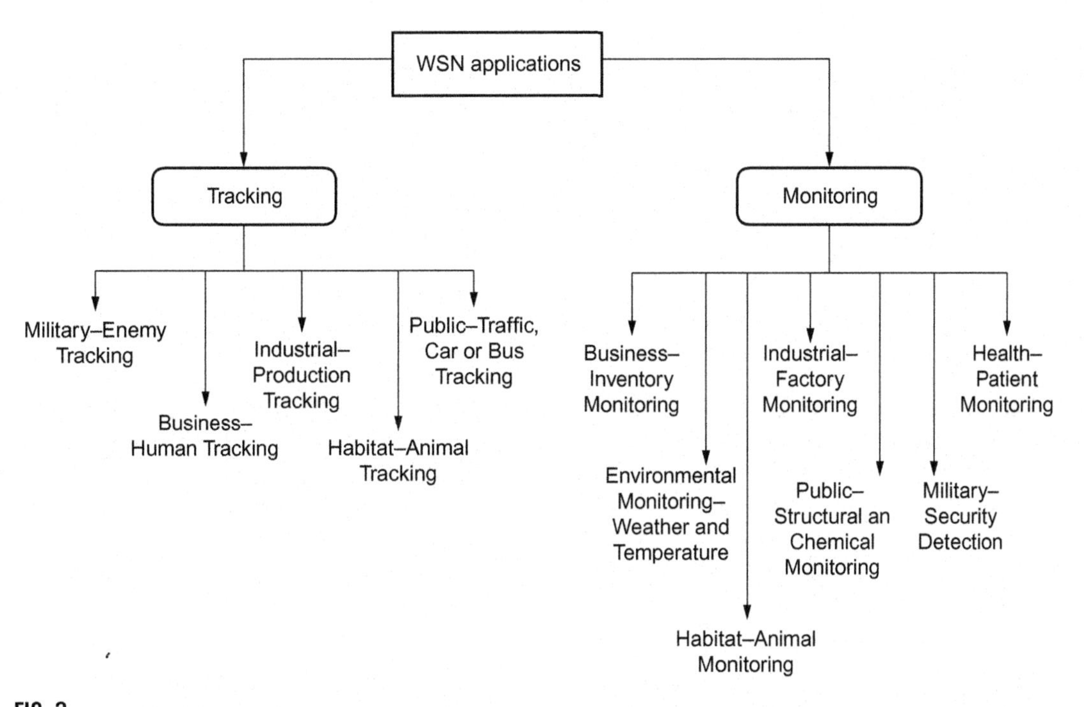


**FIG. 2**

Application area of WSN.

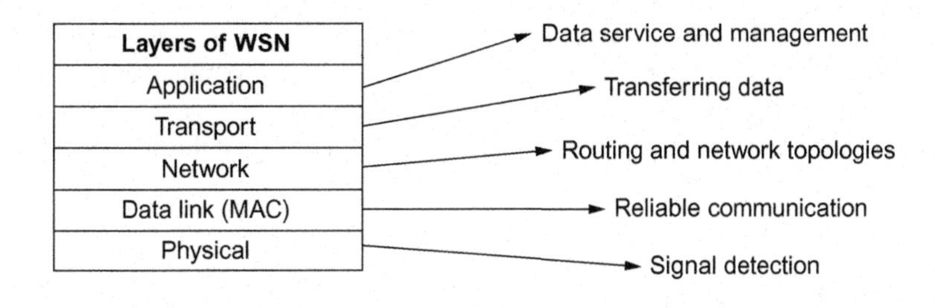


**FIG. 3**

Layers of a WSN.

the last layer, the Application layer, is responsible for data service and management (Al-Karaki and Kamal, 2004).

In addition to the layers and the types, the WSN has another important characteristic that depends on the environment.

- Terrestrial: These sensors consist of hundreds or thousands of sensors organized either structured or unstructured (Toumpis and Tassiulas, 2006). These kinds of sensors are very cheap but the difficulty is managing the battery power, because of the vast quantity of sensors much and they consume too many energies.

- Underground: These sensors are placed underground, in mine or caves. The difficulty is accessing the sensors when they are placed (Akyildiz and Stuntebeck, 2006).
- Underwater: These sensors are placed under the water. With the acoustic atmosphere of the water, underwater sensors have many difficulties, such as expense, difficult to access the sensors, high energy consumption, limited bandwidth, and long propagation delay (Heidemann et al., 2005).
- Multimedia: These sensors have a camera or microphone, and are organized in a structured type. Multimedia sensors have a high bandwidth demand and high energy consumption (Akyildiz et al., 2007).
- Mobile: These sensors have the ability to move. The key difference between these and other sensors is that mobile sensors can reposition and organize themselves in the network (Leopold et al., 2003).

Each type of sensor has a different usage type that depends on the environment, yet each WSN has similar working cycle. A WSNs working cycle contains three stages, these are the birth, life, and death stages. The birth stage is the installation and configuration stage, such as the initialization of protocols and algorithms. The life stage is the working mode. In this stage, sensors are sense, monitor, track, communicate, and report. Then, the WSN enters the last stage, the death stage, which occurs due to depleted energy, node failure or malicious attacks (Dohler, 2008).

Due to unique characteristics of the usage type mentioned previously, the WSN has some constraints. They are (Bajaber and Awan, 2011):

- Energy limitations: Depending on the size and the place of use, sensors are not rechargeable and changing batteries is difficult. Also, sensors could consume too much energy due to excessive workload. These reasons cause sensors to have a limited energy.
- Lifespan: The sensor has a short lifespan because of energy limitations.
- Storage: Due to the size of the sensors, their storage capacity is small.
- Application: Sometimes WSN are unable to support too many kinds of applications.
- Processing power: Sensors have many jobs and they try to done by a limited energy and limited storage power. So, their processing power is not enough.
- Scalable: The control and management of the sensors is difficult when the WSN contains too many sensors.

These limitations cause the WSN to have many problems. The two most important problems are security and heavy data transmission load between nodes or sink. There are many methods available to find a solution of these problems. One of the best methods to solve data transmission is balancing the load (Bajaber and Awan, 2011).

## 2 LOAD BALANCING

As mentioned previously, the WSN consists of a large number of sensors with some constraints. With these constraints, the WSN has too many difficulties. Load balancing is the most important difficulty because the main goal of the WSN is to gather information by monitoring the environment and send it to the sink (Rowstron and Druschel, 2001). Load balancing is a technique to spread out the work load between two or more network links, CPUs, computers, or other resources in order to maximize the throughput and utilization, and minimize the response time and energy consumption. In short, load balancing is a technique that can equalize the workload of each of the sensors or servers in the WSN (Akkaya and Younis, 2005).

WSNs really need load balancing techniques because of their constraints. With load balancing (Wajgi and Thakur, 2012a), WSNs can:

- increase sensors battery life
- increase sensors processing
- increase network scalability

# 3 LOAD BALANCING TECHNIQUES IN A WSN

As mentioned previously, sensors have limited resources in a WSN. The most important limited resource is energy, because processing and communication power consumes energy. If the sensor energy finishes quickly, it means that the sensor lifetime is short. It is a big problem because sometimes it is very difficult to replace or recharge the battery (Swain et al., 2010). Therefore load balancing is a technique that balances the load and extends the sensors lifetime. There are many protocols and algorithms to balance the load in a WSN (Wajgi and Thakur, 2012b).

## 3.1 LOAD BALANCING PROTOCOLS IN A WSN

Aljawawdeh and Almomani (2013) present a Dynamic Load Balancing Protocol (DLBP) for the WSN. The aim of this protocol is to equalize the load and increase the lifetime of sensors in the WSN. In this study, three comparison metrics were used: network success ratio, routing overhead, and network lifetime. The simulation results show that DLBP increase the lifetime of sensors 20%.

Ming-hao et al. (2011) proposed a multipath routing protocol with load balancing. The aim of their research is to decrease the energy consumption and increase wireless interference. With this protocol, data packages are distributed on many sensors and energy consumption is minimized. The simulation results show that this protocol decreases the total energy consumption and increases the lifetime of the network.

Ozdemir (2009) presented a Secure Load Balancing (SLB) protocol in heterogeneous sensor networks via hierarchical data aggregation. The goal of this research is to improve data accuracy and bandwidth utilization under the security constraints. The research study was simulated in terms of data accuracy, average data rate per sensor node, and data aggregation efficiency. The performance analysis and the simulation results show improve data accuracy and bandwidth utilization. Also, the SLB protocol minimizes the energy consumption and prolongs the lifetime of the WSN. According to same author (Özdemir, 2007) the SLB protocol can be implemented on a WSN via inter cluster relaying. In this paper, the author tries to decrease problems of accuracy and also ensure a secure data aggregation. The simulation performance result shows that the SLB protocol achieves this goal.

Merzoug and Boukerram (2011) proposed a cluster-based communication protocol for load balancing. This protocol is designed for homogenous nodes and energy constrained in the WSN. The aim of the protocol is to reduce and balance energy consumption. The simulation uses energy consumption, network lifetime, received data messages, and scalability as performance metrics. The simulation result shows that this protocol extends the lifetime of sensors.

Saraswat and Kumar (2013) conducted a study to design a Load Balanced and Energy Aware Routing (LEAR) protocol. This protocol is designed to prolong the lifetime of a large-scale WSN. The sensors, which are managing by LEAR protocol, try to choose lower data traffic routes and minimize energy consumption. The analysis proves that the LEAR protocol improves the lifetime of large scale WSN.

Kavitha and Pushpalatha (2015) proposed an adaptive and secure load balancing multipath routing protocol based on AODV. The protocol aims to increase the network throughput and prolong the lifetime of the sensors. Researchers analyze three features of the protocol, packet delivery scheme, authentication, and load balancing. The simulation results show that this protocol improves both the throughput of the network and the lifespan of the sensors.

Shancang et al. (2014) developed an Adaptive and Secure Load Balancing Multipath Routing (SM-AODV) protocol. This protocol contains four features: application independence, secure data delivery, adaptive congestion control and rate adjustment, and extensibility. The goal of this protocol is to improve the network performance. The simulation results show that the SM-AODV protocol increases the network performance in terms of packet delivery ratio, average delay, and throughput. However, there are still some problems, such as high percentage of end-to-end delay, packet delivery ratio, and data lost ratio. At the same time the network is not highly secure. To address these problems, Lata et al. (2015) presented a protocol, called Secure Adaptive Load-Balancing Routing (SALR), that tries to find a solution. The mathematical analyses and simulation results prove that the SALR protocol is better than SM-AODV in terms of delay, packet delivery ratio, data lost ratio, and security.

Wu et al. (2008) proposed a novel Load Balanced and Lifetime Maximization Routing (BLM) protocol to maximize the lifespan of a WSN. Researchers simultaneously distribute multiple sinks to equalize the energy consumption among the sensors. Two performance metrics are used in the simulation: the lifetime of the network and the number of packets received by sinks under a given amount of events. So, in the simulation results, BLM can prolong the lifetime of the network.

Siavoshi et al. (2014) developed and presented a load-balanced, energy-efficient clustering protocol for intra-cluster communication. In this protocol, the network is divided by many virtual circles to mitigate the energy consumption. This protocol can be compared to the LEACH, TCAC, and DSBCA protocols. This simulation results show that the proposed protocol increases the network lifetime by the rate of 73% more than LEACH, 52% more than TCAC, and 21% more than DSBCA.

Yuvaraju et al. (2014) proposed a new multipath protocol, called Secure Energy Efficient Load Balancing Multipath Routing Protocol with Transmission Power Adjustment (tSEL). This protocol uses multipath routing to get better load balancing than a single path routing and it provides security. The tSEL protocol chose a disjointed multiple paths and distribute workload among the paths effectively. The results of this simulation show that the network lifetime will increase.

Nam (2013) presented a Load Balancing Routing Protocol (LBRP) to equalize the workload traffic and prolong the lifespan of the WSN. In this protocol, each node distributed the workload through the multiple paths instead of single path. When each node uses multiple paths routing, the energy consumption decreased. Therefore, the network lifespan will increase. The result of this simulation shows that the LBRP protocol achieves the purpose.

Hongseok et al. (2010) proposed a protocol, called Gradient-based routing protocol for LOad BALancing (GLOBAL). Each routing protocol must distribute the workload with multiple sinks to extend the lifetime of the WSN. In these existing routing protocols, sensors establish their gradient. But in this case, sensors cannot prevent using a path that is the most overloaded. The GLOBAL protocol found a solution and presented a new gradient-based model. With this model, the least loaded path effectively prevents the most overloaded node. These simulation results show that the GLOBAL protocol increased the lifetime of the network when compared to other existing routing protocols, such as SPR and CPL.

Talooki et al. (2009) modified the Dynamic Source Routing (DSR) protocol to the Load Balanced Dynamic Source Routing (LBDSR) protocol. In DSR, traffic and energy consumption are unbalanced this means that the delay, blocking, and dependence on global information from all nodes will cause rates to increase. The simulation uses traffic load balance, energy consumption balance, average end to end delay, and the route's reliability as performance metrics. These simulation results show that the LBDSR will increase traffic and energy consumption balancing by 15%, decrease the average end to end delay by 10% and decrease the node failure rate.

Chen et al. (2006) proposed a Gathering-Load Balanced Tree Protocol (LBTP). The existing protocols have a broadcasting tree but they have no data gathering. So, the purpose of this protocol is to gather data efficiently and minimize the energy consumption to prolong network lifetime. The results of this simulation show that the LBTP is appropriate for data gathering.

The RAEED protocol was designed for the WSN to protect Denial of Service attacks (DoS). The research related with the RAEED protocol shows that the RAEED performs very well against the DoS attacks (Saghar, 2010), but the RAEED consumes too much energy to protect the WSN. So, Khan et al. (2016) developed a protocol to solve this problem, called the Robust Formally Analyzed Protocol for WSN Deployment with Load Balancing (RAEED-LB). A comparison of these two protocol results show that RAEED-LB increases the lifespan of network from 10% to 35%.

## 3.2 LOAD BALANCING ALGORITHMS IN WSN

Chung et al. (2011) designed an algorithm to balance load, called the Balanced Low-Latency Converge-Cast Tree (BLLCT). The BLLCT tries to achieve bottom to top load balancing. In this algorithm, the node selects one of the candidate parents according to specific properties. As a result of this study, researchers have determined that this algorithm, while not overly complex, still consumed much energy.

Laszlo et al. (2011) proposed a novel Load Balancing Scheduling algorithm. The main purpose of this algorithm is to provide an optimal scheduling for packet forwarding. By using a mathematical approach, the algorithm achieves their purpose.

Na et al. (2007) conducted a load balancing algorithm. Researchers proposed an Energy-Aware Multilevel Clustering Tree with Gateway (EAMCT-G) algorithm. The goal of this algorithm is to set remaining energy as a priority in the selection of cluster head and to select a gateway for the members of a cluster set. The algorithm avoids energy consumption and prolongs the sensor's lifetime. After the simulation, EMGCT-G had a problem with the lifetime of sensors. It was determined that if the sensor node that has the greatest energy becomes a cluster head, then all of the nodes in one hop will join that cluster set. So this will increase the data load and energy consumption causing the sensors life to become short. But according to Zhang et al. (2011), the Load Balancing Cluster Algorithm for Data Gathering (LCA-DG) can solve this problem. LCA-DG tries to balance the load of each cluster head and prevent the overload of cluster heads in EAMCT-G. The LCA-DG uses two factors to improve a cluster member's strategy: the distance between the cluster head and the members, and residual energy. The results of this simulation shoe that this algorithm extends the network lifetime and is suitable for large-scale WSN.

Dai and Han (2003) designed a node-centric load balancing algorithm. This algorithm uses a load balanced tree to prolong the lifetime of the sensors. The researchers use some metrics with the simulation and find that the algorithm's routing tree is more effectively balanced than the others.

Israr and Awan (2007) presented a multihop clustering algorithm for load balancing. This algorithm tries to find a solution of load balancing and energy efficiency. The results of this simulation show that this algorithm can balance the load and decrease energy consumption.

Levendovszky et al. (2011) proposed a novel load balancing algorithm. The aim of their algorithm is to balance the load to minimize energy consumption and prevent packet loss probability. This mathematical approach and the accompanying simulation results shows that this algorithm saves energy and decreases packet loss in the network.

Gupta and Younis (2003) conducted study with the load balance clustering algorithm. The goal of this algorithm is to maximize the system stability and minimize the energy consumption, in order to extend the sensor's lifetime. The result of this simulation shows that the presented algorithm improves the lifetime of sensors in terms of time to network partition, average lifetime of sensors, average delay per packet, network throughput, average energy consumed per packet, average power consumed, and standard deviation of load per cluster.

Karger and Ruhl (2003) presented a dynamically load balancing algorithm among peers. The authors analyzed the distribution of the workload by moving data from higher loaded to lower loaded nodes. In conclusion, the researchers discovered that this algorithm maximized node utilization.

Zhang et al. (2009) designed a Load-balanced Algorithm in WSN Based on Pruning Mechanism. In a WSN, the nodes that are close to the sink carry more load than others. These kinds of overload nodes become a hot point and consume too much energy. So the authors designed this algorithm to find a solution to this hot point problem and to balance the energy. This algorithm uses nodes location, residual energy, and a count of cluster nodes to solve the hot point problem and balance the workload. The performance evaluation results show that the proposed algorithm achieves this purpose and extends the network lifetime.

Low et al. (2007) proposed a Load-Balanced Clustering Algorithms for the cluster head in a WSN. The proposed algorithm considers workload traffic to enhance the overall scalability of the network. The algorithm selects nodes to use as gateways; these gateways act like cluster heads to balance their workload. The results of this simulation shows that the proposed algorithm increases the system scalability and balances the workload.

Deng and Hu (2010) proposed a Load Balancing Group Clustering (LBGC) algorithm to balance the energy consumption based on SEP, LEACH, DEEC, and SGCH in a heterogeneous WSN. The proposed algorithm uses a multihop group formation method to balance energy consumption. The results of this simulation analysis shows that LBGC balances the energy consumption and effectively improve the network lifetime.

Touray et al. (2012) presented an algorithm, called the Biased Random Algorithm for Load Balancing (BRALB) in Wireless Sensor Networks. It is based on energy biased random walk for environmental monitoring. The main purpose of the BRALB algorithm is to mitigate energy consumption to prolong the lifespan of the network. The researchers used probability theory based on energy resources in each node to analyze energy consumption. The results of these analyses show that the BRALB algorithm decreased energy consumption and prolonged the lifespan of the network.

Kim et al. (2008) proposed a load balancing scheduling algorithm. The aim of this algorithm is to minimize the packet loss probability and balance the load. The proposed algorithm uses the optimal scheduling algorithm for packet forwarding to achieve this goal. The results of this analyses show that the proposed algorithm minimized the packet loss and balanced the load.

Mathapati and Salotagi (2016) presented a load balancing and RSA security algorithm. Their algorithm attempts to find better solution to balance load in a secure way. The results of this simulation shows that the presented algorithm provides better performance of load balancing and security than the existing algorithms in terms of throughput, end-to-end delay, and packet delivery.

Researchers have tried to find a solution to workload constraint with load balancing protocols and algorithms. But in a WSN that has too many constraints, such as those mentioned previously, researchers must propose a solution to provide quality service and satisfy the users.

## 4 QUALITY OF SERVICE

With the rapid development of technology, networks are changing affecting people's lives in many different ways. In the last decade, the network has become wireless, which consists of a large number of sensors. In this new technology, many new concepts have come to life. However only one concept is never changing: quality of service (QoS) to satisfy the users demand (Martínez et al., 2007).

Quality of Service (QoS) is a service characteristic of a network that must be met when transporting a packet from the source to the destination (Crawley et al., 1998). In general, QoS has two perspectives: the network and the application/users. From a network perspective, the goal is to provide the best service to maximize the resource utilization and satisfy the users demand. From the application/users perspective, users are never concerned about resources like protocols and algorithms. The users concern only the service quality (Chen and Varshney, 2004).

In a WSN, QoS has many challenges that differ from the traditional network because of its unique characteristics. They are (Yuanli et al., 2006; Bhuyan et al., 2010; Chen and Varshney, 2004):

- Limited energy: The battery power is the most important limitation in sensors, because its affect on the lifespan of the sensors.
- Limited resources: Each sensor has limited resources, such as bandwidth, memory, processing power, and buffer size. These limitations affect the service quality.
- Geographical deployment: Sensors can deploy in geographically-challenging places, such as underwater and underground. And sensors can move. This can pose problems for QoS.
- Dynamic topology: a WSN must have dynamic topology because of node failure, link failure, and node mobility. This dynamic property can cause a problem for QoS.
- Scalability: The increase or decrease of the number of sensors in the WSN can affect the service quality.
- Various types of service: Due to widespread usage, a WSN can suggest too many services. QoS should not be affected by these different types of services.
- Unbalanced mixture traffic: Data traffic that contains a large number of sensors can be have high or mixed traffic. So the QoS method must be aware of data traffic.
- Physical arrangement: As with the geographical regions, the environment can damage to the sensors, such as rain, waves, and wind. These damages can cause damage to the network.
- Data redundancy: WSN has a data redundancy property that can affect the service quality.
- Data importance: All data differ in that they contain a variety of information. So, data must get priority treatment when it is transported from source to destination.

- Security: In any network type, security is the most important thing. So, security can affect the service quality easily.

In other words, QoS is directly related with the performance of any network, including the WSN. There are parameters to check the performance of a WSN depending on the challenges mentioned previously (Mbowe and Oreku, 2014).

## 5 PERFORMANCE ISSUES IN WSN

Performance is the most important issue in any network because high performance rate means high service quality. Researchers use parameters to find out the performance rate of a network. In the WSN, important performance parameters are (Gamal et al., 2004; Chipara et al., 2006; Mbowe and Oreku, 2014):

- Latency: Latency or delay is a passing time when the packets transfer from source to destination point. Latency is categorized as: network delay, processing delay, destination processing delay, propagation delay, and end-to-end delay.
- Throughput: Throughput is a measurement of the data rate or packets (per second) that are reached at the destination.
- Average jitter: The jitter is the time difference in packet-interarrival time, which also means difference between packet delays.
- Network lifetime: The Network is a time until the packet loss rate is above a threshold value.
- Packet loss or corruption rate: This is the rate of corruption packets when transferring from source to destination.
- Packet delivery: Packet delivery is a number of packets to be delivered to the destination.
- Packet generation rate: This is a number of packets that the sensor nodes generate.
- Energy consumption: This is a consuming energy rate of all the sensors or other devices in the network.

By considering both performance parameters and QoS challenge factors, researchers can use performance analysis to find a better solution for the WSN constraints (Alazzawi and Elkateeb, 2008).

## 6 SECURITY IN WSN

Security can be defined as the careful analysis used to protect the network against threats. In short, security means, to "protect right things in a right way" (Anderson, 2008). For the WSN, security is a crucial concept because of the vast usage area of the WSN, which includes the military, medical, and public safety arenas. For example, WSNs can be used by the military on the battlefield in a mine site operation or in a quick intelligence application. In the field of medicine, WSNs can be used to expedite data transfers from patient to doctor, and can also be used as a public safety application, such as with a burglar alarm (Nigam et al., 2014).

Due to the customization afforded individual WSNs, they are vulnerable to security attacks and cannot use traditional defense mechanisms because of specific constraints and requirements (Oreku, 2013).

## 6.1 VULNERABILITIES IN THE WSN

The WSN has different vulnerabilities than those found in other networks. They are

- Limited resources: All security applications need resources for implementation, such as energy, memory, and processing power. In WSNs, these are very limited resources (Hill et al., 2000).
- Unattended deployment: Due to the specific usage area of a WSN, sensors can be deployed in an extreme area, for example, an enemy territory on the battlefield. On the other hand, natural disasters can affect the sensors, like earthquakes or hurricanes. These scenarios afford the WSN little physical protection (Li and Fu, 2015).
- Unreliable wireless channel: WSNs use a wireless radio communication channel, which is an open communication system. This means that anyone who can uses wireless communication devices with the same frequency band will capture the signal easily and can damage the network. With these open channel communications, data conflicts or delays can occur. Therefore open communication channels are another big vulnerability for the WSN (Xing et al., 2006; Li and Fu, 2015).

## 6.2 SECURITY REQUIREMENTS IN THE WSN

The WSN is a special type of ad hoc network, so it has specific vulnerabilities and requirements. To secure the network, the WSN requires the following (Carman et al., 2000):

- Confidentiality: One of the most important requirements of the WSN, it is vital that no sensor, other than the recipient sensor, should read data. The network must protect the privacy of all data. For example, the military intelligence service must protect its data.
- Secure location: In the WSN, each sensor must be located precisely, because it is important to identifying the faults.
- Integrity: Packets that are transported from source to destination must reach their destination safely. Sometimes a packet does not lose any data, however, data can be changed during transportation. So data integrity is an important concept in the WSN.
- Time synchronization: Any data must synchronize with the time.
- Availability: All networks incur some costs, like energy. Security approaches can consume too much energy making sensors unavailable.
- Data freshness: Data must reach the destination point on time. Data must be up-to-date, otherwise it is not useful. For example, if the alarm sensors deployed around a building to detect a burglar experience a one-hour delay, it will be too late to save the building.
- Self-organization: Each sensor in the WSN must not be a fixed infrastructure. It must be flexible and independent to be self-organizing.
- Authentication: An important concept for the WSN, authentication is a code or mechanism that confirms that the received data is coming from the actual sensor.

## 6.3 ATTACKS AND COUNTERMEASURES IN WSN

Most of the attacks against security in a WSN are different from wired networks because of the WSNs vulnerabilities, requirements, and constraints. So WSNs are vulnerable to many attacks. According to literature, WSN attacks can be classified according to the OSI layers protocol, which is the application layer attack, the transport layer attack, the network layer attack, the link layer attack, and the physical layer attack. With the exception of OSI layers, an attack can be internal or external, according to the legitimate status of a sensor. External attacks are done by a node that is not a member of the network, whereas internal attacks are done by a node that is a member of the network. This kind of node, which is committing internal attacks, is called a selfish node. In the WSN internal attacks are more important than external attacks (Humphries and Carlisle, 2002; Al-Sakib et al., 2006).

Depending on the interaction, attacks can be either active or passive. Passive attacks never damage the communication. Passive attacks only capture data and read information. Active attacks, however, damage the network with no mercy. Active attacks can inject unnecessary packets on to the network and overload the traffic (Karlof and Wagner, 2003).

### 6.3.1 Types of attacks in the WSN

As mentioned previously, attacks are classified in two groups, active and passive. Eavesdropping and traffic analysis are considered passive attaches, whereas DoS or DDoS attacks, and routing attacks are classified as active attacks, as shown in Fig. 4.

As shown in Fig. 4, active attacks divided by five groups:

**(a)** DoS or DDoS attacks
**(b)** Routing attacks
**(c)** Digital Signature attacks

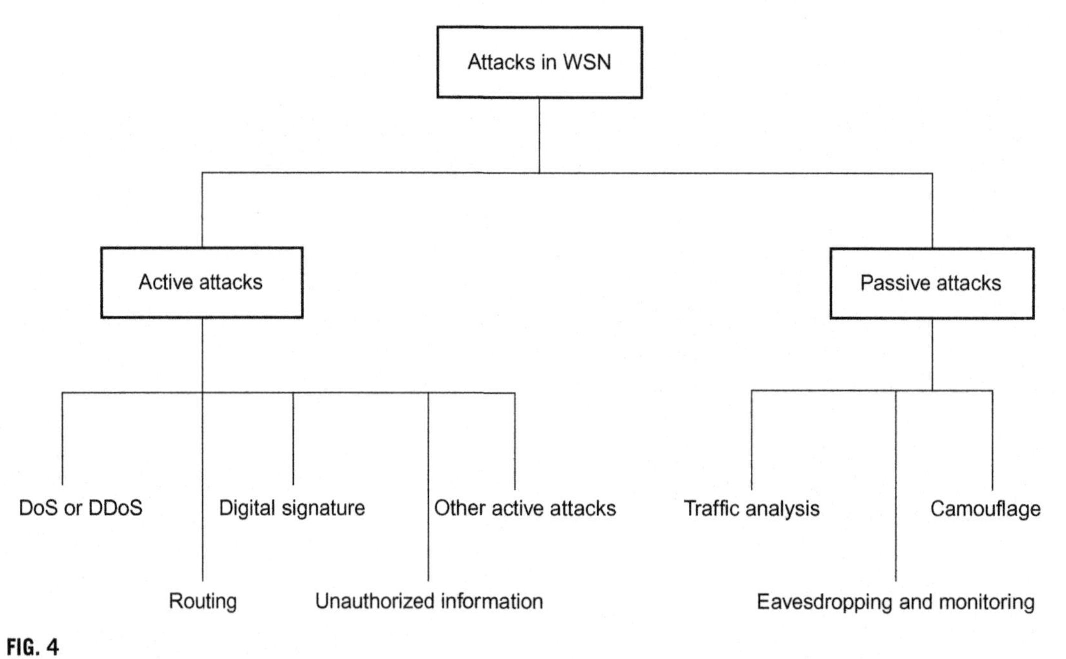


**FIG. 4**

Active and passive attacks.

**(d)** Unauthorized Information attacks
**(e)** Other active attacks

(a) DoS or DDoS Attacks: DoS or DDoS attacks are the most serious type of attack in the WSN. Researchers are still trying to create a defense mechanism to minimize the effect of a DoS or DDoS attack (Blackert et al., 2003).

According to Walters and Liang (2006) a DoS attack is an event that can minimize or eliminate a network's capacity to slow down traffic and cause improper functionality. On the other hand, Raymond and Midkiff (2008) define a DoS attack as an attack that is produced by the unintentional failure of nodes or malicious action. Research shows that a DoS attack tries to consume the resources that are available in the victim node by flooding the system with unnecessary packets. So, users have a difficulty getting timely and efficient data, and experience the loss of an effective network service.

Sari (2015) defines DoS and DDoS attack succinctly, as an attack that paralyzes a node or network with unnecessary packets in order to consume the resources of the network.

According to the researchers' definitions, there is an only one difference between a DoS and a DDoS attack. A DoS attack is an attack done by a single Internet connection. Whereas a DDoS attack is done by multiple devices that are distributed across the Internet (Vijay and Nikhil, 2015).

In a WSN, several kinds of DoS or DDoS attacks are categorized by the layers, as shown in Table 1. Jamming and tampering attacks occur in the physical layer:

- Jamming: Radio frequencies are used in a jaming attach to jam or set a node. The aim of the jamming attack is to transmit or block the message so the message cannot reach its destination, as shown in Fig. 5. When a jamming attack is successful, no messages are able to be sent or received between node to node or node to sink. If the jamming attack has an enough power, it can damage the whole network (Shi and Perrig, 2004; Chen et al., 2003).

**Table 1 DoS Attacks Categorized by Layers**

| Layers of WSN | Attacks |
|---|---|
| Physical | Jamming<br>Tampering |
| Data Link | Collision<br>Exhaustion<br>Unfairness |
| Network | Neglect and Greed<br>Homing<br>Misdirection<br>Black Hole<br>Smurf |
| Transport | Malicious Flooding or Flooding<br>De-synchronization |
| Application | Sending Large Amount of Stimuli<br>Path-based DoS<br>Network programming |

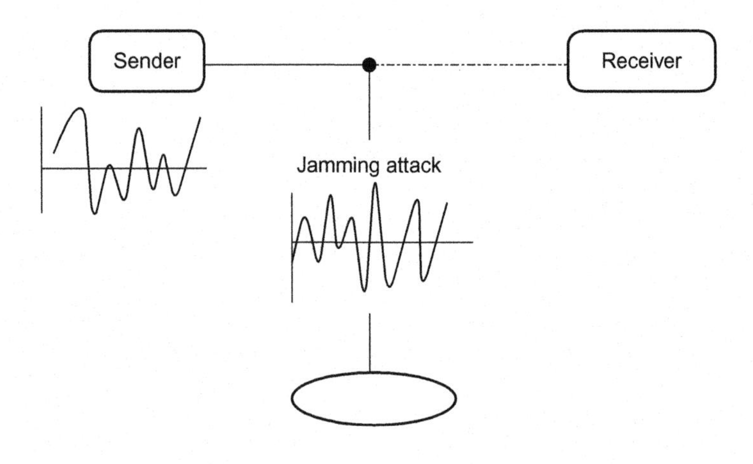


**FIG. 5**

Jamming attack.

- Tampering: In a WSN, sensors are deployed in hostile environment, such as underwater, or in a cave or war zone. Because of these environmental conditions, a tampering attack is the way to attack physically. With a tampering attack, the attacker can damage, modify, or even replace the sensor with a malicious sensor (Becher et al., 2006; Wang and Bhargava, 2004).

The link layer contains collision, exhaustion and unfairness attacks:

- Collision: When the communication protocol is disrupted by two sensors transmitting packets on the same frequency, it is called collision. to attempt collision. When comparing collision attack with jamming attacks, a collision attack has few advantages. A collision attack consumes less energy power and is difficult to detect (Wood and Stankovic, 2002).
- Exhaustion: A repeated collision attack is called an exhaustion attack. Exhausting is the continuously transmitting of packets that are affected by collision. With this retransmitting, an exhausting attack consumes sensor electric and processing power (Sahu and Pandey, 2014).
- Unfairness: This weak type of attack on the DoS or DDoS attacks is done for the priority mechanism and intermittently disrupts the frame transmission (Alquraishee and Kar, 2014; Wood and Stankovic, 2002).

Neglect and greed, homing, misdirection, black hole, and smurf attacks are categorized as network layer attacks (Sangwan, 2016; Padmavathi and Shanmugapriya, 2009):

- Neglect and Greed: The attacker creates alternative routing paths when the packets start to transmit from source to destination. With this attack, the behaviors of adjoining nodes are damaged.
- Homing: An attacker investigates the network traffic to find the base station or cluster head's geographic location. Then, the attacker begins to attack crucial nodes.
- Misdirection: An attacker replays or changes the routing information, as shown in Fig. 6.
- Black Hole: A black hole attack is the simplest DoS or DDoS attack. As shown in Fig. 7, a black hole attack establishes the shortest route to a destination point by using a malicious node as an intermediate node. When the malicious node becomes a communication node, it is able to do

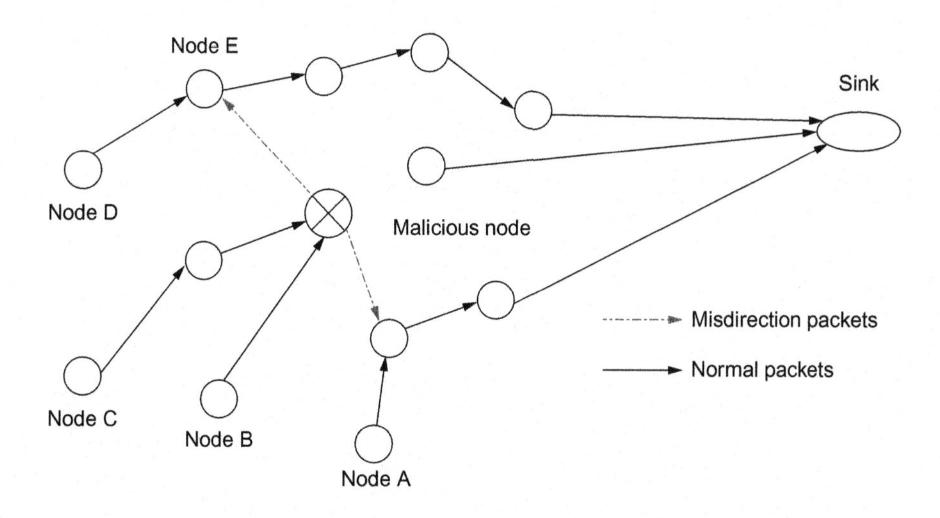


**FIG. 6**

Misdirection attack.

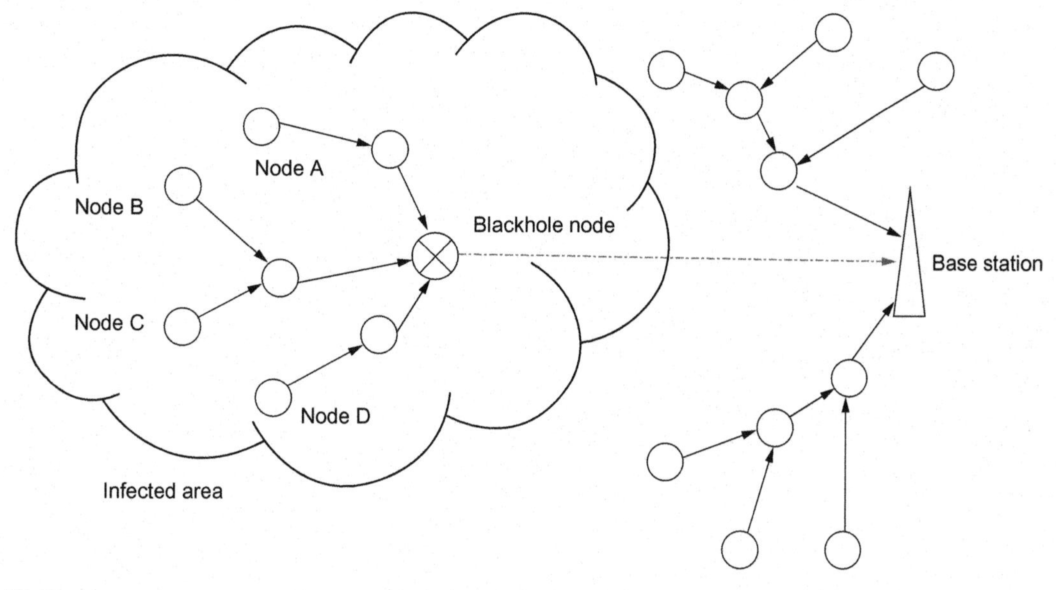


**FIG. 7**

Blackhole attack.

anything with the passing packets. This process creates a black hole to swallows or absorbs all the packets (Culpepper and Tseng, 2004; Al-Shurman et al., 2004).

- Smurf: A smurf attack is called amplification attack. As shown in Fig. 8, an attacker attacks the IP (Internet Protocol) and ICMP (Internet Control Message Protocol) and sends slews of ICMP Echo request packets to create busy network traffic (Charles, 2005).

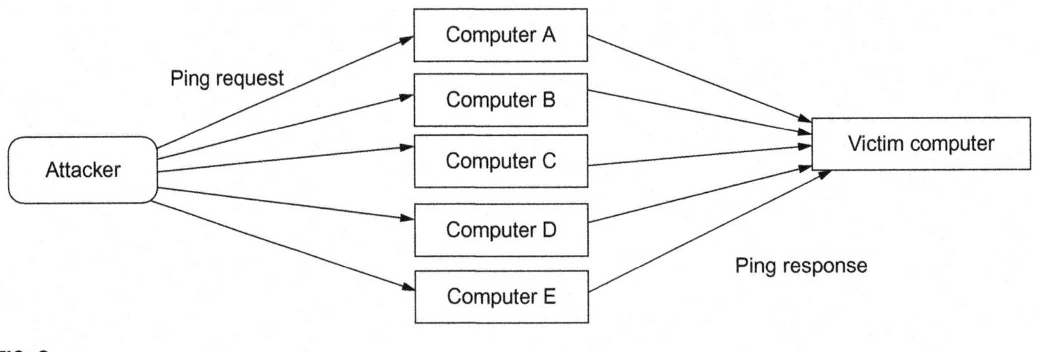


**FIG. 8**

Smurf attack.

Malicious flooding and de-synchronization attacks occur in the transport layer (Benenson et al., 2008; Wood and Stankovic, 2002):

- Malicious Flooding or Flooding: The WSN is vulnerable to a flooding attack at any time during the communication stream. An attacker begins to ask a new communication request repeatedly to the targeted node. A flooding attack consumes the node's energy and processing power with this request. It does not stop stops until node death.
- De-synchronization: The interruption or breaking of the existing communication between two nodes by continuously forging messages to these nodes is called de-synchronization. As a result of this attack, the node's energy is wasted, which degrades the performance of the whole network because of unnecessary traffic.

At lastly, sending a large number of stimuli, path-based DoS, and network programming attacks occur in the application layer (Mpitziopoulos et al., 2009; Kaushal and Sahni, 2015):

- Sending Large Number of Stimuli: In this type of attack, stimuli are controlled by the application and alerts for motion detection are continuously sent, causing large amounts of network traffic.
- Path-based DoS: An attacker sends all packets to the base station directly, which requires too much network bandwidth, energy, and processing power.
- Network programming: An attacker reprograms the nodes with a false program and this affects the whole network.

(b) Routing Attacks:

- Spoofed, Altered and Replayed Routing Information: An attacker directly attacks the routing information as shown in Fig. 9. The attacker spoofs, alters, or replays the routing information to interrupt the whole network. As a result of this attack, routing loops are created, service routes are extended or shortened, false error messages are generated and increase end-to-end latency (Jaydip, 2009; Karlof and Wagner, 2003).
- Selective Forwarding: This type of attack occurs in a network layer and includes two cases. The attacker selectively sends the information to the particular sensor that is message selective forwarding. On the other hand, the attacker sends the information to the selected sensor that is

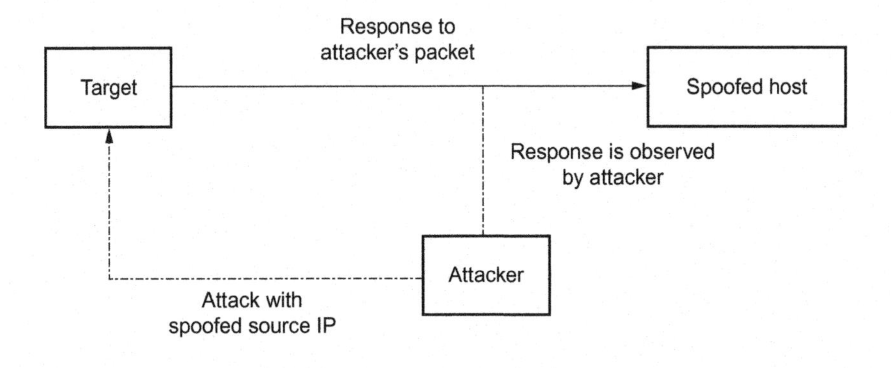


**FIG. 9**

Spoofing attack.

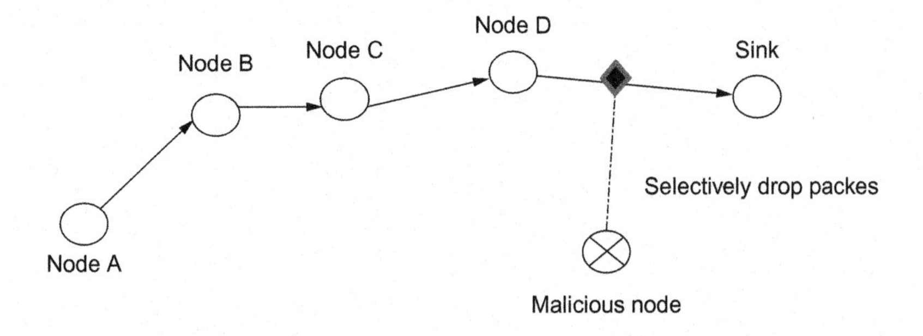


**FIG. 10**

Selective forwarding attack.

senor selective forwarding. Then, the attacker drops a few packets that satisfy a particular condition, as shown in Fig. 10 (Ganesan et al., 2002; Karlof and Wagner, 2003).

- Sybil: This is a network or application layer attack. As shown in Fig. 11, Sybil attack is a malicious node that has multiple identities and other nodes have the possibility of choosing this node as a routing node. The aim of this attack is to fill a neighbor node's memory by sending incorrect information. So, this attack is a serious threat for a geographic routing protocol (Arora and Gupta, 2014; Braginsky and Estrin, 2002).
- Wormhole: This is a critical attack for the WSN and is a network or link layer attack. In this attack, attackers need two or more adversaries and these adversaries create better communication resources and better communication channel called tunnels. The explanation of an attacking procedure is shown in Fig. 12. The sender uses a node to send a message to the other node, which is the receiver and the receiver node sends that message to its neighbors. The neighbor node thinks the incoming message is coming from the sender, but usually the sender is out of range. So, the neighbor node tries to send the message to the originating node, but it is never reached (Perrig et al., 2004).

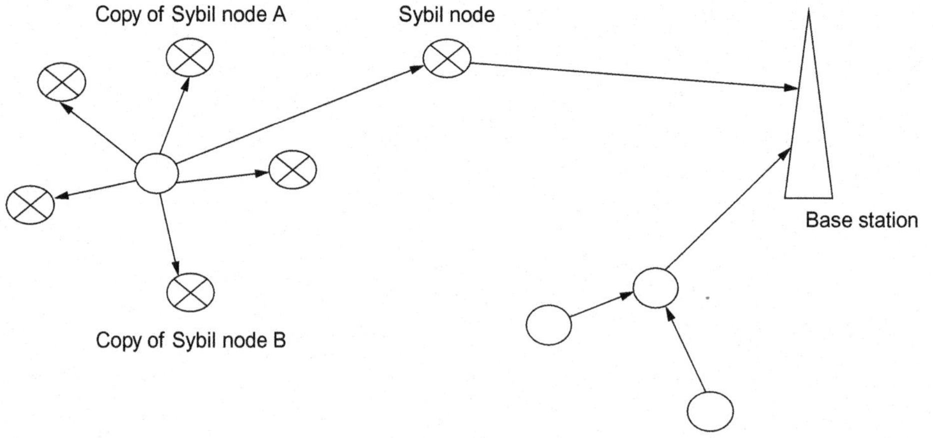

**FIG. 11**

Sybil attack.

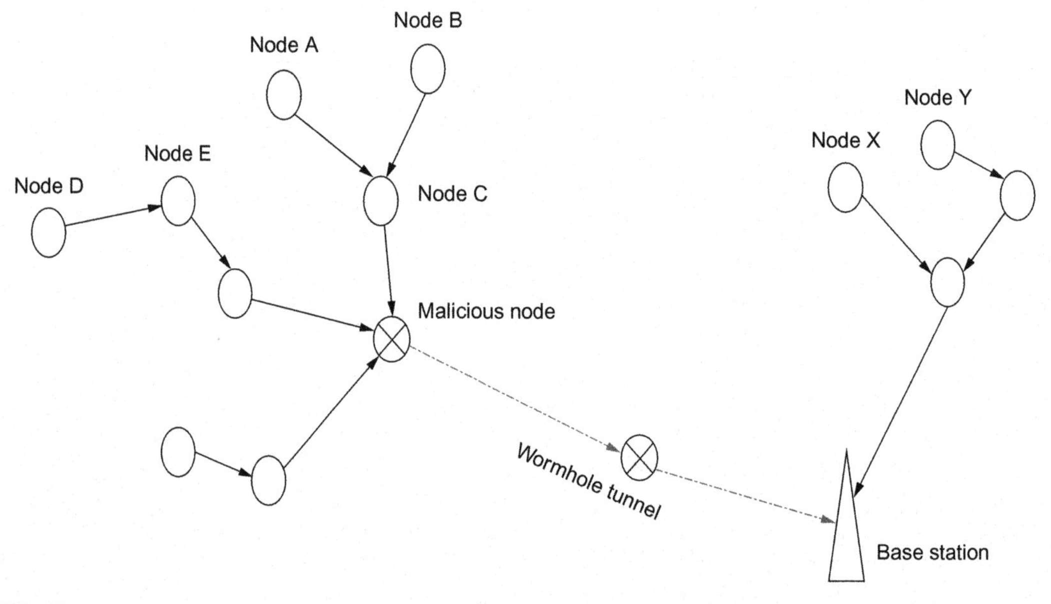


**FIG. 12**

Wormhole attack.

- Syn Flood: This attack uses a hello message packet as a weapon to convince the sensor to accept the communication. As shown in Fig. 13, a malicious node sends a hello message with high radio transmission power and processing power and convinces the receiver node that the malicious node is a neighbor. When the receiver node forwards the message to the malicious node, the message will be lost and disrupt whole network (Lazos et al., 2005).

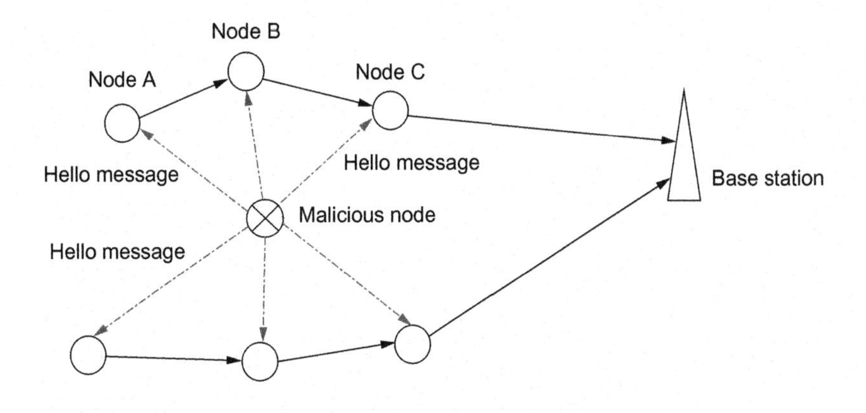


**FIG. 13**

Syn flood attack.

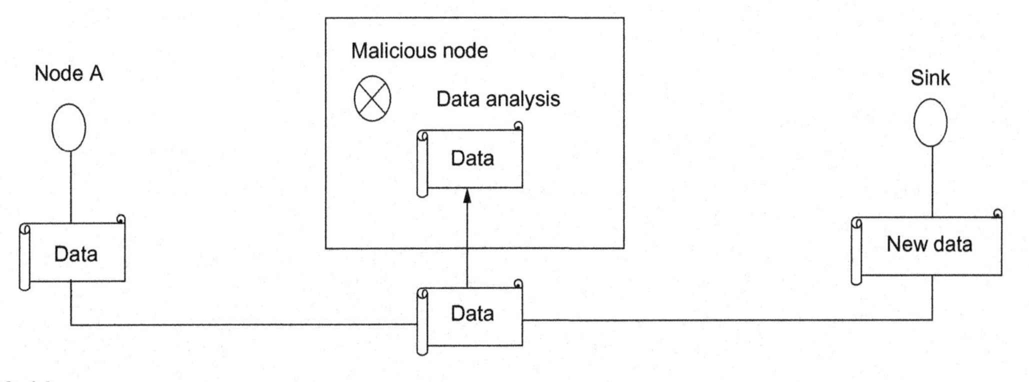


**FIG. 14**

Known message attack.

(c) Digital Signature Attacks (Min, 2004; Sari, 2015):

- Known Message: As shown in Fig. 14, an attacker knows a set of messages that are signed by the victim and has an access to signatures for a list of messages.
- Chosen Message: The attacker chooses a specific message from a list of messages that he wants to victim to sign.
- Key Only: The key only attack, as shown in Fig. 15, the attacker only knows the real signer's public verification algorithm or public key.

(d) Unauthorized Information Attacks (Luo et al., 2000; Sari, 2015):

- Unauthorized Disclosure Information: In this type of attack, the attacker tries to learn the contents of data.

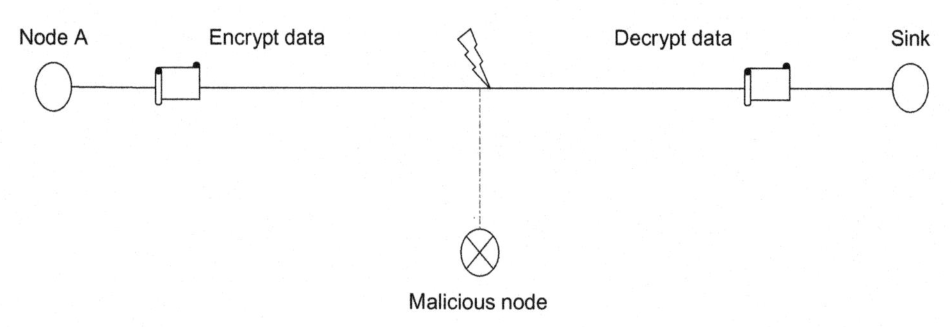


**FIG. 15**

Key only attack.

- Unauthorized Modification Information: The attacker modifies, alters, reorders, or delays the message to unauthorized effect.
- Unauthorized Access to Information: The attacker has unauthorized access to the user's resources or systems.

(e) Other Active Attacks:

- False or Malicious Node: This is the most popular method used by attackers against the security in a WSN. In this attack, the attacker inserts false information to the system (Pathan et al., 2006).
- Node Capturing or Node Subversion: In this attack, the attacker can capture the node and store malicious information on it (Pathan et al., 2006).
- Node Malfunction: An attacker uses the malfunction node to generate unnecessary packets and these unnecessary packets damage the whole network (Naeem and Kok-Keong, 2009).
- Node Replication: This type of attack is very simple. An attacker copies a node in an existing network. These replicating nodes are disrupting the network performance (Naeem and Kok-Keong, 2009).
- Passive Information Gathering: The attacker uses powerful resources to collect information. This information contains specific contents IDs, timestamps, and location of nodes. With this information, the attacker can destroy the node and network (Undercoffer et al., 2002).
- Byzantine: In this attack, the attacker creates forwarding packets through nonoptimal paths, selectively dropping packets, and routing loops. With these creations, the attacker damages the routing service of the whole network (Molva and Michiardi, 2002).
- Resource Consumption or Sleep Deprivation: Attackers try to consume the power of the nodes by using useless packets and requesting route discovery excessively (Conti et al., 2003).
- Packet Dropping: The attacker directly tries to damage the communication (Lawson, 2005).
- Rushing: In this attack, the attacker exploits the duplicate suppression mechanism to raise the speed of routing process (Graf, 2005).
- Man-in-the-Middle (MitM): The attacker inserts a malicious node into a conversation between two nodes to intercept communications. With this malicious node, attacker can gain access to the communication and learn everything, as shown in Fig. 16 (Gasser et al., 1989).

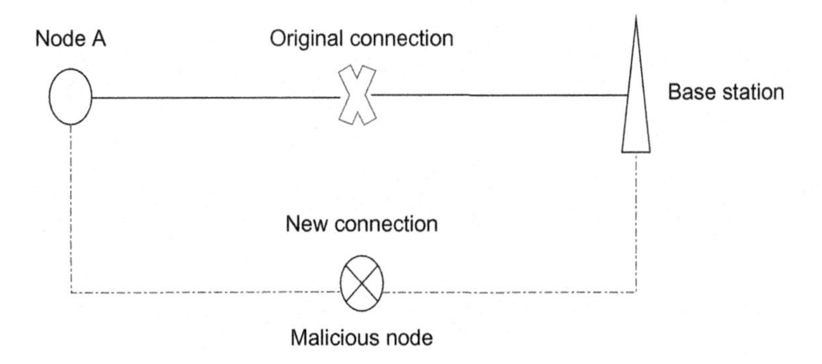


**FIG. 16**

Man-in-the-middle attack.

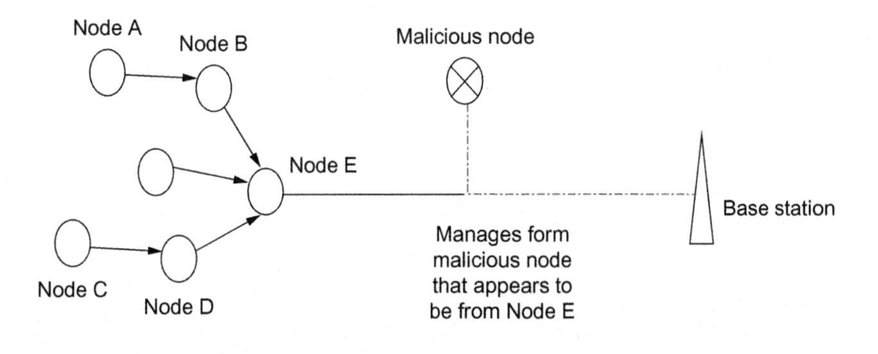


**FIG. 17**

Masquerading attack.

- Masquerading: In a masquerading attack, the attacker steals passwords and logons to gain access to network information. After the attacker gets the passwords, he can create a new connection with the receiver, as shown in Fig. 17 (Zhang and Lee, 2000; Sari, 2015).
- Repudiation of Action: This attack is against accountability. An attacker generates incorrect log information to change the authoring information of the actions executed (Kong et al., 2001).
- Pseudorandom Number Attack: The attacker uses a public key mechanism to generate the pseudorandom number to break cryptography in the communication (Kaufman et al., 2002).

As shown in Fig. 4, passive attacks are divided into three groups (Gruteser et al., 2003; Chan and Perrig, 2006):

- Eavesdropping and Monitoring, or Sniffing: These are the most common and the easiest type of passive attack. As shown in Fig. 18, if the data is not encrypted, an attacker can easily listen to the contents of the data. So, these types of attacks are only related to data.

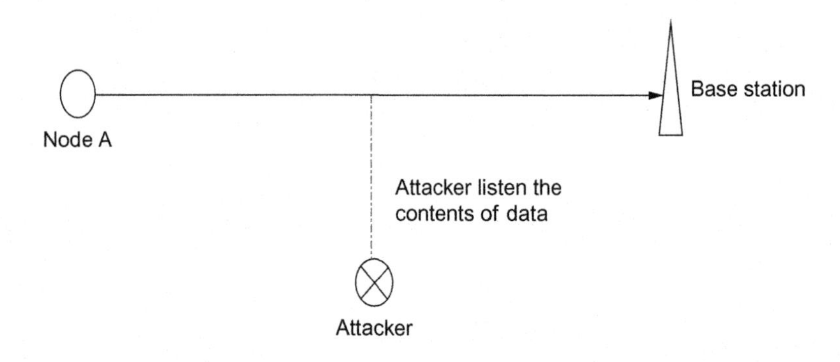


**FIG. 18**

Eavesdropping and monitoring attack.

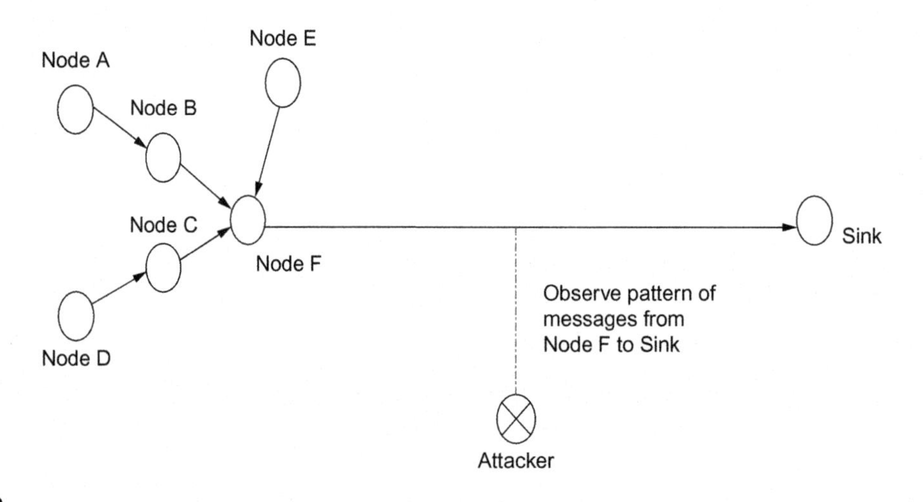


**FIG. 19**

Traffic analysis attack.

- Traffic Analysis: Traffic analysis attacks are combined with eavesdropping and monitoring to be effective. When combined with other attacks, a traffic analysis attack not only has access to the contents of data, but also can learn the sensors specialties and roles in the WSN, as shown in Fig. 19.
- Camouflage: In this attack, the attacker inserts and hides their nodes in the network. These hide nodes behave as a normal node and attack to the network traffic for learning the information.

### 6.3.2 Existing defense mechanism in WSN

According to the authors mentioned previously, WSNs are too vulnerable against different types of attack. Defense mechanisms can be used to recover, protect, and detect from the security attacks. In the WSN, defense mechanisms can protect from more than one type of attack but sometimes each attack requires a special defense mechanism.

- Defense mechanism for wormhole:

Kaissi et al. (2005) proposed a proactive routing protocol combat mechanism for a wormhole attack, which is called Dawwsen. This routing protocol is based on the hierarchical tree where the base station is the root and the sensors are the leaves. Dawwsen defense mechanism has two advantages. One advantage is that the geographical location of the sensors is that the not important and the second is, time stamp is not necessary to detect wormhole.

Hu and Evans (2004) presented a mechanism to combat a wormhole attack, called the directional antenna mechanism. This mechanism detects the malicious nodes and prevents them from establishing a wormhole tunnel.

Wang et al. (2004) proposed another approach to protect the network against a wormhole attack, called the visualization approach. In this approach, multiple dimension scaling is used for calculating the estimated distances between all sensors and neighbors. With these calculations wormhole tunnels can be found and eliminated easily.

Sebastian (2011) studied a security protocol, called the Secure Routing protocol against a Wormhole Attack (SeRWA) to defend network against wormhole attack. The SeRWA protocol created secure routing with a false positive. But the author uses a mobile agent to minimize the false positivity. With the mobile agent system, the SeRWA protocol defends the network effectively against a wormhole attack. On the other hand, Madria and Yin (2009) proposed another protocol with a same name, but this protocol has a different working mechanism. This SeRWA protocol uses symmetric key cryptography to defend the network against a wormhole.

Farooq et al. (2014) proposed a defense mechanism to detect a wormhole attack based on the medical application environment. This defense mechanism detects and eliminates the injected nodes that create a wormhole tunnel.

- Defense mechanism for sybil:

Douceur (2002) used identity certificates to prevent sybil attack. In this method, each node has unique information and the server uses this unique information to create unique identity certificate. After the certificate is created and each node is set up, nodes check the information with the server to see if it is a match. This process requires the exchange of a few messages, so it increases the network traffic but it protects the network against a sybil attack.

Newsome et al. (2004) presented a detection mechanism to detect sybil nodes in the network. This detection mechanism uses radio resource testing to detect sybil nodes. With this mechanism, it is easy to find and eliminate the sybil nodes and network work properly.

Eschenauer and Gligor (2002) proposed another method for sybil attack. The authors use a random key predistribution technique. In this technique, a set of random keys is assigned to each sensors. Each sensor can discover the common key with a neighbor node at the set-up phase. This technique minimizes a sybil attack.

- Defense mechanism for selective forwarding:

According to Ganesan et al. (2001) multipath routing technique can be used to prevent a selective forwarding attack. In this technique, nodes choose to next hop in a set of possible candidate hops dynamically. So this minimizes the selective forwarding attack to gain the control of data transmit.

- Defensive mechanism for sinkhole:

Zorzi and Rao (2003) proposed a geographic routing protocol. This type of protocol constructs a topology-based on demand and only uses localized interaction and information. According to the analysis, this protocol minimizes the threats of sinkhole.

- Defensive mechanism for jamming:

Wood et al. (2003) studied two methods for a jamming attack. The first study proposes a mapping protocol to detect jamming regions and save the communication. In some analysis, this protocol easily finds jamming regions. The second method is Frequency-hopping Spread Spectrum (FHSS). FHSS uses the pseudo-random sequence for the transmitter and the receiver. These sequence rapidly and continuously change the signals of transmitting. So attackers cannot catch the frequency.

Cagalj et al. (2007) presented a defense mechanism that uses wormholes to avoid jamming attacks. As a result, the help of a wormhole prevents jamming attacks.

- Defensive mechanism for spoof and alter:

Ye et al. (2005) proposed a mechanism to protect a WSN against the spoofing attack, called statistical en-route filtering (SEF). The SEF mechanism is used to find and drop injected data. Their analysis resulted in the SEF minimizing the effect of a spoofing attack.

Slijepcevic et al. (2002) proposed a security mechanism to prevent information spoofing. If parts of the network are affected by spoofing, this security mechanism still protect the rest of the world, because this security scheme protects various sensitivity levels separately. This property maximizes the security percentage against the spoofing attack.

Yin et al. (2003) studied the Secure Routing on the Diameter (SRD) mechanism to detect spoofing. This mechanism is a token-based mechanism and prevents a spoofing attack when data is transmitted from the source to the destination.

- Defensive mechanism for syn flood:

Hamid et al. (2006a) presented a protocol called probabilistic secret sharing. This protocol defends a network against to syn flood attacks. This protocol uses a bidirectional verification technique and multipath, multibase station routing.

- Defensive mechanism for blackhole:

Karakehayov (2005) proposed a REWARD routing algorithm against to blackhole attacks. REWARD uses geographic routing and broadcasts interradio behavior to watch neighbor transmissions to detect and eliminate blackhole attacks. The results of this analysis show that the REWARD algorithm protects the network against the blackhole attack.

- Defensive mechanism for digital signature attacks:

Freier et al. (2011) proposed a presented RSA-based signature algorithm. This algorithm needs two parties to work, like a handshake process. Two parties verify each other's certificate and negotiate the public key cryptography in the communication between them. Shortly, the client set-up communication and server respond the communication. On the other hand, Hankerson et al. (2004) proposed an Elliptic Curve Digital Signature Algorithm (ECDSA) that uses an ECC-based signature public key cryptography. This algorithm also uses a handshake process like the RSA-based signature. As a result of this analysis, ECDSA and the RSA-based signatures consume almost the same energy at the client side when transferring the public key cryptography. But at the server side, there is a difference between the ECDSA and the RSA-based algorithm. The ECDSA is cheaper than the RSA-based signature algorithm.

- Defensive mechanism for node capture:

Chan et al. (2003) analyzed the vulnerability of the node capture. The results of their analysis determined that the increasing amount of key overlap in the ring would increase the node capture possibility. The authors proposed a q-composite random key predistribution scheme to solve this vulnerability. This scheme required at least q common keys shared before they set-up the communication. These q common keys establish a secure communication between two neighbor nodes.

- Defensive mechanism for collision, exhaustion, and unfairness:

Jaydip (2009) tried to detect and find a solution for collision, exhaustion and unfairness attacks. For collision attack, the author uses an error correction code to prevent collision. This correction code can detect a malicious node and prevent collision but it is still not enough to protect the network 100% from a collision attack. On the other hand, this method consumes too much processing and communication power. For an exhaustion attack, the author studied the MAC admission control technique. This technique checks all requests and ignores suspicious requests that are intended to cause damage with an exhaustion attack. For an unfairness attack, the author studied in small frames. The author used small frames to reduce the amount of time, so an attacker cannot easily capture the communication channel. So, this small frame technique minimizes the possibility of an unfairness attack.

Liu et al. (1997) presented an Error Correction Codes to defend the network against a collision attack. This defense mechanism finds the collision and eliminates the attack but it needs expensive communication and consumes extra processing power.

Znaidi et al. (2008) proposed a defense mechanism for an exhaustion attacks. With this defense mechanism, the network ignores the extreme request because the defense mechanism limits the MAC admission control rate. The same authors proposed a different solution for an exhaustion attack that limits the usage of the MAC channel. With this solution, each sensor in the network should not use the MAC channel too long for transmitting packets. This limited usage minimizes the effect of an exhaustion attack.

- Defensive mechanism for flood attack:

Aura et al. (2001) presented a mechanism to detect a flood attack, called client puzzle. The idea is, the client demonstrates its commitment to the connection by solving puzzles. With this technique, the attacker does not have an infinite source to quickly create a new connection.

- Defensive mechanism for node replication:

Parno et al. (2005) proposed a mechanism that uses two algorithm to prevent node replication. One of the algorithms randomly distributes the location information to each nodes and the second one detects node replication and eliminates it. The old mechanisms have a single point of failure problem but with these two algorithms there is no single point of failure.

Lemos et al. (2010) presented a Collaborative IDS scheme for a node replication attack. In this scheme, some nodes are assigned as a monitoring node, called a supervisor node. These supervisor nodes monitor the whole network behavior based on rules that detect the node replication attack.

- Defensive mechanism for traffic analysis:

Deng et al. (2004) proposed a mechanism to defend the network against traffic analysis. This mechanism contains four strategies. First, the parent routing tree deployed to the sensors and sensor can easily send packets to the multiple parents. Second, a controlled random walk is deployed. Third, a random fake path is deployed to confuse the attackers. Forth, multiple high communication random areas are created. These four strategies prevent the analysis of traffic.

- Defensive mechanism for eavesdrop:

Xi et al. (2006) and Özturk et al. (2004) proposed a Greedy Random Walk (GROW) protocol to protect the network against an eavesdrop attack. In this protocol, the sink and sensors separately create an N or M hop random walk. When the transmitted packet comes to the intersection of hops, it is directly forwarded to the hop created by sink. This proposed protocol minimizes an eavesdropping attack.

- Defensive mechanism for more than one attack:

Perrig et al. (2002) presented two protocols, called SNEP and μTESLA. These two protocols prevent information spoofing and replay attacks, and provide data confidentiality, freshness, and broadcast authentication. On the other hand, the same authors presented another protocol to prevent passive attacks, called the SPINs protocol. The SPINS protocol changes the data traffic pattern randomly, so it is hard to listen to the communication.

Karlof et al. (2004) proposed a TinySec security mechanism. TinySec uses an efficient symmetric key encryption protocol to protect the network against the information spoofing and replay attack. The TinySec focuses on providing message authentication, confidentiality, and integrity.

Hu et al. (2003) presented a symmetric key cryptographic algorithm, called TIK, to defend against to a wormhole and spoofing attack. This algorithm requires accurate time synchronization and creates a secure communication between all nodes.

Zhu et al. (2003) presented a protocol, called a Localized Encryption and Authentication Protocol (LEAP). LEAP is a key management protocol based on a symmetric key algorithm. LEAP uses four different types of symmetric key mechanisms for each packet depending on the security requirements. LEAP is a popular protocol in the security of the WSN but it still has a weakness in that it is impossible to inject false data or decrypt earlier messages.

Lai et al. (2002) proposed a Broadcast Session Key (BROSK) negotiation protocol. BROSK is a key management protocol that uses a session key to protect the network. The BROSK protocol shares a master key with all the nodes to establish a session key and negotiate. On the other hand, BROSK has other advantages; it is a scalable and energy efficient protocol.

Deng et al. (2002) presented a protocol, called the Intrusion Tolerant Routing Protocol (INSENS). This protocol uses a routing-based approach to detect a DoS attack. The INSENS constructs routing tables and deploys to each of the nodes, bypassing the malicious nodes. So, a malicious node cannot attack the network. This protocol detects and minimizes the damage of a DoS attack at the same time.

Brutch and Ko (2003) presented a defense mechanism with three architectures to defend the network against to various attacks. In the first architecture, each node has its own intrusion detection system to detect the attack directly toward them. Second, the architecture is responsible for detecting local attacks on the sensors. Third, the architecture is responsible for routing attacks.

Albers et al. (2002) studied a local intrusion detection system (LIDS). In this system, networks must collaborate with another network. Networks exchange security date and alert one another of an intrusion. So, with these exchanges each node's vision has extended. This system can easily detect and eliminate various attack.

Karp and Kung (2000) proposed a Greedy Perimeter Stateless Routing (GPSR) security mechanism for defending the network against sinkhole and wormhole attacks. This mechanism detects a broadcast probe request to find and eliminate attacks effectively.

Shiva et al. (2012) presented a protocol for selective forwarding, sinkhole, and altering attacks, called the Energy Efficient Node Disjoint Multipath Routing Protocol (EENDMRP). When data start to transmit, EENDMRP uses the digital signature cryptography system to reach its destination point effectively. With this protocol, selective forwarding, sinkhole, and altering attacks are effectively minimized.

El-Bendary et al. (2011) proposed the Secure Directed Diffusion Routing (SDDR) protocol. This protocol uses the μTESLA algorithm to protect the network against a blackhole and spoofing attack.

Hamid et al. (2006b) studied a multipath and multibase routing mechanism to prevent syn flood and replay attack. This mechanism uses a key and one-way hash key chain for assigning to a multitree key protocol. With this procedure, the routing mechanism effectively defends the network.

Wei et al. (2009) added RSA public key encryption, AES, and digital signature algorithm to the AODV protocol and it is called SAODV. The SAODV protocol defends the network against various attacks such as eavesdrop, DoS, and syn flooding. The results of this analysis show that the SAODV protects the network effectively, and other properties like energy consumption, are similar to the AODV protocol.

Xin-sheng et al. (2013) proposed Load-Balanced Secure Routing Protocol (LSRP) to protect the network from various attacks and balance the load. This protocol uses one-way hash key chain and symmetric key technology to defend the network.

Babu et al. (2013) prevented a security mechanism for the transport and network layer of DoS and DDoS attacks. This mechanism is based on an address registration protocol. This mechanism analyzes and examines network traffic to detect unwanted flow. However, this mechanism is sometimes not efficient because several of attacks reside in the transport and network layers of the DoS and the DDoS.

Stetsko et al. (2010) proposed a neighbor-based intrusion detection system. The detection system is based on the Collaboration Tree Protocol (CTP). This system detects syn flood, selective forwarding, and jamming attacks. However, two main problems occur with this system, the communication overhead and the incidence of false alarm for detection.

Kline et al. (2011) proposed a DoS defense system, called the Shield. The Shield controls and monitors the whole network traffic. The aim of the Shield is to identify the malicious node and detect the network. If there is a malicious node, the Shield directly blocks and identifies the malicious node and prevents the DoS attack. On the other hand, Ranjan et al. (2009) modified the Shield for the DDoS attack. In DDoS, the Shield has a suspicious assignment mechanism and DDoS-resilient scheduler, with the exception of the Shield of the DoS. However, the working mechanism is exactly the same as with the Shield of the DoS.

Chaudhary and Thanvi (2015) presented an Ad-hoc on-Demand Distance Vector (AODV) protocol. The AODV protocol analyzes and detects malicious nodes or DoS attacks in the whole network. After detection, the AODV protocol applies the prevention scheme to the injected area and eliminates the attack.

Kaushal and Sahni (2016) presented a security scheme against a DoS (SSAD) attack. This mechanism is based on cluster-based WSNs. This mechanism uses a trusted management to detect and eliminate the DoS attacks in cluster nodes.

## REFERENCES

Akkaya, K., Younis, M., 2005. A survey on routing protocols for wireless sensor networks. Ad Hoc Netw. 3 (3), 325–349.

Akyildiz, I.F., Stuntebeck, E.P., 2006. Wireless underground sensor networks: research challenges. Ad Hoc Netw. 4, 669–686.

Akyildiz, I.F., Melodia, T., Chowdhury, K.R., 2007. A survey on wireless multimedia sensor networks. Comput. Netw. 51, 921–960.

Al-Karaki, J.N., Kamal, A.E., 2004. Routing techniques in wireless sensor networks: a survey. IEEE Wirel. Commun. 11 (6), 12–19.

Al-Sakib, K.P., Hyung-Woo, L., Choong Seon, H., 2006. Security in wireless sensor networks: issues and challenges. In: The 8th International Conference Advanced Communication Technology, ICACT'06, pp. 1043–1048.

Al-Shurman, M., Yoo, S., Park, S., 2004. Black hole attack in mobile ad hoc networks. In: In the 42nd Annual Southeast Regional Conference. ACM, Huntsville, AB, pp. 536–548. doi:10.1145/986537.986560.

Alazzawi, L., Elkateeb, A., 2008. Performance evaluation of the WSN routing protocols scalability. J. Comput. Syst. Netw. Commun. 9, 1–9.

Albers, P., Camp, O., Percher, J.M., Jouga, B., Puttini, R., 2002. Security in ad hoc networks: a general intrusion detection architecture enhancing trust based approaches. In: First International Workshop on Wireless Information Systems, pp. 1–12.

Aljawawdeh, H., Almomani, I., 2013. Dynamic load balancing protocol (DLBP) for wireless sensor networks. In: IEEE Jordan Conference on Applied Electrical Engineering and Computing Technologies (AEECT).

Almomani, I., Saadeh, M., Jawawdeh, H.A., Al-Akhras, M., 2011. Energy awareness tree-based routing protocol for wireless sensor networks. In: The 10th WSEAS International Conference on Applied Computer and Applied Computational Science (ACACOS '11), pp. 26–30.

Alquraishee, A.G.A., Kar, J., 2014. A survey on security mechanisms and attacks in wireless sensor networks. Contemp. Eng. Sci. 7 (3), 135–147.

Amrinder, K., Sunil, S., 2013. Simulation of low energy adaptive clustering hierarchy protocol for wireless sensor network. Int. J. Adv. Res. Comput. Sci. Softw. Eng. 3 (7).

Anderson, R.J., 2008. Security Engineering: A Guide to Building Dependable Distributed Systems, second ed. John Wiley & Sons, New York. ISBN 978-0-470-06852-6.

Arora, P., Gupta, A., 2014. A survey on wireless sensor network security. Int. J. Comput. Sci. Inform. Technol. Res. 2 (2), 67–76.

Aura, T., Nikander, P., Leiwo, J., 2001. DOS-resistant authentication with client puzzles. In: Revised Papers From the 8th International Workshop on Security Protocols. Springer-Verlag, Cambridge, UK, pp. 170–177.

Babu, C.M., Lanjewar, A.U., Manisha, C.N., 2013. Network intrusion detection system on wireless mobile ad hoc networks. Int. J. Adv. Res. Comput. Commun. Eng. 2 (3), 1495–1500.

Bajaber, F., Awan, I., 2011. Adaptive decentralized re-clustering protocol for wireless sensor networks. J. Comput. Syst. Sci. 77 (2), 282–292.

Becher, A., Benenson, Z., Dornseif, M., 2006. Tampering with motes: real-world physical attacks on wireless sensor networks. Security Pervasive Comput. 3934, 114–118.

Benenson, Z., Cholewinski, P.M., Freiling, F.C., 2008. Vulnerabilities and attacks in wireless sensor networks. In: Wireless Sensor Network Security. IOS Press, Amsterdam, pp. 22–44.

Bhuyan, B., Kumar, H., Sarma, D., Sarma, N., Kar, A., Mall, R., 2010. Quality of service (QoS) provisions in wireless sensor networks and related challenges. Wirel. Sens. Netw. 2, 861–868.

Blackert, W.J., Gregg, D.M., Castner, A.K., Kyle, E.M., Hom, R.L., Jokerst, R.M., 2003. Analyzing interaction between distributed denial of service attacks and mitigation technologies In: Proceeding DARPA Information Survivability Conference and Exposition, vol. 1 (22), 26–36.

Braginsky, D., Estrin, D., 2002. Rumor routing algorithm for sensor networks. In: Inproceedings of the 1st ACM International Workshop on Wireless Sensor Networks and Applications. ACM Press, New York, USA, pp. 22–31.

Brutch, P., Ko, C., 2003. Challenges in intrusion detection for wireless ad-hoc networks. In: Inproceedings of the Symposium on Applications and the Internet Workshops, pp. 1–6.

Cagalj, M., Capkun, S., Hubaux, J.P., 2007. Wormhole-based anti-jamming techniques in sensor networks. IEEE Trans. Mobile Comput. 6 (1), 1–15.

Carman, D.W., Krus, P.S., Matt, B.J., 2000. Constraints and approaches for distributed sensor network security. Technical Report 00-010. NAI Labs, Network Associates Inc., Glenwood, MD.

Chan, H., Perrig, A., 2006. Security and privacy in sensor network. In: IEEE Communications Surveys & Tutorials, IEEE Computer Magazine, pp. 103–105.

Chan, H., Perrig, A., Song, D., 2003. Random key pre-distribution schemes for sensor networks. In: Proceedings of the IEEE Symposium on Security and Privacy, pp. 197–203.

Charles, C.T., 2005. Security review of the light-weight access point protocol. In: IETF CAPWAP Working Group, pp. 1–27.

Chaudhary, S., Thanvi, P., 2015. Performance analysis of modified AODV protocol in context of denial of service (DOS) attack in wireless sensor networks. Int. J. Eng. Res. Gen. Sci. 3 (3), 1–6.

Chen, D., Varshney, P.K., 2004. Qos support in wireless sensor network: a survey. In: Proceedings of the International Conference on Wireless Networks (ICWN'04), Las Vegas, Nevada, USA, vol. 1, pp. 304–316.

Chen, D., Deng, J., Varshney, P.K., 2003. Protecting wireless networks against a denial of service attack based on virtual jamming. In: Proceedings of the Ninth Annual International Conference on Mobile Computing and Networking. ACM, New York, USA, pp. 548–561.

Chen, T.-S., Tsai, H.-W., Chu, C.-P., 2006. Gathering-load-balanced tree protocol for wireless sensor networks. In: Proceedings of the IEEE International Conference on Sensor Networks, Ubiquitous, and Trustworthy Computing (SUTC'06), pp. 32–35.

Chipara, O., He, Z., Xing, G., Chen, Q., Wang, X., 2006. Real-time power-aware routing in sensor networks. In: Proceedings of the 14th IEEE International Workshop on Quality of Service, pp. 83–92.

Chung, T.-P., Lin, T.-S., Zheng, X.-Y., Yen, P.-L., Jiang, J.-A., 2011. A load balancing algorithm based on probabilistic multi-tree for wireless sensor networks. In: Fifth International Conference on Sensing Technology (ICST). IEEE, New York, pp. 527–532.

Conti, M., Gregori, E., Maselli, G., 2003. Towards reliable forwarding for ad hoc networks. In: Personal Wireless Communications, 8th International Conference, Venice, Italy. Springer, Berlin, Heidelberg, pp. 790–804. doi: 10.1007/978-3-540-39867-7_71.

Crawley, E., Nair, R., Rajagopalan, B., Sandick, H., 1998. A Framework for QoS-Based Routing in the Internet, pp. 138–144.

Culpepper, B.J., Tseng, H.C., 2004. Sinkhole intrusion indicators in DSR MANETs. In: Proceedings First International Conference on Broad Band Networks, pp. 681–688.

Dai, H., Han, R., 2003. A node-centric load balancing algorithm for wireless sensor networks. In: Global Telecommunications Conference, GLOBECOM'03. IEEE, vol. 1. IEEE, New York.

Deng, Y., Hu, Y., 2010. A Load Balanced Clustering Algorithm for Heterogeneous Wireless Sensor Networks. ISBN 978-1-4244-7159-1, pp. 1–4.

Deng, J., Han, R., Mishra, S., 2002. INSENS: Intrusion-tolerant routing in wireless sensor networks. Technical Report CU-CS-939-02. Department of Computer Science, University of Colorado at Boulder.

Deng, J., Han, R., Mishra, S., 2004. Countermeasures against traffic analysis in wireless sensor networks. Technical Report CU-CS-987-04. University of Colorado at Boulder, pp. 1–13.

Dohler, M., 2008. Wireless sensor networks: the biggest cross-community design exercise to-date. Recent Pat. Comput. Sci., 1 (1), pp. 9–25.

Douceur, J.R., 2002. The Sybil attack. In: 1st International Workshop on Peer-to-Peer Systems (IPTPS '02), pp. 1–6.

El-Bendary, N., Soliman, O.S., Ghali, N.I., Hassanien, A.E., Palade, V., Liu, H., 2011. A secure directed diffusion routing protocol for wireless sensor networks. In: Proceedings of the 2nd International Conference on Next Generation Information Technology (ICNIT '11), pp. 149–152.

Eschenauer, L., Gligor, V.D., 2002. A key-management scheme for distributed sensor networks. In: Inproceedings of the 9th ACM Conference on Computer and Networking, pp. 41–47.

Farooq, N., Zahoor, I., Mandal, S., Gulzar, T., 2014. Systematic analysis of dos attacks in wireless sensor networks with wormhole injection. Int. J. Inform. Comput. Technol. 4 (2), 173–182.

Freier, A., Karlton, P., Kocher, P., 2011. The Secure Sockets Layer (SSL) Protocol Version 3.0. 1–67.

Gamal, A.E., Mammen, J., Prabhakar, B., Shah, D., 2004. Throughput-delay trade-off in wireless networks. In: Proceedings of the 23rd Annual Joint Conference of IEEE Computer and Communications Societies, vol. 1, pp. 464–475.

Ganesan, D., Govindan, R., Shenker, S., Estrin, D., 2001. Highly-resilient, energy-efficient multipath routing in wireless sensor networks. Mobile Comput. Commun. Rev. 4 (5), 1–13.

Ganesan, D., Krishnamachari, B., Woo, A., Culler, D., Estrin, D., Wicker, S., 2002. An Empirical Study of Epidemic Algorithms in Large Scale Multihop Wireless Networks, pp. 1–15.

Ganesan, P., Venugopalan, R., Pedddabachagari, P., Dean, A., Mueller, F., Sichitiu, M., 2003. Analyzing and modeling encryption overhead for sensor network nodes. In: The 2nd ACM International Conference on Wireless Sensor Networks and Applications (WSNA'03) Washington, vol. 9, pp. 3–5.

Gasser, M., Goldstein, A., Kaufman, C., Lampson, B., 1989. The digital distributed system security architecture. In: Inproceedings of the National Computer Security Conference, pp. 305–319.

Graf, K., 2005. Addressing Challenges in Application Security. Watchfire White Paper, pp. 1–26. Retrieved from http://www.watchfire.com.

Gruteser, M., Schelle, G., Jain, A., Han, R., Grunwald, D., 2003. Privacy-aware location sensor networks. In: Inproceedings of the 9th USENIX Workshop on Hot Topics in Operating Systems, (HotOSIX), pp. 163–167.

Gupta, G., Younis, M., 2003. Performance evaluation of load-balanced clustering of wireless sensor networks. In: 10th International Conference on Telecommunications, pp. 1–7.

Haenggi, M., 2005. Opportunities and challenges in wireless sensor network. In: Handbook of Sensor Networks Compact wireless and Wired Sensing Systems. CRC Press, Boca Raton, FL, pp. 21–34.

Hamid, M.A., Rashid, M.M., Choong, S.H., 2006a, Defense against lap-top class attacker in wireless sensor network. In: Proceedings of the 8th International Conference Advanced Communication Technology (ICACT '06), pp. 314–318.

Hamid, M.A., Rashid, M.O., Hong, C.S., 2006b. Routing security in sensor network: hello flood attack and defense. In: IEEE ICNEWS, pp. 77–81.

Hankerson, D., Menezes, A., Vanstone, S., 2004. Guide to Elliptic Curve Cryptography. Springer-Verlag, New York. ISBN 0-387-95273-X.

Heidemann, J., Li, Y., Syed, A., Wills, J., Ye, W., 2005. Underwater sensor networking: research challenges and potential applications. In: Proceedings of the Technical Report ISI-TR-2005-603, USC/Information Sciences Institute.

Hill, J., Szewczyk, R., Woo, A., Hollar, Z., Culler, D.E., Pister, K., 2000. System architecture directions for networked sensors. In: Architectural Support for Programming Languages and Operating Systems, pp. 93–104.

Ho, C.K., Robinson, A., Miller, D.R., Davis, M.J., 2005. Overview of sensors and needs for environmental monitoring. Sensors 5 (1), 4–37.

Hongseok, Y., Moonjoo, S., Dongkyun, K., Kyu, H.K., 2010. Global: a gradient-based routing protocol for load-balancing in large-scale wireless sensor networks with multiple sinks. In: International Symposium Computers and Communications (ISCC), pp. 556–562.

Hu, L., Evans, D., 2004. Using directional antennas to prevent wormhole attacks. In: Inproceedings of the 11th Annual Network and Distributed System Security Symposium, pp. 1–11.

Hu, Y.C., Perrig, A., Johnson, D.B., 2003. Packet leashes: a defense against wormhole attacks in wireless networks. In: Twenty-Second Annual Joint Conference of the IEEE Computer and Communications Societies. IEEE INFOCOM, vol. 3, pp. 1976–1986.

Humphries, J.W., Carlisle, M.C., 2002. Introduction to cryptography. ACM J. Educ. Resour. Comput. 2 (3), 2–7.

Israr, N. and I. Awan 2007. Multihop clustering algorithm for load balancing in wireless sensor networks. Int. J. Simul. 8 (3), 13–25.

Jabbar, S., Butt, A.E., Sahar, N., Minhas, A.A., 2011. Threshold based load balancing protocol for energy efficient routing in WSN. In: 2011 13th International Conference on Advanced Communication Technology (ICACT), pp. 196–201.

Jaydip, S., 2009. A survey on wireless sensor network security. Int. J. Commun. Netw. Inf. Security 1 (2), 55–78.

Kaissi, R.E., Kayssi, A., Chehab, A., Dawy, Z., 2005. DAWWSEN: a defense mechanism against wormhole attack in wireless sensor network. In: Proceedings of the Second International Conference on Innovations in Information Technology (IIT'05), pp. 1–10.

Karakehayov, Z., 2005. Using reward to detect team black-hole attacks in wireless sensor networks. In: Workshop on Real-World Wireless Sensor Networks (REALWSN'05), Stockholm, Sweden, pp. 1–5.

Karger, D., Ruhl, M., 2003. New algorithms for load balancing in peer-to-peer systems. Tech. Rep. MIT-LCS-TR-911, MIT LCS.

Karlof, C., Wagner, D., 2003. Secure routing in wireless sensor networks: attacks and countermeasures. Ad Hoc Netw. 1, (2–3), 299–302. Special Issue on Sensor Network Applications and Protocols.

Karlof, C., Sastry, N., Wagner, D., 2004. TinySec: a link layer security architecture for wireless sensor networks. In: Proceedings of the 2nd International Conference on Embedded Networked Sensor systems, Baltimore, MD, USA, pp. 162–175.

Karp, B., Kung, H.T., 2000. GPSR: greedy perimeter stateless routing for wireless networks. In: Proceedings of the 6th Annual International Conference on Mobile Computing and Networking, pp. 243–254.

Kaufman, C., Perlman, R., Speciner, M., 2002. Network Security Private Communication in a Public World. Prentice Hall PTR, Upper Saddle River, NJ, pp. 752–761 .

Kaushal, K., Sahni, V., 2015. Dos attacks on different layers of WSN: a review. Int. J. Comput. Appl. 130 (17), 8–11.

Kaushal, K., Sahni, V., 2016. Early detection of ddos attack in wsn. Int. J. Comput. Appl. 134 (13), 14–18.

Kavitha, H.L., Pushpalatha, K.N., 2015. Secure authentication technique for wireless sensor networks with load-balancing routing protocol. Int. J. Innov. Res. Comput. Commun. Eng. 3 (6), 5164–5169.

Khan, N.A., Saghar, K., Ahmad, R., Kiani, A.K., 2016. Achieving energy efficiency through load balancing: a comparison through formal verification of two WSN routing protocols. In: 13th International Bhurban Conference on Applied Sciences & Technology (IBCAST), pp. 350–354.

Kim, N., Heo, J., Kim, H.S., Kwon, W.H., 2008. Reconfiguration of cluster head for load balancing in wireless sensor networks. Comput. Commun. 153–159.

Kline, E., Afanasyev, A., Reiher, P., 2011. Shield: DOS filtering using traffic deflecting. In: 19th IEEE International Conference on Network Protocols, pp. 37–42.

Kong, J., Zerfos, P., Luo, H., Lu, S., Zhang, L., 2001. Providing robust and ubiquitous security support for mobile ad hoc networks. In: 9th International Conference on Network Protocols (ICNP), pp. 251–260.

Kun, S., Pai, P., Peng, N., 2006. Secure distributed cluster formation in wireless sensor networks. In: Proceedings of the 22nd Annual Computer Security Applications Conference (ACSAC'06), pp. 131–140.

Lai, B., Kim, S., Verbauwhede, I., 2002. Scalable session key construction protocols for wireless sensor networks. In: In IEEE Workshop on Large Scale Real Time and Embedded Systems, pp. 1–6.

Laszlo, E., Tornai, K., Treplan, G., Levendovszky, J., 2011. Novel load balancing scheduling algorithm for wireless sensor networks. In: The Fourth International Conference on Communication Theory, Reliability, and Quality of Service, pp. 112–117.

Lata, B.T., Sumukha, T.V., Suhas, H., Tejaswi, V., Shaila, K., Venugopal, K.R., Dinesh, A., Patnaik, L.M., 2015. SALR: secure adaptive load-balancing routing in service oriented wireless sensor networks. In: Signal Processing, Informatics, Communication and Energy Systems (SPICES).

Lawson, L., 2005. Session hijacking packet analysis. SecurityDocs.com Report, pp. 1–8.

Lazos, L., Poovendran, R., Meadows, C., Syverson, P., Chang, L.W., 2005. Preventing wormhole attacks on wireless ad hoc networks: a graph theoretic approach. In IEEE Wireless Communications and Networking Conference, pp. 1193–1199. doi:10.1109/WCNC.2005.1424678.

Lemos, M.V.S., Leal, L.B., Filho, R.H., 2010. A new collaborative approach for intrusion detection system on wireless sensor networks. In: Novel Algorithms Techniques Telecommunication, Network, pp. 239–244. doi:10.1007/978-90-481-3662-9_41.

Leopold, M., Dydensborg, M.B., Bonnet, P., 2003. Bluetooth and sensor networks: a reality check. In: Proceedings of the Sensys'03, Los Angeles, CA, pp. 65–72.

Levendovszky, J., Tornai, K., Treplan, G., Olah, A., 2011. Novel load balancing algorithms ensuring uniform packet loss probabilities for WSN. In: Vehicular Technology Conference 73rd, pp. 88–93.

Li, Y., Fu, Z., 2015. The research of security threat and corresponding defense strategy for WSN. In: Seventh International Conference on Measuring Technology and Mechatronics Automation, pp. 1274–1277.

Li, Z., Gong, G., 2008. A Survey on Security in Wireless Sensor Networks. Department of Electrical and Computer Engineering, University of Waterloo, Canada.

Liu, H., Ma, H., Zarki, M.E., Gupta, S., 1997. Error control schemes for networks: an overview. Mobile Netw. Appl. 2 (2), 167–182.

Low, C.P., Fang, C., Mee, J., Hang, Y.H., 2007. Load Balanced Clustering Algorithm for Wireless Sensor Networks. In: IEEE International Conference on Communications, 2007. ICC '07. IEEE Communications Society.

Luo, H., Kong, J., Zerfos, P., Lu, S., Zhang, L., 2000. Self securing ad-hoc wireless networks. In: IEEE Symposium on Computers and Communications (ISCC'02), pp. 1–17.

Mathapati, M.I., Salotagi, S., 2016. Load balancing and providing security using RSA in wireless sensor networks. Int. J. Adv. Res. Ideas Innov. Technol. 2 (2), 1–7.

Martínez, J.-F., García, A.-B., Corredor, I., López, L., Hernández, V., Dasilva, A., 2007. Qos in wireless sensor networks: survey and approach. In: Proceedings of the 2007 Euro American Conference on Telematics and Information Systems, p. 20.

Madria, S., Yin, J., 2009. SERWA: a secure routing protocol against wormhole attacks in sensor networks. Ad Hoc Netw. 7 (6), 1051–1063.

Mbowe, J.E., Oreku, G.S., 2014. Quality of service in wireless sensor networks. Wirel. Sens. Netw. 6, 19–26.

Merzoug, M.A., Boukerram, A., 2011. Cluster-based communication protocol for load-balancing in wireless sensor networks. Int. J. Adv. Comput. Sci. Appl. 3 (6), 105–112.

Min, S., 2004. A Study on the Security of NTRU Sign Digital Signature Scheme. Master Thesis in Information and Communications University, Korea.

Min, R., Bhardwaj, M., Cho, S.H., Shih, E., Sinha, A., Wang, A., Chandrakasan, A., 2001. Low-power wireless sensor networks. In: Proceedings of 14th International Conference on VLSI Design (VLSI DESIGN 2001), pp. 205–210.

Ming-hao, T., Ren-lai, Y., Shu-jiang, L., Xiang-dong, W., 2011. Multipath routing protocol with load balancing in WSN considering interference. In: 6th IEEE Conference on Industrial Electronics and Applications, pp. 1062–1067.

Molva, R., Michiardi, P., 2002. Security in ad hoc networks. In: Personal Wireless Communications, 8th International Conference, Venice, Italy. Springer, Berlin, Heidelberg, pp. 756–776. doi: 10.1007/978-3-540-39867-7_69.

Mpitziopoulos, A., Gavalas, D., Konstantopoulos, C., Pantziou, G., 2009. A survey on jamming attacks and countermeasures in WSNS. IEEE Commun. Surv. Tut. 11 (4).

Na, A., Xinfang, Y., Yufang, Z., Lei, D., 2007. A virtual backbone network algorithm based on the multilevel cluster tree with gateway for wireless sensor networks. In: Proceedings The IET International Communication Conference on Wireless Mobile and Sensor Networks. Shanghai, China, pp. 462–465.

Naeem, T., Kok-Keong, L., 2009. Common security issues and challenges in wireless sensor networks ieee 802.11 wireless mesh networks. Int. J. Digital Content Technol. Appl. 3 (1), 89–90.

Nam, J., 2013. Load balancing routing protocol for considering energy efficiency in wireless sensor network. Adv. Sci. Technol. Lett. 44, 28–31.

Newsome, J., Shi, E., Song, D., Perrig, A., 2004. The Sybil attack in sensor networks: analysis and defenses. In: Proceedings of the Third International Symposium on Information Processing in Sensor Networks. ACM, New York, USA, pp. 259–268.

Nigam, V., Jain, S., Burse, K., 2014. Profile based scheme against DDoS attack in WSN. In: Fourth International Conference on Communication Systems and Network Technologies, pp. 112–116.

Oreku, G.S., 2013. Reliability in WSN for security: mathematical approach. In: Computer International Conference on Applications Technology, pp. 82–86.

Özdemir, S., 2007. Secure load balancing for wireless sensor networks via inter cluster relaying. In: Information Security and Cryptology Conference with International Participation, pp. 249–253.

Ozdemir, S., 2009. Secure load balancing via hierarchical data aggregation in heterogeneous sensor networks. J. Inform. Sci. Eng. 25, 1691–1705.

Ozturk, C., Zhang, Y., Trappe, W., 2004. Source-location privacy in energy-constrained sensor network routing. In: Proceedings of the 2nd ACM Workshop on Security of Ad Hoc and Sensor Networks. ACM, New York, NY, USA.

Padmavathi, G., Shanmugapriya, D., 2009. A survey of attacks, security mechanisms and challenges in wireless sensor networks. Int. J. Comput. Sci. 4 (1), 1–9.

Parno, B., Perrig, A., Gligor, V., 2005. Distributed detection of node replication attacks in sensor networks. In: Proceedings of IEEE Symposium on Security and Privacy, pp. 1–15.

Pathan, A.S.K., Hyung-Woo, L., Hong, C.S., 2006. Security in wireless sensor networks: issues and challenges. In: Advanced Communication Technology, pp. 1–6.

Perrig, A., Szewczyk, R., Wen, V., Culler, D., Tygar, J.D., 2002. SPINS: security protocols for sensor networks. Wirel. Netw. 8 (5), 521–534.

Perrig, A., Stankovic, J., Wagner, D., 2004. Security in wireless sensor networks. Commun. ACM 47 (6), 53–57.

Ranjan, S., Swaminathan, R., Uysal, M., Nucci, A., Knightly, E., 2009. DDoS-shield: DDoS resilient scheduling to counter application layer attacks. IEEE/ACM Trans. Netw. 17 (1), 26–39.

Raymond, D.R., Midkiff, S.F., 2008. Denial-of-service in wireless sensor networks: Attacks and defenses. IEEE Pervasive Comput. 7 (1), 74–81.

Rowstron, A., Druschel, P., 2001. Pastry: scalable, distributed object location and routing for large-scale peer-to-peer systems. In: Proceeding in Middleware.

Saghar, K., 2010. Formal modelling and analysis of denial of services attacks in wireless sensor networks. PhD dissertation, Northumbria University, Newcastle upon Tyne.
Sahu, S.S., Pandey, M., 2014. Distributed denial of service attacks: a review. Int. J. Mod. Educ. Comput. Sci. 1, 65–71.
Sangwan, A., 2016. Evaluation of threats and issues in wireless sensor networks. Int. J. Adv. Res. Comput. Sci. Manag. Stud. 4 (2), 6–13.
Saraswat, L., Kumar, S., 2013. Comparative study of load balancing techniques for optimization of network lifetime in wireless sensor networks. Int. J. Comput. Electron. Res. 2 (2), 189–193.
Sari, A., 2015. Security issues in mobile wireless ad hoc networks: a comparative survey of methods and techniques to provide security in wireless ad hoc networks, In: New Threats and Countermeasures in Digital Crime and Cyber Terrorism, Advances in Digital Crime, Forensics, and Cyber Terrorism (ADCFCT) Book Series (Chapter 5).
Sebastian, T.J., 2011. Secure route discovery against wormhole attacks in sensor networks using mobile agents. In: Inproceedings of the 3rd International Conference on Trends in Information Sciences and Computing (TISC '11), pp. 110–115.
Shancang, L., Shanshan, Z., Xinheng, W., Kewang, Z., Ling, L., 2014. Adaptive and secure load-balancing routing protocol for service-oriented wireless sensor networks. IEEE Syst. J. 8 (3), 858–867.
Sharma, K., Ghose, M.K., 2010. Wireless sensor networks: an overview on its security threats. Int. J. Comput. Appl. (Special Issue Mobile Ad-hoc Networks) (1), 42–45.
Sharma, R., Jain, G., 2015. Adaptive clustering using round robin technique in WSN. Int. J. Comput. Sci. Commun. 6 (2), 32–35.
Shi, E., Perrig, A., 2004. Designing secure sensor networks. Wirel. Commun. Mag. 11 (6), 38–43.
Shiva, M.G., D'Souza, R.J., Varaprasad, G., 2012. Digital signature-based secure node disjoint multipath routing protocol for wireless sensor networks source. IEEE Sens. J. 12 (10), 2941–2949.
Siavoshi, S., Kavian, Y.S., Sharif, H., 2014. Load-balanced energy efficient clustering protocol for wireless sensor networks. IET Wirel. Sens. Syst. 6 (3), 67–73. Special Issue: Selected Papers from the 9th International Symposium on Communications Systems, Networks and Digital Signal Processing.
Slijepcevic, S., Potkonjak, M., Tsiatsis, V., Zimbeck, S., Srivastava, M.B., 2002. On communication security in wireless ad-hoc sensor networks. 11th IEEE International Workshops on Enabling Technologies: Infrastructure for Collaborative Enterprises, vol. 10 (12), pp. 139–144.
Stetsko, A., Folkman, L., Matyáš, V., 2010. Neighbor-based intrusion detection for wireless sensor networks. In: Proceedings of the 6th International Conference on Wireless and Mobile Communications (ICWMC). IEEE Xplore Press, Valencia, pp. 420–425.
Swain, A.R., Hansdah, R.C., Chouhan, V.K., 2010. An energy aware routing protocol with sleep scheduling for wireless sensor networks. In: Advanced Information Networking and Applications (AINA), 24th IEEE International Conference, Perth, WA, pp. 933–940.
Talooki, V.N., Rodriguez, J., Sadeghi, R., 2009. A load balanced aware routing protocol for wireless ad hoc networks. In: International Conference on Telecommunications, ICT '09, pp. 25–30.
Toumpis, S., Tassiulas, T., 2006. Optimal deployment of large wireless sensor networks. IEEE Trans. Inform. Theory 52, 2935–2953.
Touray, B., Shim, J., Johnson, P., 2012. Biased random algorithm for load balancing in wireless sensor networks (BRALB). In: 15th International Power Electronics and Motion Control Conference, pp. 1–5.
Undercoffer, J., Avancha, S., Joshi, A., Pinkston, J., 2002. Security for sensor networks. In: Inproceedings of the CADIP Research Symposium. University of Maryland, Baltimore County, USA, pp. 1–11.
Vijay, U.E., Nikhil, S., 2015. Study of various kinds of attacks and prevention measures in WSN. Int. J. Adv. Res. Trends Eng. Technol 2 (10), 1223–1235.
Wajgi, D., Thakur, N.V., 2012a. Load balancing algorithms in wireless sensor network: A survey. Int. J. Comput. Netw. Wirel. Commun.

Wajgi, D., Thakur, V.N., 2012b. Load balancing based approach to improve lifetime of wireless sensor network. Int. J. Wirel. Mobile Netw. 4 (4), 155–167.

Walters, J.P., Liang, Z., 2006. Wireless Sensor Network Security: A Survey, Security in Distributed, Grid, and Pervasive Computing. Auerbach Publications, CRC Press, Boca Raton, FL, pp. 1–5.

Wang, W., Bhargava, B., 2004. Visualization of wormholes in sensor networks. In: Inproceedings of the ACM Workshop on Wireless Security. ACM Press, New York, USA, pp. 51–60.

Wang, X., Gu, W., Schosek, K., Chellappan, S., Xuan, D., 2004. Sensor network configuration under physical attacks. Technical report (OSU-CISRC-7/04-TR45). Department of Computer Science and Engineering, Ohio State University.

Wei, L., Ming, C., Mingming, L., 2009. Information security routing protocol in the WSN. In: Fifth International Conference on Information Assurance and Security, pp. 651–656.

Werff, T.J.V.D., 2003. Global future report in technology review. Available at http://www.globalfuture.com/mit-trends2003.html.

Wood, A.D., Stankovic, J.A., 2002. Denial of service in sensor networks. IEEE Comput. 35 (10), 54–62.

Wood, A.D., Stankovic, J.A., Son, S.H., 2003. JAM: a jammed-area mapping service for sensor networks. In: 24th IEEE Real-Time Systems Symposium, pp. 286–297.

Wu, C., Yuan, R., Zhou, R., 2008. A novel load balanced and lifetime maximization routing protocol in wireless sensor networks. In: Vehicular Technology Conference, 2008. VTC, IEEE.

Xi, Y., Schwiebert, L., Shi, W., 2006. Preserving privacy in monitoring-based wireless sensor networks. In: Inproceedings of the 2nd International Workshop on Security in Systems and Networks (SSN '06). IEEE Computer society, pp. 1–8.

Xiangyu, J., Chao, W., 2006. The security routing research for WSN in the application of intelligent transport system. In: International Conference on Mechatronics and Automation, Luoyang, China, pp. 2319–2323.

Xin-sheng, W., Yong-zhao, Z., Liang-min, W., 2013. Load-balanced secure routing protocol for wireless sensor networks. Int. J. Distrib. Sens. Netw., 1–13.

Xing, K., Srinivasan, S.S.R., Rivera, M., Li, J., Cheng, X., 2006. Attacks and countermeasures in sensor networks: a survey. In: Network Security. Springer, New York, pp. 1–28.

Ye, F., Luo, H., Lu, S., Zhang, L., 2005. Statistical en-route filtering of injected false data in sensor networks. IEEE J. Selected Areas Commun. 23 (4), 839–850.

Yick, J., Mukherjee, B., Ghosal, D., 2008. Wireless sensor network survey. Comput. Netw. 52, 2292–2330.

Yin, C., Huang, S., Su, P., Gao, C., 2003. Secure routing for large scale wireless sensor networks. In: Proceedings of the International Conference on Communication Technology, pp. 1282–1286.

Yu, Y., Prasanna, V.K., Krishnamachari, B., 2006. Information Processing and Routing in Wireless Sensor Networks. World Scientific Publishing Co. Pte. Ltd, Singapore, pp. 1–21. ISBN 978-981-4476-70-6..

Yuanli, W., Xianghui, L., Jianping, Y., 2006. Requirements of quality of service in wireless sensor network. In: International Conference on Networking, International Conference on Systems and International Conference on Mobile Communications and Learning Technologies (ICNICONSMCL'06), pp. 256–269.

Yuvaraju, M., Sheela, K., Rani, S., 2014. Secure energy efficient load balancing multipath routing protocol with power management for wireless sensor networks. In: International Conference on Control, Instrumentation, Communication and Computational Technologies (ICCICCT), pp. 331–335.

Zhang, Y., Lee, W., 2000. Intrusion detection in wireless ad-hoc networks. In: The 6th Annual International Conference on Mobile Computing and Networking, pp. 275–283.

Zhang, Y., Zheng, Z., Jin, Y., Wang, X., 2009. Load balanced algorithm in wireless sensor networks based on pruning mechanism. In: International Conference on Communication Software and Networks, pp. 57–61.

Zhang, H., Li, L., Yan, X., Li, X., 2011. A load balancing clustering algorithm of WSN for data gathering. In: 2011 2nd International Conference on Artificial Intelligence, Management Science and Electronic Commerce (AIMSEC), pp. 915–918.

Zhu, S., Setia, S., Jajodia, S., 2003. Leap: efficient security mechanism for large-scale distributed sensor networks. In: Inproceedings of the 10th ACM Conference on Computer and Communications Security. ACM Press, New York, USA, pp. 62–72.

Znaidi, W., Minier, M., Babau, J.P., 2008. An ontology for attacks in wireless sensor networks. Research Report RR-6704. INRIA, pp. 1–13.

Zorzi, M., Rao, R.R., 2003. Geographic random forwarding (GERAF) for ad hoc and sensor networks: multihop performance. IEEE Trans. Mobile Comput. 2 (4), 337–348.

## FURTHER READING

Balen, J., Zagar, D., Martinovic, G., 2011. Quality of service in wireless sensor networks: a survey and related patents. Recent Pat. Comput. Sci. 4, 188–202.

Ganzc, A., Ganz, Z., Wongthavarawat, K., 2004. Multimedia Wireless Networks: Technologies, Standards, and QoS. Prentice Hall, Upper Saddle River, NJ, pp. 267–279.

Ranjan, N., Krishna, G., 2013. Wireless sensor network: quality of services parameters and analysis. In: Conference on Advances in Communication and Control Systems, pp. 332–337.

Zhang, L., Chen, F., 2014. A round-robin scheduling algorithm of relay nodes in WSN based on self-adaptive weighted learning for environment monitoring. J. Comput. 9 (4), 830–835.

CHAPTER

# MACHINE LEARNING TECHNIQUES FOR THREAT MODELING AND DETECTION

# 8

**Michał Choraś*, Rafał Kozik***

*UTP University of Science and Technology in Bydgoszcz, Bydgoszcz, Poland**

## 1 INTRODUCTION

In this chapter our goal is to prove that the bio-inspired solutions can be successfully applied to cyber security. The motivation of our work and results, presented in this chapter, come from the current needs to protect ICT systems and computer networks from cyberattacks, cybercrime, and cyber terrorism. The major contribution is the set of novel practical implementations of the bio-inspired solutions used for computer networks protection. In Section 2, we provide the general overview of the current challenges cyber security. In Section 3, we present our own novel developments and implementations of the bio-inspired techniques for cyber security, namely:

- bio-inspired optimization techniques (genetic algorithms for SQL injection attacks detection and for detection of anomalies in HTTP requests),
- techniques mimicking the behavior of living organisms (e.g., correlation approach and ensemble of classifiers),
- collective intelligence and distributed computation. Conclusions and references are given thereafter.

## 2 CYBERSECURITY: A CHALLENGE

As always in the history of the world, when technology is created and evolves, it can be used for good and criminal purposes. The same happens with the quick evolution of the communication networks and software applications that are now often the source or the target of so-called cybercrime and cyberattacks. What is more, currently it is very difficult, even for large and wealthy organizations (such as big industrial companies, banks, public administration), to counter and eliminate cyberattacks. Of course, the same situation (even in greater extent) applies to smaller organizations (e.g., SMEs) or citizens, where fewer resources are available for security. The important question in our civilization has always been, "What can be done to assure effective security?" The general (independent on application or domain) view on the security chain is as follows: analyze and understand the context and situation; look for vulnerabilities; analyze threats and manage risks; observe the situation using sensors and monitoring capabilities (also humans); collect data; analyze data and, on the basis of the data processing and analysis; either detect danger and attacks, or decide that there are no attacks happening at a given

Security and Resilience in Intelligent Data-Centric Systems and Communication Networks. https://doi.org/10.1016/B978-0-12-811373-8.00008-2

Copyright © 2018 Elsevier Inc. All rights reserved.

moment. The last stage is the reaction and remediation, and of course, there are different reactions available depending on situation, capabilities, and legal aspects. There are plenty of methods to analyze the collected data. Hereby, we will focus on the bio-inspired techniques for application layer attacks detection. As for decision-making, there are two possible approaches based on detecting certain patterns in the observed data. Let us assume that all the required and desired sensors, probes, monitoring devices are installed and operational. Then, what can be done with the collected data? The first approach is to learn patterns of evil (e.g., cyberattacks or terrorists attacks) and detect those patterns. The second approach is to learn the pattern of a normal and safe state, and detect abnormalities (also called anomalies or outliers) that do not fit into the normal and typical patterns. Those two approaches apply not only to computer networks, but also to security in general, e.g., in domains such as counter terrorism, analysis of bank transactions, and urban safety. If we knew the modus operandi and patterns of the terrorists, they would be easier stopped once effective monitoring is applied. However, the terrorists never promised to use the same ways and modus operandi, and they always want to outsmart those responsible for security and safety. Law enforcement agencies even train officers not to look for the biased patterns (e.g., white vans) but to analyze context, look for anomalies and think out-of-the-box. As for the networked systems, current cyber security solutions can be also classified as signature-based and anomaly-based. Typically, signature-based solutions are installed and widely used on personal computers and intrusion detection systems. For deterministic attacks it is fairly easy to develop patterns that will clearly identify a particular attack. The drawback of signature-based solutions is that, since there are no signatures (patterns) of the future attacks, new cyberattacks and so called zero-day exploits cannot be detected and mitigated. The reality is that cyber hackers and cyber terrorists never promised to use the same attacks, tools, means, and worms. For instance, top-ranked application layer attacks such as SQL Injection attacks or XSS (Cross Site Scripting) are changing all the time due to their diversity, complexity, and availability of obfuscation techniques Therefore signature-based approaches are not efficient and anomaly-based solutions are needed. Of course, the typical drawback of an anomaly-based approach is that such solutions produce a significant number of false alarms. In other words, not all of the detected anomalies are the signs of terrorist or cyberattacks, and the context needs to be understood while making decisions (e.g., rapid growth on network traffic to certain service might not be the sign of a DDoS (Distributed Denial of Service) attack, it can be the start of selling tickets to important sports events or concerts). The evolution of species is based on the ongoing battle between the predator and the victim in such a battle the victim learns to avoid and protect from predator (either biologically or in behavior), while the predator has to improve (skills, behavior or biological characteristics) to catch the victim. In this chapter, we focus on data analysis, processing, attacks detection, and decision-making for cyber security. As for the bio-inspired techniques, in the following part of the chapter, we will present how the bio-inspired optimization, collective intelligence, as well as mimicking the behavior of the living organisms, can be practically realized to enhance cyber security of computer networks.

## 3 BIO-INSPIRED METHODS FOR CYBER SECURITY PRACTICAL EXAMPLES AND IMPLEMENTATIONS

We live in a world of information that is ruled by the information theory; on the other hand we also live in a natural world bounded by the laws of physics. These two worlds present common analogies and similarities visible at macro and micro levels. For instance, the particles and data have similar statistical properties (i.e., uncertainty and entropy) that can be measured using common tools. This fact is heavily exploited by a variety of optimization techniques like simulated annealing, stochastic climbing, and particle filtering. Also the macro scale of our physical world (interaction between organisms, complex mammals brains capable of multimodal perception or evolution of species) inspires a variety of large-scale genetic and evolutionary-based optimization or swarm/ant colony optimization techniques. We can also observe many similarities between computer networks and biological organisms, especially when it comes to communication and security of telecommunication systems (Sharp, 2001). Even the term viruses has been borrowed from life sciences to highlight the behavior resemblance. As for the cyber defense and protection, there are also examples of solutions that are inspired by biology (Mazurczyk and Rzeszutko, 2015; Rzeszutko and Mazurczyk, 2015). Some of the methods include artificial neural networks, swarm optimization methods, ant colonies, collective intelligence, artificial immune systems, and the genetic algorithms. In this section, we analyze different bio-inspired techniques applied for cyber security domain. During the analysis we make several references to our past projects related to cyber security.

## 3.1 PRACTICAL REALIZATIONS OF THE BIO-INSPIRED OPTIMIZATION TECHNIQUES APPLIED TO CYBER SECURITY IN THE APPLICATION LAYER

Is this subsection we will present two practical implementations of the bio-inspire techniques:

- the genetic algorithm for SQL injection attacks detection, and
- the genetic algorithm for detection of anomalies in HTTP requests.

The growing popularity of publicly available Web services is one of the driving forces for so-called "web hacking" activities. According to the Symantec report the number of web attacks blocked per day has increased by 23% in comparison to previous years (Symantec, 2014). Protection from the cyberattacks targeting the application layer is difficult. According to White Hat Security Report, the Cross-Site Scripting remains the number one in 2014 (WhiteHat, 2014). The authors of this report also stated that such security exploits may allow the attacker to access sensitive data or overtake the site and user's accounts. According to this report, it may take 180 days (on average) to have security vulnerability patched. Therefore it is important to have effective and efficient tools to counter such attacks. Among all the attacks targeted at web-servers, the SQLIA (SQL Injection Attack) still remains one of the most important network threat that is ranked on top of the OWASP list (OWASP, 2013). SQL injection and other similar exploits are the results of interfacing a scripting language by directly passing information through another language and are ultimately caused by insufficient input validation. SQL Injection Attacks (SQLIA) refer to a code injection attacks category in which part of the user's input is treated as SQL code. Such code, if executed on the database, may change, erase, or expose sensitive data stored in the database. The SQL Injection Attacks are relatively easy to perform and hard to detect or prevent. In order to perform injection attack, the attacker sends a text, which exploits the syntax of the targeted interpreter, therefore almost any source of data can be an injection attack vector. As a result, injection can cause serious consequences including data loss, corruption, and lack of accountability, or denial of access.

### 3.1.1 Genetic algorithm to generate regular expressions and to detect SQL injection attacks

A variety of Evolutionary Algorithms (EAs) that mimic the biological evolution or social behavior of the living organisms have been successfully used over the last decades to find near-optimum solutions to large-scale optimization problems. Most commonly used ones are: genetic algorithms, particle swarm, ant-colony, firefly algorithm, or shuffled frog leaping. A practical implementation is proposed in Bankovic et al. (2007), where the authors proposed a Genetic Algorithm (GA) based technique to learn IF-THEN rules of the firewall from the historical data. Authors first extracted the relevant features describing TCP/IP connections using the PCA technique, and then they encoded the rules as chromosomes within the typical GA framework. In Bin Ahmad et al. (2014) the authors used the genetic algorithm to enhance the effectiveness of the fuzzy-classifier for detecting the insider threats. In our recently finalized research project SECOR, Choraś et al. (2012) proposed innovative anomaly detection methods, and we proposed a novel method for an SQL Injection Attack detection based on the Genetic Algorithm (GA) for determining anomalous queries. Our proposed solution exploited genetic algorithm implementing a variant of social behavior of species, where the individuals in the population explored the lines in the log-files that were generated by the SQL database. In our model, each individual delivers a generic rule (which was a regular expression) that describes the visited line of the log. The proposed algorithm is divided into the following steps:

- Initialization. The line from the log file is assigned to each individual. Each newly selected individual is compared to the previously selected in order to avoid duplicates.
- Adaptation phase. Each individual explores the fixed number of lines in the log file (the number is predefined and adjusted to obtain reasonable processing time of this phase).
- Fitness evaluation. The fitness of each individual is evaluated. The global population fitness as well as rule level of specificity is taken into consideration, because we want to obtain the set of rules that describe the lines in the log file.
- Crossover. Two randomly selected individuals are crossed over using an algorithm for string alignment. If the newly created rule is too specific or too general, it is dropped in order to keep low false positives and false negatives.

In our work, we used the modified version of the Neddleman-Wunsch algorithm (Needleman and Wunsch, 1970), originally invented to find the best match between DNA sequences. In order to find correspondence between those two sequences, but also for any text strings such as the logs analyzed here, it is allowed to modify the sequences by inserting the gaps. For each gap (and for mismatch) there is a penalty while the award is given for genuine matches. For the Needleman and Wunsch algorithm, the most important aspect is to find the best alignment between two sequences (the one with highest award). From an anomaly detection point of view the parts where gaps are inserted are also important, because they are the points of injections. These parts are described with regular expressions using the following heuristics:

- The character sequences are encoded as [a-zA-Z]+ whenever those contain only letters
- The character sequences are encoded as [0-9]+ whenever those contain only numbers
- The sequences can also be encoded using both of the mentioned above sequences (e.g., [a-zA-Z0-9]+
- Whenever encoded character sequences contain special characters (e.g., whitespace and brackets), those are included into the regular expression accordingly.

The fitness function that is used to evaluate each individual takes into account the effectiveness of the particular regular expression (number of times it fires), the level of specificity of such a rule, and the overall effectiveness of the whole population. The level of specificity indicates the balance between number of matches and number of gaps. This parameter enables the algorithm to penalize these individuals that try to find general rule for significantly different queries like SELECT and INSERT. The fitness is computed using following cost function:

$$E(i) = \alpha \sum_{j \in (I-i)} E_f(j) + \beta E_f(i) + \gamma E_s(i). \tag{1}$$

In this formula the $\alpha$, $\beta$, $\gamma$ are constants that normalize the overall score and balance the each coefficient importance. $E_f$ indicates of regular expression (number of times the rule fires) and $E_s$ indicates the level of specificity. The level of specificity indicates balance between number of matches and number of gaps. The SQL injection attacks detection results for our method are comparable to those obtained with standard signature-based solutions like SNORT , Apache SCALP, PHP-IDS, and ICD (see Table 2). The Apache SCALP is an analyzer of the Apache server access log files. It is able to detect several types of attacks targeted at web application. The detection is a signature-based one. The signatures have a form of regular expressions that are borrowed from the PHP-IDS project. SNORT is a deployed IDS system that uses the set of rules that are used for detecting web application attacks. However, most of the available rules are intended to detect very specific type of attacks that usually exploit very specific web-based application vulnerabilities. PHP-IDS is a simple to use, well-structured, fast and state-of-the-art security layer for PHP based web applications. It is based on a set of approved and heavily tested filter rules. Each attack is given a numerical impact rating that makes it easy to decide what kind of action should follow the hacking attempt. This could range from simple logging to sending out an emergency mail to the development team, displaying a warning message for the attacker, or even ending the users session. The ICD (Idealized Character Distribution) method is based on a character distribution model for describing the genuine traffic generated to web application. The Idealized Character Distribution (ICD) is obtained during the training phase from perfectly normal requests sent to the web application. The ICD is calculated as mean value of all character distributions. During the detection phase the probability that the character distribution of a query is an actual sample drawn from its ICD is evaluated. For that purpose Chi-Square metric is used. In our experiments, we aimed at:

- Evaluating each method for SQL injection attack separately (Table 2).
- Investigating whenever combining above methods together can additionally improve overall effectiveness of injection attack detection (Table 3). The set of tools combined with each other are indicated in Table 1.

In Table 1 the configuration for different experiments are shown. We have investigated different configurations of scenarios for effectiveness evaluation. These scenarios try to reflect the deployment aspects connected to the mentioned above injection attack detection tools. For example, PHP-IDS requires changes in the HTTP server configuration in order to be operational. Moreover, PHP-IDS can only be deployed when the HTTP server supports PHP technology. When it comes to the Apache SCALP tool, it is capable of analyzing a HTTP-GET requests by default. Therefore to increase the detection effectiveness it is required to modify the server configuration.

For evaluation purposes a 10-fold approach is used. The information obtained from our proposed method, PHP-IDS, SCALP, ICD, and SNORT are used to build classifier for attack detection. In order

**Table 1 Experimental Setup Sensors Used in a Given Experiment (E1, . . . , E6) Are Indicated With the Letter X**

| | Our Method | SNORT | ICD | SCALP | PHP-IDS |
|---|---|---|---|---|---|
| E1 | | | X | X | |
| E2 | | | X | X | X |
| E3 | X | | X | X | |
| E4 | X | | X | X | X |
| E5 | | X | X | X | X |
| E6 | X | X | X | X | X |

**Table 2 Effectiveness of the Different Tools**

| | Our Method (%) | SNORT (%) | ICD (%) | SCALP (%) | PHP-IDS (%) |
|---|---|---|---|---|---|
| TP | 87.8 | 66.3 | 97.9 | 50.9 | 93.5 |
| TN | 97.7 | 80.5 | 94.5 | 96.1 | 98.1 |
| Weighted Average | 96.2 | 78.3 | 95.0 | 89.0 | 98.1 |

to conduct the experiments the WEKA toolkit is used. As it is shown in Table 2, the PHP-IDS algorithm slightly outperforms other approaches (but as mentioned above it cannot be used in all the situations/configurations). When it comes to modeling the genuine queries, our proposed method is almost as good as the PHP-IDS. However, for queries having the symptoms of an attack, the proposed method is about 10% worse when compared to the ICD, which has the best performance of attacks detection.

The effectiveness for different configuration setups (experiments from 1 to 6) is shown in Table 3. Notice that the highest effectiveness is reported for experiment 6, when all detectors are used. The effectiveness is increased by 1.5% when compared to the PHP-IDS and is reported to be 99.67%. Experiments 1 and 3 have no impact on the server configuration. All detectors used in these scenarios (ICD, SCALP, and the proposed method) are transparent for web servers since they process log files generated by HTTP and DB daemons. Noticed that when the proposed algorithm is added, the effectiveness increases by more than 2%. Without the proposed method the performance is slightly

**Table 3 Effectiveness of the Different Tools Combined With Each Other in Different Experimental Setups and Various Classifiers**

| | NaiveBayes | PART | Ridor | J48 | REPTree | AdaBoost |
|---|---|---|---|---|---|---|
| E1 | 96.73 | 96.84 | 96.38 | 96.87 | 96.83 | 96.03 |
| E2 | 97.54 | 97.06 | 96.99 | 97.02 | 97.12 | 96.07 |
| E3 | 96.91 | 99.00 | 98.93 | 99.10 | 98.97 | 98.80 |
| E4 | 98.35 | 99.03 | 99.02 | 99.12 | 98.99 | 98.89 |
| E5 | 98.83 | 99.24 | 99.27 | 99.24 | 99.26 | 99.30 |
| E6 | 99.08 | 99.51 | 99.40 | 99.54 | 99.37 | 99.67 |

worse, but it is still better than each detector's individual effectiveness (which is 95% for the ICD). When experiments 3 and 4 are compared, notice that the PHP-IDS does not increase the effectiveness which is 99.1%. Therefore the proposed algorithm, ICD, and SCALP can be a good alternative in situation when it is impossible to deploy the PHP-IDS.

### *3.1.2 Using genetic algorithm to identify structure in raw packets in order to detect anomalous HTTP requests*

When it comes to communication in the application layer, the Hypertext Transfer Protocol (HTTP) is one of the most frequently used protocols. This is because nowadays the significant part of ICT solutions rely on Web servers or Web services, and the HTTP protocol is a reliable mean enabling communication between computers in distributed networks. The HTTP is a request-response plain-text protocol. Basically, the request is a type of method that can be called (executed) on a resource uniquely identified by a URL address. There are different types of methods that (among others) allow performing CRUD (Create, Read, Update, and Delete) operations on a resource. Some of the methods are accompanied with the request payload (e.g., POST, PUT). Different protocols that use HTTP for transportation exhibit different structures of the payload. For example, the structure sent via plain HTML form will be different from a GWT-RPC or SOAP call. In this approach, we take advantage of the request-response nature of the HTTP protocol and we use the preclassification approach shown in Fig. 1. Thus instead of a whole packet analysis, we focus on the request payload in order to extract the structure from the consecutive calls. The procedure is as follows:

- Once the HTTP request is delivered to our proposed anomaly detection system it is separated into three streams. It means that we extract the information about the HTTP method and the object it is called, the URL arguments (if they exist), and the payload of the request.
- Afterwards, we take advantage of the fact that the number of objects (URL addresses) is limited and can be used to build a whitelist. It is based on the assumption that unknown methods called on existing objects, or methods called on unknown objects, may be considered anomalous.
- Finally, we take feedback from the whitelist, the data contained in the URL and the payload to validate the request structure and content.

Our method works as an additional cyber security measure protecting the WWW server from cyberattacks. The current implementation works as a passive analyzer that analyzes the HTTP streams.

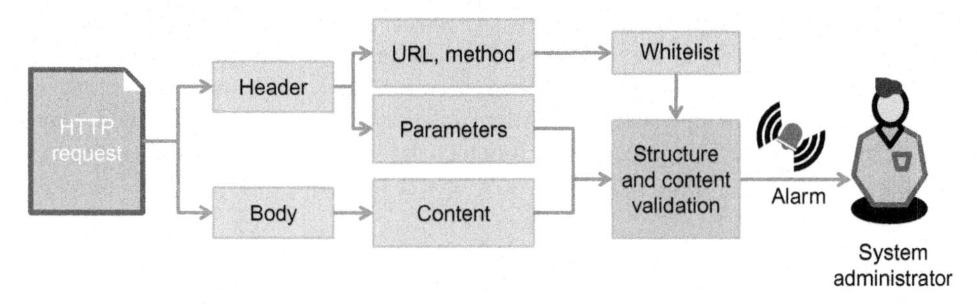


**FIG. 1**

Information flow in the proposed method for detecting anomalous HTTP requests.

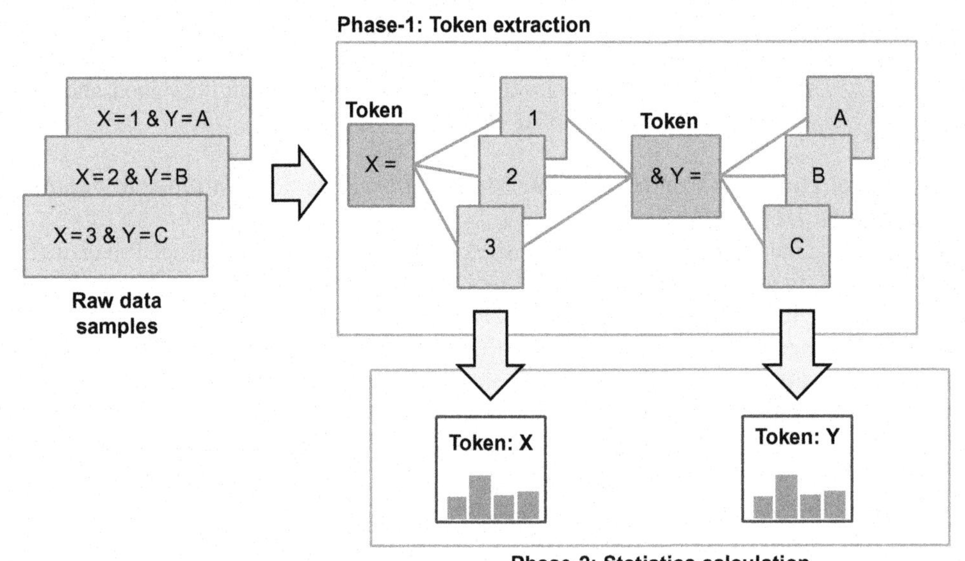


**FIG. 2**

Overview of the method for the structure extraction from raw packets.

Therefore the proposed algorithm operates on a server side where the web application is deployed. It intercepts the HTTP(S) traffic generated by the web browser of the client. Through the proxy server, it is possible to split the HTTP streams (in order to process them simultaneously) without affecting the quality of the Web service. In our work, we analyze and classify the content of HTTP requests. The general overview of our method is presented in Fig. 2.

We represent the structure of the payload by means of tokens. The token of an HTTP request is defined as the sequence of bytes that are common for all requests sent to the same resource. There could be several tokens identified for one request. Tokens are used to identify delimiters of those regions of the requests sequences, which are likely to be related to the data provided by the client sending the request. Hence, it allows us to identify possible points where malicious code can be injected. It is out of the scope of this chapter to show how tokens are generated. However, once we have tokens, we apply a genetic algorithm to align them for further processing and decision-making. In order to build the HTTP request model, we need to identify the right subset of tokens and their order. In our case a single token represents an item. We assign the value to each token (in current implementation we favor longer tokens over shorter) and mass, which represents the position of the token in a sequence. The limit, in our case, is determined by the analyzed sequences. More formally, the problem is constrained by the following optimization problem:

$$\begin{aligned} \underset{x}{\text{maximize}} \quad & C(x) = \sum_{i=0}^{n} v_i x_i \\ \text{subject to} \quad & \sum_{i=0}^{n} w_i x_i \leq W, \; x_i \in \{0, 1\}. \end{aligned} \tag{2}$$

To solve this optimization problem, we proposed to use genetic algorithm with the classical binary chromosome encoding schema and one point crossover. The chromosome in our algorithm represents a candidate solution, and it is a string of bits (1 indicates that the given token is taken to build the structure of request, while 0 is used to reject a given token). The genetic algorithm is used in the following manner: The population is initialized randomly. The chromosome length is determined by the number of tokens identified during the extraction procedure. The fitness of each chromosome is measured. Individuals are ordered by fitness values. Two chromosomes are selected randomly from the population. Selected chromosomes are subjected to the crossover procedure. The procedure is terminated (and the individual with the best fitness is selected) if a maximal number of iterations is exceeded, otherwise it goes back to step 2. Further in our system, once the tokens are identified, we describe the sequences between tokens using their statistical properties and apply machine-learning algorithms to decide if the requests represented by tokens are anomalous or not. We have proposed to build histograms of characters. However, instead of a bin-per-character approach, we have calculated the number of characters for which the decimal value in ASCII table belongs to the following ranges: [0, 31], [32,47], [48,57], [58,64], [65,90], [91,96], [97,122], [123,127], [128,255]. It may look heuristic, but different ranges represent different type of symbols like numbers, quotes, letters, or special characters. This approach allowed us to reduce the length of the feature vectors while obtaining satisfactory results. In our experiments, we used two approaches to classify given request as normal or anomalous, namely statistical $\chi^2$ test and a variety of ensemble classifiers.

The proposed method containing practical realization of the genetic algorithm achieves satisfactory results, better than state-of-the-art methods (e.g., signature-based approaches like the PHP-IDS or Apache ModSecurity), on a benchmark CSIC10 database (Torrano-Gimnez et al., 2010).

The results have been presented in Table 4 and Fig. 3.

## 3.2 PRACTICAL REALIZATIONS OF THE TECHNIQUES MIMICKING BEHAVIOR OF LIVING ORGANISMS

The second group of bio-inspired methods include mechanisms that mimic the defense techniques adapted by living organisms. One of these techniques is called the Moving Target (MT) strategy and aims at providing security through the system diversity. It is achieved by changing various system properties (system configuration). For instance, in Lucas et al. (2014) authors used genetic algorithms to address the problem of uniform and deterministic configuration (e.g., of computing clusters and databases farms). Authors modeled the configuration of a single computer as a chromosome and used

**Table 4 Effectiveness of Different Methods**

| | True Positives (%) | False Positives (%) |
|---|---|---|
| PHP-IDS | 20.4 | 1.25 |
| Apache ModSecurity | 26.3 | 0.34 |
| $\chi^2$ calculated on whole packet (no structure extracted) | 33.2 | 0.1 |
| $\chi^2$ with our proposed method | 91.1 | 0.7 |
| Our proposed method with Random Forest Classifier | 91.90 | 0.7 |

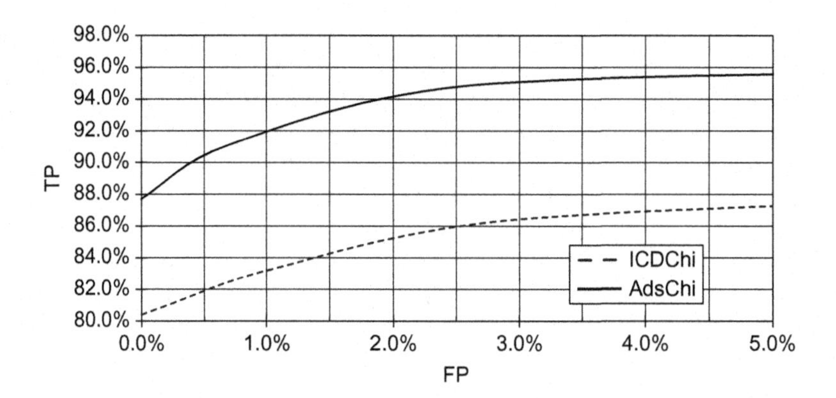


**FIG. 3**

Receiver Operating Characteristic curves showing the accuracy of our proposed method (AdsChi) with respect to an approach using $\chi^2$ calculated for the whole packet (ICDChi).

the evolutionary approach to identify new possible configurations. Other MT strategies may include dynamic IP addresses translation or techniques to fool the network scanners (Kewley et al., 2001).

## 3.3 ENSEMBLE OF CLASSIFIERS

We followed such a bio-inspired approach in practice in one of our previous works (Kozik and Choraś, 2015), where we used several techniques adapting the idea of ensemble learning. Our main goal was to increase the effectiveness of attack detection by adapting an ensemble of one-class classifiers. One of the challenges of producing the ensemble of classifiers is the diversity problem. Although a formal definition does not exist, it can be intuitively perceived as correlation and similarity of classifiers results. For instance, if the outputs produced by the pool of classifiers are similar, those will have poor diversity, thus we may not expect performance improvement. According to Wozniak (2014), the following methods improve diversity, namely:

- to use different partition of the data to train the classifiers,
- to exploit local specialization of given classifiers, and
- to use different subset of features.

In order to address the first and the second approach, we applied boosting and bagging techniques. For the last one, we applied a random selection of features subspace. In our approach, we selected two types of classifiers that build the ensemble, namely:

- Decision Stump (DS)—a machine learning model that is a decision tree with the single level. For example, if the subsequent features are considered in a one-class classification problem, this machine learning technique will produce a threshold.
- Reduces Error Pruning Tree (REPTree)—a machine learning technique that uses the pruned decision tree. The REP Tree algorithm generates multiple regression trees in each iteration. Afterwards, it chooses the best one. It uses a regression tree adapting variance and information

**Table 5 Comparison of the Different Hybridization Techniques**

| | True Positives (%) | False Positives (%) |
|---|---|---|
| DS Bagging | 99.3 | 4.6 |
| Boosting | 93.5 | 0.1 |
| RS | 94.4 | 0.3 |
| RepTree Bagging | 98.0 | 1.7 |
| Boosting | 94.0 | 0.4 |
| RS | 95.6 | 0.4 |

gain (by measuring the entropy). The algorithm prunes the tree using a back fitting method. Our experiments were conducted on a publicly available benchmark database to show that ensembles of weak classifiers can achieve better results than the classical approach using a single classifier. Some of the results are presented in Table 5.

### *3.3.1 Heterogeneous data sources correlation*

Another recent strategy inspired by nature is to use heterogeneous multimodal sources of information and to correlate them for improved decisions. The multimodal perception of the physical world that is exhibited by mammals' brains is also used as guidance when prototyping machine learning algorithms. For instance, living organisms use different heterogeneous sources of information (touch and smell), in order to reduce the uncertainty of single source and to better identify objects, threats, or to estimate more accurately the position with respect to the environment. The same phenomenon also applies when it comes to pattern matching, object detection or identification, data mining, and machine learning. In fact, there is no single pattern recognition algorithm that is suitable for all the problems. Each classifier has its own domain of competence. The reason why the researchers are focusing on the correlation approach and ensemble of classifiers is the fact that the combined classifiers: (1) can improve the overall effectiveness of recognition, (2) can be easier deployed in distributed systems, (3) allow overcoming the initialization problem of many machine- learning methods (e.g., k-means, tree learner, GMM). Therefore we have also explored the advantages of the data heterogeneity and multimodality in order to detect the cyberattacks conducted in the application layer. In the same way that a living organisms uses different senses to identify and avoid threats, we use different sensors to detect a wide variety attack targeting web applications. For instance, we deploy sensors and firewalls in different layers of the TCP/IP protocol stack, but we also deploy different detection techniques at the same layer (e.g., we combine anomaly-based attacks detection with signature-based detection. Our experiments (see Table 3) showed that this technique leads to significant improvement of the detection effectiveness.

## 3.4 PRACTICAL REALIZATION OF THE COLLECTIVE INTELLIGENCE AND DISTRIBUTED COMPUTATION

The third group of the bio-inspired methods include techniques that mimic the collaborative strategies of social insects such as bees, ants, and fireflies. In this section, we present the practical realization of those ideas. However, in contrast to the previous subsections, our proposal is not algorithmic, but it is realized on a conceptual, and later on the deployment, level. Previously, for instance in McKinnon et al. (2013), the authors proposed a system for adapting an ant colony to identify a potential cyber security attack against smart meter deployments. In Chhikara and Patel (2013) the authors combined ant colony optimization with cyber security scanners to identify vulnerabilities in the networks in a more effective way. We have applied such a bio-inspired approach to design and develop the Federated Networks Protection System (Choraś et al., 2012). Our motivation was that the successful cyberattacks are considered to be a threat for military networks and public administration computer systems. Therefore the goal of the Federated Networks Protection System, developed in the SOPAS project, is to protect public administration and military networks which are often connected into Federations of Systems. While adopting the concept of the federation of networks and collective intelligence, the synergy effect for security can be achieved. In our approach, we use the capability of the federated networks and systems to share and exchange information about events in the network, detected attacks, and the proposed countermeasures. Also in our case, the collective intelligence concept refers to a set of different independent systems, which are not centrally managed, but cooperate in order to share knowledge and increase their security. Of course, as in nature, an important factor for implementation of an approach such as this is trust. Trust of the networked systems has to be managed by administrators and decision-makers following certain procedures. The general architecture of the Federated Networks Protection System is presented in Fig. 4. It consists of several interconnected domains, which exchange information in order to increase their security level and the security of the whole federation. Different subnets are arranged in domains, according to the purpose they serve (e.g., WWW, FTP, or SQL

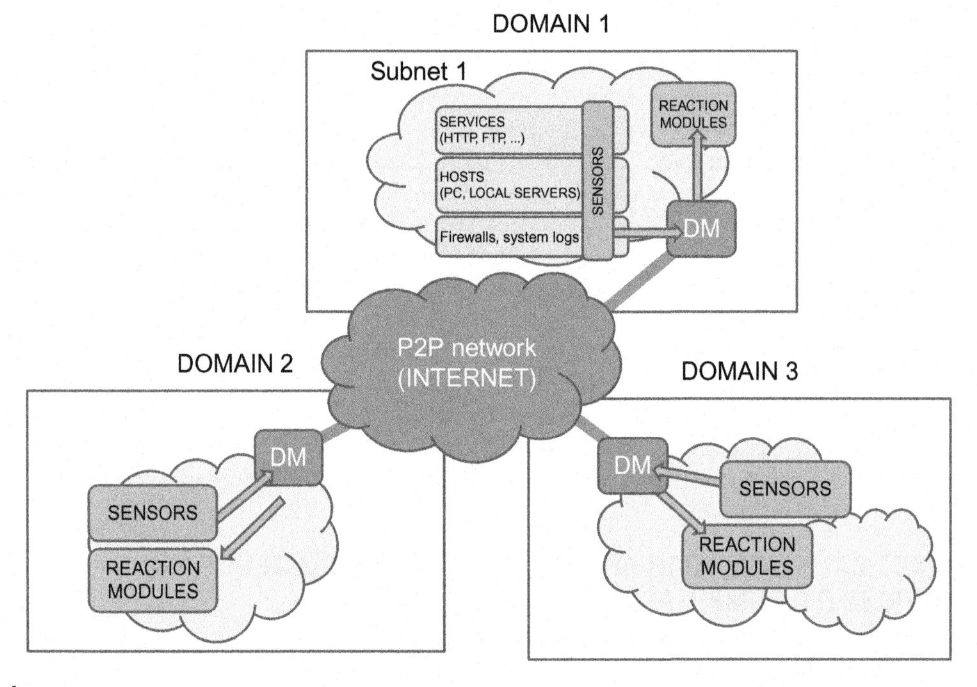


**FIG. 4**

The concept of the multidomain collective intelligence system for networks protection.

servers) or according to their logical proximity (two networks closely cooperating with each other). In each of the domains, a Decision Module (marked as MD in Fig. 4) is deployed. Each MD is responsible for acquiring and processing network events coming from sensors distributed within the domain.

If the attack or its symptoms are detected in one domain, the relevant information is disseminated to other cooperating domains so that the appropriate countermeasures can be applied. All Decision Modules within the federation can also interact with each other and exchange the security-related information. The information about the network incidents, such as attacks in one domain, may be sent to different Decisions Modules in order to block the attacker before the attack takes place in another domain. Communication between domains and Decision Modules is based on a P2P (Peer-to-Peer) protocol in order to increase communication resiliency and enable data replication. Moreover, for decision-making, we proposed a semantic approach to network event correlation for large-scale federated intrusion detection system. In our experiments, we showed that the proposed system can, for example, correlate various network events from different layers and domains (traffic observations and application logs analysis) in order to detect injection attacks on the public administration Web service. As a result of attack detection, the Decision Module creates reaction rules and sends it to MD in another domain. Therefore the same injection attack targeted at the other network can be prevented. It is worthwhile to notice, that Decision Modules are central units in their own domains, but are treated as advanced cooperating sensors by other domains. Of course, the decision and reaction can be different in each domain (even for the same attack or event) depending on internal policies and legal requirements.

## 4 CONCLUSIONS

In this chapter, we summarized our results and presented implementations related to cyber security solutions that exploit techniques inspired by nature. We showed how those techniques can be practically applied for cyberattacks detection, anomaly detection, and protection of computer networks. Particularly, we have investigated and presented the practical solutions for evolutionary-based optimization techniques, collective intelligence, and techniques that mimic social behavior of species. The proposed genetic algorithms improve the detection of SQL injection attacks and anomalies within HTTP requests. Similarly, the proposed ensemble of classifiers and correlation techniques allow for improved networks protection. Furthermore, the collective intelligence concept has been successfully implemented in the Federated Networks Protection System. We believe that the bio-inspired techniques will further find many applications in cyber security domain since, as proven, the readiness of such technology has increased and practical implementations are possible.

## REFERENCES

Bankovic, Z., Stepanovic, D., Bojanic, S., Nieto-Taladriz, O., 2007. Improving network security using genetic algorithm approach. Comput. Electr. Eng. 33 (5–6), 438–451.

Bin Ahmad, M., Akram, A., Asif, M., Ur-Rehman, S., 2014. Using genetic algorithm to minimize false alarms in insider threats detection of information misuse in windows environment. Math. Prob. Eng. 2014.

Chhikara, P., Patel, A.K., 2013. Enhancing network security using ant colony optimization. Global J. Comput. Sci. Technol. Netw. Web Security 13 (4).
Choraś, M., Kozik, R., Puchalski, D., Houbowicz, W., 2012. Correlation approach for SQL injection attacks detection. Adv. Intell. Soft Comput. 189, 177–186.
Choraś, M., Kozik, R., Renk, R., Houbowicz, W., 2012. Information exchange mechanism between federated domains: P2P approach. Adv. Intell. Soft Comput. 189, 187–196.
Kewley, D., Fink, R., Lowry, J., Dean, M., 2001. Dynamic approaches to thwart adversary intelligence gathering. In: Proceedings of the DARPA Information Survivability Conference & Exposition II (DISCEX '01), vol. 1, pp. 176–185.
Kozik, R., Choraś, M., 2015. Adapting an ensemble of one-class classifiers for web-layer anomaly detection systems. In: Proceedings of 3GPCIC, Cracow, pp. 724–729.
Lucas, B., Fulp, E.W., John, D.J., Canas, D., 2014. An initial framework for evolving computer configurations as a moving target defense. In: Proceedings of the 9th Annual Cyber and Information Security Research Conference (CISRC).
Mazurczyk, W., Rzeszutko, E., 2015. Security—a perpetual war: lessons from nature. IEEE IT Prof. 17 (1), 16–22.
McKinnon, A.D., Thompson, S.R., Doroshchuk, R.A., Fink, G.A., Fulp, E.W., 2013. Bio-inspired cyber security for smart grid deployments. In: Innovative Smart Grid Technologies (ISGT), 2013 IEEE PES, pp. 1–6.
Needleman, S.B., Wunsch, C.D., 1970. A general method applicable to the search for similarities in the amino acid sequence of two proteins. J. Mol. Biol. 48, 443–453.
OWASP, 2013. OWASP Top 10 (2013). https://www.owasp.org/index.php/Top_10_2013-Top_10.
Rzeszutko, E., Mazurczyk, W., 2015. Insights from nature for cybersecurity. Biosecur. Bioterror. 13 (2), 82–87.
Sharp, A.T., 2001. A novel telecommunications-based approach to mathematical modeling of HIV infection. Dissertations and Student Research in Computer Electronics and Engineering, University of Nebraska.
Symantec, 2014. Internet Security Threat Report (2014). http://www.symantec.com/security_response/publications/threatreport.jsp.
Torrano-Gimnez, C., Prez-Villegas, A., Alvarez, G., 2010. The HTTP dataset CSIC 2010. http://www.isi.csic.es/dataset/.
WhiteHat, 2014. WhiteHat Website Security Statistics Report. https://www.whitehatsec.com/resource/stats.html.
Wozniak, M., 2014. Hybrid classifiers. In: Studies in Computational Intelligence. Springer, Berlin, Heidelberg.

CHAPTER

# COGNITIVE DISTRIBUTED APPLICATION AREA NETWORKS

9

**Gianni D'Angelo*, Salvatore Rampone***
*University of Sannio, Benevento, Italy**

## 1 INTRODUCTION

We live in an era in which machine intelligence has become more "*pervasive*." The machine intelligence spreads as a virus, and almost without realizing it, now it appears in every process of human life, embedded, in various forms, in many tools that we use every day. The finance, transports, health, justice, and even our free time are only some examples of the using of this machine intelligence. We are witnessing a digital transformation of our society. The size and impact, at various levels, of this "new digital world," made of increasingly interconnected technology, are huge, and the most important feature that distinguishes it is the immense amount of data created every moment. It is estimated 2.5 exabytes every day. The 90% of the data produced in the world today has been created only in the last two years. There are an incredible amount of information, which are derived from the digitization process, and that is growing exponentially year by year. Some studies estimate that the total volume will reach 35,000 exabytes in 2020. In recent years, this phenomenon is known as Big Data (Esposito et al., 2015).

Unfortunately, in most cases, data present themselves ever more in unstructured form, and in countless formats, including text, audio, images, and video. Furthermore, they are increasingly becoming a product separated from the machines themselves in an autonomous way, and stored in multiple places, from traditional databases to social networks. Also, due to constantly production of such a data, they can appear in both static and dynamic form. As an example, data provided by sensors, RFID tracking systems, video surveillance cameras, and smart metering systems can change rapidly over time.

Therefore the machines must become more and more intelligent in order to manage such a huge amount of data, and mainly to extract "*value*" (D'Angelo and Rampone, 2014) from them. The researchers' effort of implementing the advanced analytic tools was performed just to respond to the need to address the complexities that the phenomenon of digitization intrinsically brings with it. Nevertheless, this has required the introduction of new methods and approaches to effectively extract value from data, and imitate the decision-making of the human being (D'Angelo and Rampone, 2016). The resulting scenario is known as: "*Cognitive Computing Systems*" (Banavar, 2015).

Traditional IT systems available nowadays have greatly facilitated the development of business and society, but they are no longer useful to respond the new digitization phenomenon. On the other hand, cognitive systems are changing forever the way people interact with information systems and

**Security and Resilience in Intelligent Data-Centric Systems and Communication Networks. https://doi.org/10.1016/B978-0-12-811373-8.00009-4**

Copyright © 2018 Elsevier Inc. All rights reserved.

technological systems in general. Indeed, they are never programmed to perform tasks, but they are being trained through examples coming from real world situations. This approach allows man to extend his skills on any domain of knowledge and to be able to take complex decisions requiring the evaluation of an extraordinary amount of data very quickly (D'Angelo et al., 2015).

In recent years, the widespread use of machine intelligence and the huge amount of produced data represents a useful combination to help humans to find solutions to everyday problems.

The cognitive systems represent a new category of technology that uses automatic processing of natural language and machine learning to enable people and machines to interact in a more natural way, by extending and enhancing the skills and abilities of human cognition. The mental actions of human cognition for acquiring knowledge and for understanding the environment through senses, experiences, attention, memory, judgment, and reasoning are shared with, and influenced from, the technological artifacts of the environment itself. This leads to an extension of individual cognitive resources in order to accomplish something that an individual could not achieve alone.

Fjeld et al. (2002) (referring to Nardi, 1996) argued "distributed cognition puts people and things at the same level; they are both 'agents' in a system."

In such a view, the achievements of an individual are based on a procedure in which the human cognitive processes, the artifacts, and the constraints of the world reciprocally affect each other.

Such a cognitive process can be distributed between humans and machines (known as physically distributed cognition; Norman, 1991, 1993; Perkins, 1993) or between cognitive agents (socially distributed cognition; Rogers and Ellis, 1994). Salomon in (Salomon (1997)) has stressed that the distributed cognition creates systems composed by an individual agent, his or her peers, teachers, and all the socio-culturally cognitive tools.

Distributed cognition means that the understanding of an individual, and its actions in the world, are not merely a product of the decisions or desires of that one person, but are influenced by nonhuman agents (Fig. 1).

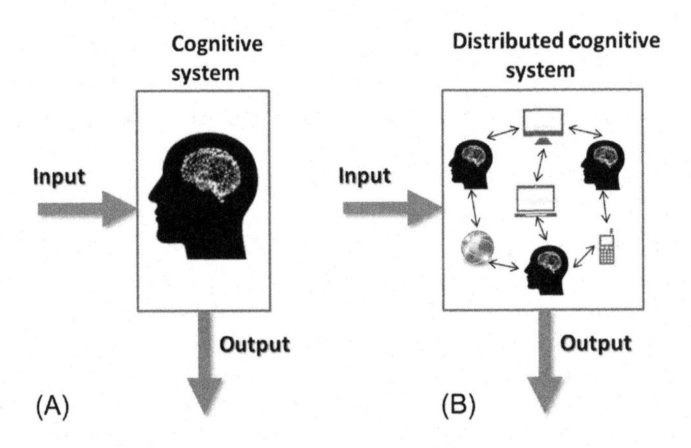


**FIG. 1**

The traditional theory on human cognitive is based on an internal model that ignores the role of external world (A). In distributed cognitive system the human's actions are influenced by nonhuman agents that act materially and temporally (B).

Professor David Stark at Columbia University during a sociology lesson showed to the students the notion of a socio-technical system (Baxter and Sommerville, 2011). He used a simple demonstration by using a certain degree of fantasy, typical of children. Professor Stark walked toward the door of the classroom, opened it, and walked away from the door, which closed itself.

Then, he asked the students what had happened. After some confused answers, one student, first reminded that the professor had opened the door, and then emphasized that the door had closed itself, without external human intervention.

The main point of the proof was that the actions performed were not just the effects of human manipulation of the environment, but that the mechanism of the door had caused the closing of the door.

The action of crossing a closed door by only opening it, and walking through it, is one of the most basic examples of distributed cognition. The person who opened the door does not worry about closing it, because the designer had already thought that the door should have been closed automatically. Thus, he included an appropriate mechanism in the door, a simple spring.

This simple example is also a demonstration of how the human behavior is influenced by nonhuman agent (the closing mechanism) whose behavior is designed in a preview time. The automatism of the door from closing, after opening, was designed previously but acts in the present.

When we start to recognize the presence of nonhuman agents in our lives, and we notice their influence on the our daily choices, we can find the socio-technical systems everywhere. In any place where there is even the simpler technology, how it could be at a store, restaurant, or home, it can be viewed as an component of a great socio-technical network capable to distribute cognition.

In the following, we outline the theory of distributed cognition, and will investigate on some most common application fields. Finally, we will examine one case study that attempts to apply the theory in a modern real-world context, that is social network security.

## 2 THEORY AND BACKGROUND

The spread of distributed cognition was motivated by the observation of the definition of cognition from a new perspective (Boden, 2006). In the traditional cognitive science, cognition is a human internal model which does not consider the role of the external world. The insight, which led to distributed cognition, was to consider the cognition as a socially distributed phenomenon. Starting from here, subsequent theories, by criticizing the traditional ones, have emphasized the role of the social and technological contexts in the cognition theory. Ultimately, the right definition of cognition must emphasize the agent-environment interaction.

The role of socio-cultural and technological contexts in the cognitive theory is emphasized by an approach called activity-theory (Engestrom, 2000).

Indeed, theorists of such a field have emphasized how the traditional studies of the human intellect reveals shortages of theories considering the human-machine interaction (HCI) (Lew et al., 2007). Distributed cognition theory offers, to the traditional definition, a similar critique, but at the same time does not reject its wisdom. Consequently, from the new perspective, in modern complex cooperative work environments, humans cognition and technological artifacts cooperate together by manipulating representational states in a different way, but they are capable of solving complex tasks. Accordingly, the cognition of the outcoming distributed system differs from the cognition of the single individuals that act in it.

The theory of distributed cognition was originally developed by Hutchins (Hutchins, 1995a) in order to provide a new and more balanced approach to problem-solving in real-work environments. More, his intent was to develop a new general framework for cognitive science.

Hutchins et al. originally developed the theory of distributed cognition by studying how a Navy crew works on a ship.

In modern society, this task is carried out by teams of individuals working with various types of artifacts.

Hutchins concentrates on studying the typical surfing phase in which the crew is employed in rectifying the ship's position. He describes how these individuals use tools to generate and maintain representational states, and how these are then propagated through the ship. In particular, he investigated on the navigation practice of two very different cultural traditions. He showed how different strategies were capable of solving the same task through different representational assumptions and implementation means. He examined that the cognitive processes required to manage the tools differ from the cognitive processes required to solve a task in which that tools are required. The outcome is that the solution of the task is not only determined by the cognitive capability of any single individual, but the solution is the result of the interactions of all individuals with each other and with a complex suite of tools. Hutchins, farther, documented the social organization of work on the ship and showed how the learning occurred both in individuals and at the organizational level. Therefore the crew, together with the tools, are viewed as a system with an its own cognition, and capable to solve specific tasks.

In another study (Hutchins and Klausen, 1996) Hutchins moves his theory from the bridges of a ship to the airline cockpit that led him to design a graphical interface for the auto-flight functions of the Boing 747-400 (Hutchins, 1996). The entire airline cockpit is considered as a single unit to be analyzed. The focus is on the functionality of the involved components (human and not) and on the interactions among their cognitive processes, rather than on the individual minds. He explains how much of the cognitive work required for flying an aircraft is done by the board instrumentation. These, along with the communication technologies, represent the nonhuman agents, which affect the human choices. Using the past cognitive efforts of other people and tools, these agents are able to take advantage of past experiences and knowledge. In such a way, the various components of the cockpit make the pilot free from many stressful tasks that the pilots could not be able to solve alone. The outcoming functional system from this combination of human and nonhuman actors is the above cited "socio-technical" system. In Wright et al. (2000) the authors show how the workload of the pilots can be different depending on the instrumental layout used. As reported in Hutchins (1995), the airspeed needs to be brought to specific values, depending on weight, for maneuvering the landing phase. The pilots have to make a comparison between a target speed and the current one. In order to do this, the target and current state resources must be brought into coordination, and the way this happens is highly dependent on how the resources are represented in the interaction. Fig. 2 shows three examples of how the target and current state may be represented to support this activity; in each case the process of coordinating the tools to make the comparison is quite different:

1. As reported by Hutchins (1995) the instrument (see Fig. 2A) has a moveable pointer known as bugs which may be preset by the pilot to indicate appropriate target speeds for the current operating conditions. On the other hand, the actual speed is represented by a pointer. The perception of the relative locations of the pointer and bug becomes a guide for pilot to accomplish the comparison between the target speed and current one, and for coordinating the right maneuvers.

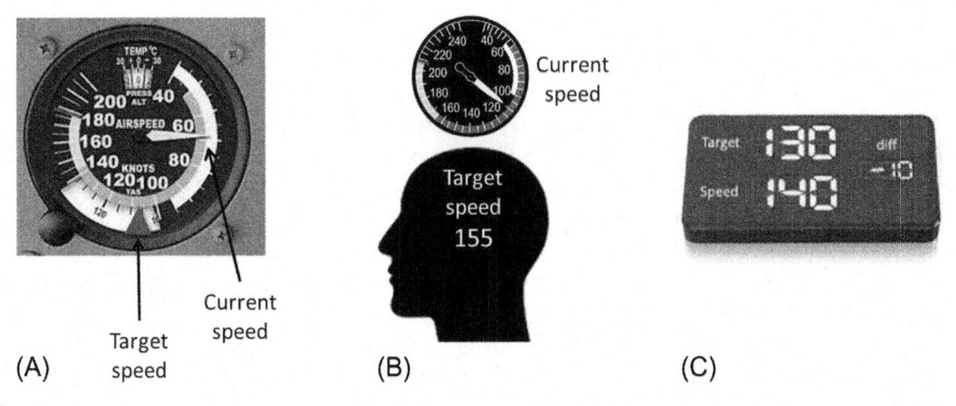


**FIG. 2**

Three examples of how the target and current state may be represented to support the speed control activity.

2. In Fig. 2B, the current speed is represented by the airspeed indicator instrument, and the target one is represented internally (i.e., remembered) by the pilot. In this case the process of coordination involves the ability of the pilot to read the display, interpret it (employing knowledge about decimal numerals), and make the comparison with the remembered target speed to determine whether the remembered speed is higher or lower.
3. In Fig. 2C, both current and target speed values are represented in numerical form in the flight deck display, allowing the pilot to make an explicit co-ordination similar to the previous case, with the difference that both values are read and interpreted, and none of them need to be kept in the pilot's memory. In addition, the flight deck computer system also makes a comparison and displays the result as a number on the display.

More, in Hollan et al. (2000) the authors list three criteria of distributed cognition:

1. *Distributed cognition in social interactions.*
The social organization is viewed as a form of cognitive architecture because the social organization determines the way the information is propagated through the group. The implication is that the models of a social group can be used to describe what is happening in an individual mind. Accordingly with Minsky (1988), such a vision gives a revolutionary answer to the age old question: "how does the mind work?" Minsky argued that each brain can be considered as an entity formed by hundreds of different machines, whose interconnections lead the brain to become as a large society made of specialized agencies. The coordination of their activities affects the goals. The concept of distributed cognition applied to phenomena belonging to the social interactions, together with the interaction between people and media, can be used to answer to typical questions in social interactions, such as: "How do the cognitive processes, associated to an individual mind, become when it is put in a group?" "How do the cognitive of the group is different from the cognitive properties of the group member?" and "How to become the cognitive properties of the single minds when they are put to participate in a group?"
Distributed cognition in social aspects have been widely studied and applied in many social environments, such as the market and jurisprudence.

2. *Distributed cognition is embodied.*
   This point of view sees the cognition of the distributed cognition approach as embodied. That is, the minds are not only passive engines capable of creating internal models which represent the external world, but they interact with the external means (artifacts and objects that are constantly surrounding us) in a complex way involving the internal resources, such as attention, individual cognition, memory, and other functions. In this perspective, the materials of work are no longer only a mere stimuli for the brain, but they become elements of the cognitive system itself. As well as the cane represents the main instrument of a blind person to perceive the external environment, the work materials become integral parts of the way people act in the distributed system.
3. *Distributed cognition is culturally embodied.*
   The study of cognition cannot ignore the culture people live in, because the people live and work in intricate cultural environments.
   Culture comes from the accumulation of solutions of past activities used to solve typical human problems, and the successes help individuals to not start from zero.
   Accordingly, culture provides us materials, mental and social structures that enable us to perform things that we could not do alone.
   Hence, the cognition is no longer separated from culture, rather the culture shapes the cognitive processes of the systems by going beyond the boundaries of individuals, as Hutchins argued (Hutchins, 1995a). Nevertheless, in order to accept the new culturally embedded cognition, it is necessary to change the actual model of the individual mind and of the ethnographical methods used to properly investigate the distributed cognitive systems.
   The ethnography of distributed cognitive systems, while retaining the model of the individual mind, needs to know how the information to be processed are arranged in the material world and social one. Based on such assumptions, Hutchins, by combining interviews with pilots, observations of pilots in flight, and studies of the operations manuals, confirmed that pilots use the airspeed indicator dial as a spatial anchor to recognize and record meaningful airspeeds, and only rarely do they think of the speed as a number. They use the spatial structure of the display in order to perceive the relation between the actual speed and target one. So, the implementation of a new digital display must include the distributed cognition theory in addition to the methods for designing digital equipment.

Hence, using the above perspectives, the distributed cognition theory can be used as an elegant and simple model to be applied for describing cognitive systems, through a way that goes beyond the traditional mind models and that generates interesting philosophical implications.

Distributed cognition theory extends the cognitive level beyond the individual cognitive process of understanding and moves toward a cognitive process at a system level. The difference between an individual who uses the pencil to write on a paper in order to remember something and an individual who relies on his memory, represents the difference between a cognitive system using an external process (pencil, paper), and an internal process (memory).

So, in the distributed cognition approach, the system is analyzed as an overall unit rather than by analyzing the single system elements. For example, in Mansour (2009), the author shows the application of the distributed cognition theory to a system made by a group of individuals by referring to it as *group intelligence*. He extends the assertion made by Argyris and Schon (1978) about the "*learning*" word, that is, this word is a term applicable to individuals within of a group, but when individuals interact one another in order to share tasks, then it is possible to identify the group itself as *learning*. The consequence, as Smith (2008) argued, is the greatest capability of the group to remember information due to the development of a particular memory created by the distributed information, which a single individual cannot possess. So, the overall group performance is improved due to the distribution of the information itself among group members.

The distribution of the knowledge between group members reveals the presence of an amplified cognition at the group level, which is not achievable by the individual mind.

The group cognition strengthens other skills and abilities in addition to the group memory, such as: group decision-making, and the group problem-solving, which form the basis for a group intelligence.

Nevertheless, group intelligence, in order to spread the cognition among the group members, needs effective communication media, by extending, then, the cognition also to the tools and efficient representational systems. Indeed, with recent advances in ubiquitous mobile computing (Ficco et al., 2007), this form of extended cognition has many consequences for HCI, and for the emerging augmented reality technology (van Krevelen and Poelman, 2010). More, the man-computer separation becomes increasingly thin.

## 3 THE SOCIAL MEDIA TECHNOLOGIES AS GLOBAL BRAIN

The World Wide Web (known also as the Web) is perhaps the mean which popularized Internet in the world, and thanks to the diffusion of numerous applications based on, it represents an important part of the cognitive revolution (Wang, 2014). Indeed, these technologies allow the connection of millions of people which can exchange any type of information among them, and thanks to the artificial intelligent systems, the Web is moving toward a massive Web made of interactions among highly intelligent entities, humans and nonhumans. So, the Web can be viewed as a techno-social system (Celina et al., 2008) able to enhance human cognition. Such a scenario can be viewed as a big and global brain made by both human individuals and nonhumans tools.

### 3.1 WEB EVOLUTION

Since the idea was introduced by Tim Berners-Lee in 1989, the Web has represented the greatest system of interaction among humans (and nonhumans) based on networks. Many advances have been made about the Web in the past decades (Aghaei et al., 2012).

*Web 1.0* is the first step for the use of Internet. In this first stage, an interconnection among Internet users is implemented through websites, portals, and platforms of Web services where users can only browse. With Web 1.0 the possibility of interaction between the company and its customers is quite limited. The unique points of contact are made by traditional means: mail, fax, telephone, and advertising. This determines a unidirectional communication flow.

*Web 2.0* is the combination of all the online applications that allow a high level of interaction between the website and the user. Such an interaction offers to everybody the opportunity to benefit, in real time, the most self-interest content or share them with other network users. In this way communication becomes participatory and, then, bi-directional, because anyone can give a contribution in the dissemination of the contents on the Internet, which thus become accessible to everyone. Tim O'Reilly and Dale Dougherty in 2004 at the O'Reilly Media Web 2.0 Conference defined the Web 2.0

as a read-write Web capable of revolutionizing the business in the computer industry due to moving the Internet as a platform (Fermentas, 2004). Web 2.0 includes main technologies and services, such as: blogs, tags, wikis, mash-up, Really Simple Syndication (RSS), folksonomy, and tag clouds.

*Web 3.0* turns the Web into an environment where the published documents (HTML pages, files, and images) become interpretable, that is they are associated with the information that specify the semantic context. In addition, they are provided in a format suitable for questions, interpretations, and generally for automatic elaborations and learning. In such a way, the Web contributes to the building and the sharing of knowledge, putting in connection the content on the Web through automated searches, and analyzes based on the meaning. Accordingly, Web 3.0 is also known as *semantic Web* (Harmelen, 2004). A typical application of the semantic Web is the ability to query a search engine by formulating questions in natural language instead of with keywords, such as Bing or Google (Hakkani-Tür et al., 2012). With the interpretation of contents of the documents, the result of a research task on the Web is much more than the mere occurrences of the content searched. In a semantic context, the research itself is able to return documents that really correspond to the thematic areas to which the content refers. In addition, by using technologies based on artificial intelligence, the semantic Web is able to make the Web readable by machines and not only by humans. In conclusion, thanks to new technologies, such as semantic Web, machine reasoning, distributed databases, and natural language processing, the Web 3.0 is an intelligent environment able to interact with the humans.

*Web 4.0* is a natural evolution of Web 3.0, and is an idea in progress. Nevertheless, it can be viewed as an alternative version of what we already have. Web 4.0 connects all devices in the real and virtual world in real-time in order to develop a ubiquitous Web, also known as *symbiotic Web* (Bernal, 2010). The Semantic Web also allows a connection to contents automatically, the applications on the Web would aim to connect people automatically based on the activities they are doing. So, people will have a valid tool to collaborate and reach common targets by putting together their resources and their skills. Web 4.0 represents the new era of the Web in which the machines, through their intelligence and automatic learning ability, are able to react to the exigencies of humans by interacting with them in symbiosis.

*Web 5.0* will be a Web oriented to emotional interaction between humans and computers (Calvo and D'Mello, 2010). Actually, the Web is emotionally neutral, which means the Web does not perceive the users feelings and emotions. In this new "*emotional Web*" (Karim et al., 2012; da Rocha Gracioso et al., 2013), the interaction between man and machine will become a daily habit for humans.

## 3.2 THE WEB AS A GLOBAL BRAIN

In 1960 James Lovelock, while working as a consultant for a team at the Institute of Technology in California that was studying methods for detecting life on Mars, hypothesized the Earth as an self-regulated living organism, in which the life is maintained thanks to complex interactions among the entire range of living organisms and geophysics components. Such a hypothesis was inspired by the observation that the life, as such, perturbs the equilibrium state of the elements that form a planet, which would remain in a fixed state in absence of life. In honor of the ancient Greek "*Earth Mother*," Lovelock called such a vision, *Gaia* (Lovelock, 2000). *Gaia* definition includes the entire biosphere, that is everything living on the planet plus the atmosphere, the oceans, and the soil. It is based on the assumption that the oceans, seas, the atmosphere, the Earth's crust, and all other geophysical components of planet Earth are maintained in a condition suitable for the presence of life, and *Gaia*'s changes are just due to the pandering of the behaviors and actions of living organisms, vegetables, and animals. So, the *Gaia* hypothesis implies that the biosphere is a single living organism.

For example, some chemical and physical parameters, such as the temperature, the oxidation state, acidity, salinity, which are fundamental for the presence of life on Earth, have constant values. This homeostasis is the effect of the active feedback carried out autonomously and from unconscious processes by the overall vegetable and animal life, the so-called *biota* (Akagi, 2006). Moreover, all these variables do not maintain a constant balance in time but evolve in sync with the *biota* itself.

Afterwards, Peter Russell, inspired by *Gaia* model, in his book titled "The Global Brain" (Russell, 2006) shows a parallelism between the brain's nervous system and the Web. He argued that, approximately, the number of nerve cells in a human brain is similar in number to the planet population. Also, there are some similarities between the growth process of the human brain and the way in which humanity is evolving.

In fact, the human brain development is based on two main phases. Firstly, the number of neuron cells increase rapidly to form the whole cerebral system. Then, the billions of isolated nerve cells connect to each other to implement functions, such as memory, attention, judgment, reasoning; namely the *cognition*.

In the same way, Earth, like a living organism according to Lovelock, has been first populated by the human society, and then, the billions of human minds have been connected by the Web into a single integrated network like a planetary nervous system. This is creating a collective consciousness (Loghry, 2013). We perceive ourselves no longer as isolated individuals, but as part of a rapidly integrating global network, namely the nerve cells of a *global brain* (Fig. 3). From Web 2.0 onwards the global brain is functioning, and its impact on the our life could be so great to go beyond our imagination. The global brain model turns the Earth in a self-conscious organism which possesses a higher cognitive performance compared to a single mind. The Web is able to store huge amount of data, because the data are distributed among billions of electronic devices in the planet. Also, their connection provides to global brain the associative memory capability, and thanks to the artificial intelligence, the interacting capability with humans. As hypothesized by Lovelock, Earth becomes a new socio-technical living organism, which self-regulates in order to safeguard life. Nevertheless, Russell warns the humans that the Earth, just like an organism, wishes to survive, so if we continue to destroy it, humanity will become a planetary cancer. Sahtouris (1999) argued: "*We are natural creatures which have evolved inside a great life system. Whatever we do that is not good for life, the rest of the system will try to undo or balance in any way it can.*"

The Web as a Global Brain is a vision that is manifested intrinsically in many applications based on Web and social media technologies. In fact, the significant impact of the above-cited Web technologies lies in connecting large numbers of human and nonhuman agents in a way that enables them to link and share their individual knowledge, which represents the individual cognition, in order to produce a group knowledge, a collective intelligence, an augmented social cognition, and, finally a group cognition (Stahl, 2006). So, the Web supports a large number of individuals in order to help them to solve problems of common interests. The Web technology within social interactions reflects a notion that is implied in the theory of distributed cognition, that is the propagation and distribution of shared knowledge.

In the following, some examples of Web-based distributed cognition are reported.

**FIG. 3**

The Earth as a global brain.

### *3.2.1 Wikipedia*

Wikipedia (Karkulahti and Kangasharju, 2012) is a Wiki-based (Raman, 2010) service offered by Wikimedia Foundation, Inc. Its goal is to encourage the growth and distribution of multilingual content, and to provide almost all contents to the world free of charge. Based on the belief that no one knows everything, but everyone knows something, the users are allowed to have access, can give their own contribution to the creation of a webpage about a specific subject. This contribution is provided in a collaborative way by creating, editing, and linking content on the webpage.

Wikipedia is based on a fundamental principle, that is: trust in users. Accordingly, any changes made by users are published immediately. Nevertheless, the authors/users could not be specialists, then they are advised about possible subsequent verification, which may modify or delete the entered contents. Indeed, each entry is subjected to a regular inspection by the site administrators who are required to make decisions, following an established editorial policy. Often the Administrative Committee discusses for a long time on the approval of certain content. Each encyclopedia entry, however, is intended to be never ending: anyone, at any time can update or modify it with insights or other content.

This process of creating knowledge in Wikipedia embodies many aspects of the distributed cognition. Indeed, the shared knowledge is a typical characteristic of the group intelligence, which is created and maintained through the interactions among components.

When an editing activity is assigned, users often proceed by interacting with one another, and the ideas emerge from discussions made by multiple perspectives. Consequently, the outcome cannot be assigned just to one person. So, the result is a group culture belonging to a *group level*. At such a level, Wikipedia can be considered as a component of the global brain devoted to the survival of the knowledge.

### 3.2.2 Bots as nonhuman agents in distributed cognition

Bots (Ferrara et al., 2016) (short for software ro*bots*) are software algorithms that are designed to mimic the way humans talk, offer news, assistance, and services. The definition is generic because in practice bots can be and do anything, ranging from responding to messages automatically to the creating a system able to help hackers in compromising sites or to sneak into remote computers.

The bots history begin when Turing (1950) theorized a test to detect if a computer was actually able to mimic human behavior: analyzing a conversation between an individual and a computer, an external person has to determine who is who. If the number of exchanges is such as to make it impossible for an answer, it is determined that the machine has passed the Turing test.

The first bot that tried to overcome the Turing test was *ELIZA* (Weizenbaum, 1966) in 1966, a software pretending to be a psychotherapist and that regulated his answers on the basis of the things written by his interlocutor. In practice, it began the conversation by asking what was the problem of the person, then analyzed the response by looking for a set of keywords and, based on this, it gave an answer. ELIZA was then able to create lifelike conversations in a rather restricted area, which was enough to pass the test.

Advances in artificial intelligence and especially the great spread of the social media applications, such as Messenger and WhatsApp, represent the main incentive for the implementation of software algorithms that exhibit human-like behavior. Thus bots are becoming more common today.

The most widely spread bot typology is the *chatbot* (Liu et al., 2015), which comes from the social chat. The main purpose of this software is to induce a user to think he is talking to a human being. They are used in dialog systems for various practical purposes, such as online help, personalized service, to get airfare, hotel pricing, and weather conditions.

Many chatbots simply scan the keywords typed in the input window and yield a response associated to the most relevant keywords. Nevertheless, some chatbots are equipped with artificial intelligence and sophisticated natural language processing systems (Tur and Mori, 2011) to respond the user like humans.

The easy implementation of the chatbots represents another motivation that contributes to their spread. Indeed, the typical chatbot's infrastructure looks like Fig. 4.

Just as in a common chat, the users through their devices add the bot to their contacts, and they use the chat interface to communicate with the bot. On the back-end, the chat platform offers some API libraries which implement a communication protocol to exchange messages with other platforms on which resides the bots intelligence. So, the desired service is implemented on a dedicated external server which is remotely connected to the chatbots through a WebHook-URL.

It follows that the bots are cheaper to develop and easier to install because they are embedded into social chat or websites which already can be used on every type of device: laptops, smartphones, and tablets.

Many big companies get benefits from the potentiality of the bots in order to extend their services. For example, Messenger in Facebook platform provides the opportunity to some companies to sell products and services within the application. Microsoft offers *Bot Framework*, a platform to build and deploy high quality bots. Motion AI, *Inc.* (Motion, 2016) offers a Web-based platform for building the

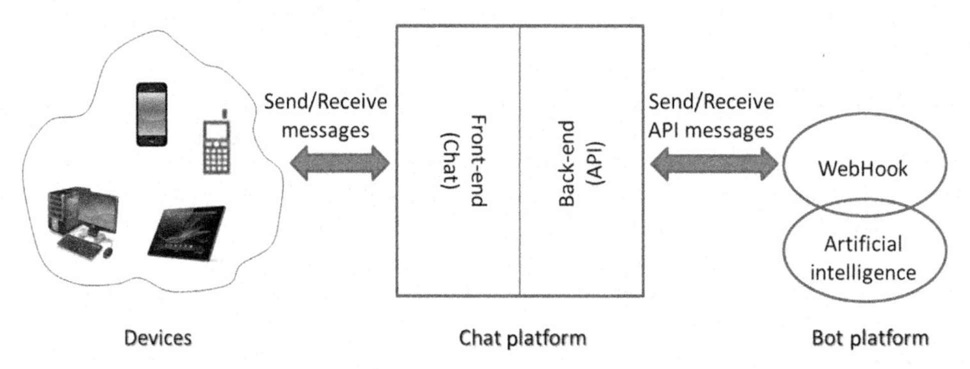


**FIG. 4**

Common ChatBot platform. The users communicate with the bot by adding it to their contacts. An API is powered to exchange messages between human interface and the bot platform. WebHook is the URL of the remote platform devoted to implement the offered service through the use of the artificial intelligence.

bots and their intelligence through graphical tools. There are also companies offering a store for bot, such as *Telegram Bot Store* (Telegram, 2016), and *BotList* (BotList, 2016).

One of the first to experience the potentiality of these bots has been *Slack*, that, as Skype program is used to coordinate work activities among people groups. Recently, many bots have been added to Slack in order to automate the workflows among people and to obtain information about some subject.

*Slackbot*, for example, appears to newcomers in a friendly way like a personal assistant which asks for information, such as name and photograph, to set more easily their personal profile, without they having to fill the classic form. The bot can also be programmed, very easily, to answer questions of various kinds, like having to remember the password for the sites used by workgroup.

There are hundreds of other bots, developed by outside companies and compatible with Slack, to do everything, such as receiving statistics on the performance of a own site, start a quick questionnaire, decide what to eat for lunch with colleagues, synchronize the calendars of the participants of a meeting, and remind future commitments.

The presence of these numerous services and the ability to create new ones have contributed to the success of Slack, which has a growing number of subscribers and collects tens of millions of dollars in the startup funding.

So, the chatbots interact with the humans to help them in solving daily common issues. Such an interaction is on a wide scale, that is it could include services and bots spread in the globe. This real-time communication leads toward an extended cognition of the human being, namely the distributed cognition.

Another example that shows the potentiality of the bots is their use in Wikipedia.

According to Wikipedia researchers, the bots are also the key to the success of the crowd-sourced encyclopedias. Indeed, in Wikipedia bots are widely used to edit articles.

Some bots accomplish simple tasks like correcting spelling and grammar errors. Others are more specialized and are able to perform tasks similar to humans, like building pages from census data or using NASA reports to create articles about asteroids.

Nevertheless, the biggest use of bots is in vandalism detecting (Halfaker and Riedl, 2012; Khoi-Nguyen Tran, 2015). Today, according to Wikimedia Foundation researcher Aaron Halfaker, the most prolific antivandal bot is: *ClueBot* (Smets et al., 2008).

It can detect and fix everything from profanity to mashed keys within seconds, and it is responsible for almost half of all edits on the English Wikipedia site.

Compared to humans, bots are highly productive, accomplishing tasks hundreds of times faster than humans. So, the bots help to coordinate human internal cognitive resources with external tools and resources. Such interaction allows both bots and humans closely connected with their environment to form a *global brain*.

# 4 A DISTRIBUTED PERSPECTIVE OF A TRUST-BASED ARCHITECTURE

In order to show the effectiveness of the distributed cognition theory, in this section we briefly discuss a distributed cognition-based architecture for pervasive computing (D'Angelo et al., 2015). Please, refer to D'Angelo et al. (2016) for details. Such an architecture makes use of a trust-model (Kagal et al., 2001), in which the network interact access rights are not static and based just on authentication and access control, but they change dynamically on the basis of trustworthiness among users. In this framework, a user must automatically check the trustworthiness of the other users before interacting with them.

The estimation of trustworthiness is made through the help of artificial intelligence-based techniques which offer a close resemblance with the human decision-making process. On this basis, the user is able to extend his cognition beyond his skills according to the distributed cognition theory. In addition, users dynamically make decisions about the trustworthiness of one another by adapting the decision to different contexts and typology of interactions.

The observation of all the users while interacting on the network leads the overall system to dynamically learn the behavioral patterns, making it able to evaluate their trustworthiness in the network.

## 4.1 PERVASIVE COMPUTING AND TRUST

The wide diffusion of small and powerful electronic devices able to provide high computing capabilities and multiple wireless communications interfaces, such as smartphones, has considerably encouraged the spread of the above-cited Web-based software. These devices offer the possibility to use any type of advanced services anywhere, anytime, and for anyone.

Such a scenario, known as *Pervasive Computing* (Weiser, 1999), can be viewed as one of the major revolutions in the world of computers, because it offers the utilization of computational services in any environment where people live and work. Despite the numerous advantages introduced by pervasive computing in daily life, such as offering useful services to people whenever and wherever they need, it has many risks connected to security and privacy too.

Indeed, communication technologies are fully integrated inside the devices, and they are driven by third-party installed software which are able to establish communications with one another, without the users conscious or explicit knowledge.

So, it is a hard task for users to know when these devices exchange personal information, such as identity, preferences, and current position.

Despite the traditional computing security, which is based on user-authentication and on access control techniques, in the pervasive computing environment the access to network must be guaranteed by autonomous access control systems (Ficco et al., 2007). To reach such a goal, these systems must be equipped with a specific intelligence.

As suggested by Kagal et al. (2001) a possible way to increase security is to use *trust* among users. In such an approach the access rights may change dynamically, depending on trust deserved by any user while interacting on the network. The trust is evaluated by a score which is calculated through specific rules and policies.

Nevertheless, trust evaluation is a hard task, which includes reasoning, perception, and many other typical abilities of human cognition.

## 4.2 THE TRUST MODEL

This section describes the proposed trust-based model. In such a model, each user, while interacting on the network, observes the behavior of each other in order to estimate their trustworthiness score.

Any user can make use of two Data Mining-based processes for evaluating the score.

First, the behavioral pattern of the user is learned through the use of association rule-based techniques (Agrawal et al., 1993), applied on the historical data. For such a goal a feature vector is employed. Such a vector represents the signature that characterizes the behavior adopted by a user in any network interaction.

The set of these interactions form a dataset of vectors, which is used as input to a Naïve Bayes classifier (Al-Aidaroos et al., 2010) for final decision. The outcome is the decision about the trustworthiness of a user explicated as a value of probability.

In order to represent the experience of the user $i$ has had with the user $j$, the feature vector is defined as follows:

$$e_{ij} = \langle EID_j, TS, ET, StDevET, LT, TC, DK, SE, HL \rangle \quad (1)$$

where,

- *(EID)—User Identification*: Each user is uniquely identified through an identity code. In most cases the EID may be resolved to a network address of the peer.
- *(TS)—Trust Score*: Any interaction between two user ends with the assigning of a score that can assume a value among: {trustworthy, dubiously, untrustworthy}.
- *(ET)—Elapsed Time*: Given a specific context of network interaction, the average value of the elapsed time between two consecutive transactions is stored. ET is updated before the interaction begins.
- *(StDevET)—Standard Deviation of ET*: The dispersion of the ET values is taken into account in order to detect malicious users that act on the network through time-based attack. That is, malicious user can have fair or unfair behavior, depending on the time.
- *(LT)—Last Time*: Given a specific context, the date of the interaction is taken in account. Old experiences act on the decision-making in similar way than new ones.

- *(TC)—Transactions Context*: It identifies the typology of transaction: e-commerce, game, social network and so on.
- *(DK)—Direct Knowledge*: This variable assumes a *true* value if the users interacting on the network are in a direct connection without intermediates, while *false* otherwise.
- *(SE)—Source Entity*: The interaction among users looks like a tree structure, in which there is a father and many sons. For example, if the user $E_1$ contacts the user $E_2$, then SE of $E_2$ assumes the EID value of the $E_1$ user.
- *(HL)—Hierarchical Level*: It represents the hierarchical level assumed by a user in the interaction tree. For example, the son of the son of the father has level 2. The level 0 is the father.

The goal of the considered feature vector is to evaluate the trustworthiness by arranging the historical interaction data that a user has had with another one. Thanks to SE, DK, and HL entries, the trust decision-making can be based on recommendations obtained from trusted third entities. The context (TC) is involved in the decision-making. In addition, the vector explicit the intransitivity of the trust. If user A trusts user B and B trusts C, this does not means that A trusts C. Also, no symmetry is included. If user A trusts user B, it does not mean that user B trusts user A. Finally, the user is uniquely identified. To summarize, the feature vector is able to take into account the reputations, recommendations, past experiences, and contexts. Decision-making is based on the use of such vectors as input to the artificial intelligence-based algorithms. As shown in Fig. 5, initially different datasets are built up. They are made of tuples belonging to the checked user (EID), one for any specific Trust Score (TS). In this case we have three sets, one for any different TS class. In order to extract associations significant among vectors parameters, an Apriori algorithm (Agrawal and Srikant, 1994) is employed. The outcoming associations represent the behavioral signature of the considered user for a specific TS class. Naturally, such a signature depends on the time the trust-evaluation is performed, and, consequently, it may change. The Apriori algorithm is also applied to the users incoming feature vector who is asking for an interaction (applicant user). Accordingly, the parameters of the feature vector are first updated, and then the signature is extracted. Soon after the extracting phase of the signatures, the trust decision-making needs to be performed. This is made by using the Naïve Bayes algorithm which discovers which TS class the incoming signature belongs to. In particular, for any TS class, the similarity between the incoming applicant user signature and the considered TS class signature is evaluated in term of probability. The final decision is expressed by considering the higher probability value.

In the described trust-based model all the users can use the historical data of third users, the so-called *recommenders*. They may inform other ones about the results of their own past interactions with a given user or about the results received by other users. This third party experience turns the single user's cognition from individual to collective. When a user receives a connection request from another one for the first time, since it has no prior experience with it, it cannot evaluate his trustworthiness. The collective knowledge of the recommenders is employed to solve this scenario by enhancing the single user's knowledge.

## 4.3 RESULTS AND DISCUSSION

As shown in D'Angelo et al. (2016), the experimental results show that the proposed trust model is able to recognize the tactics used by the malicious entities for three typical attacks: counting-based, time-based, and context-based. Moreover, the proposed trust model learns such tactics as soon as they

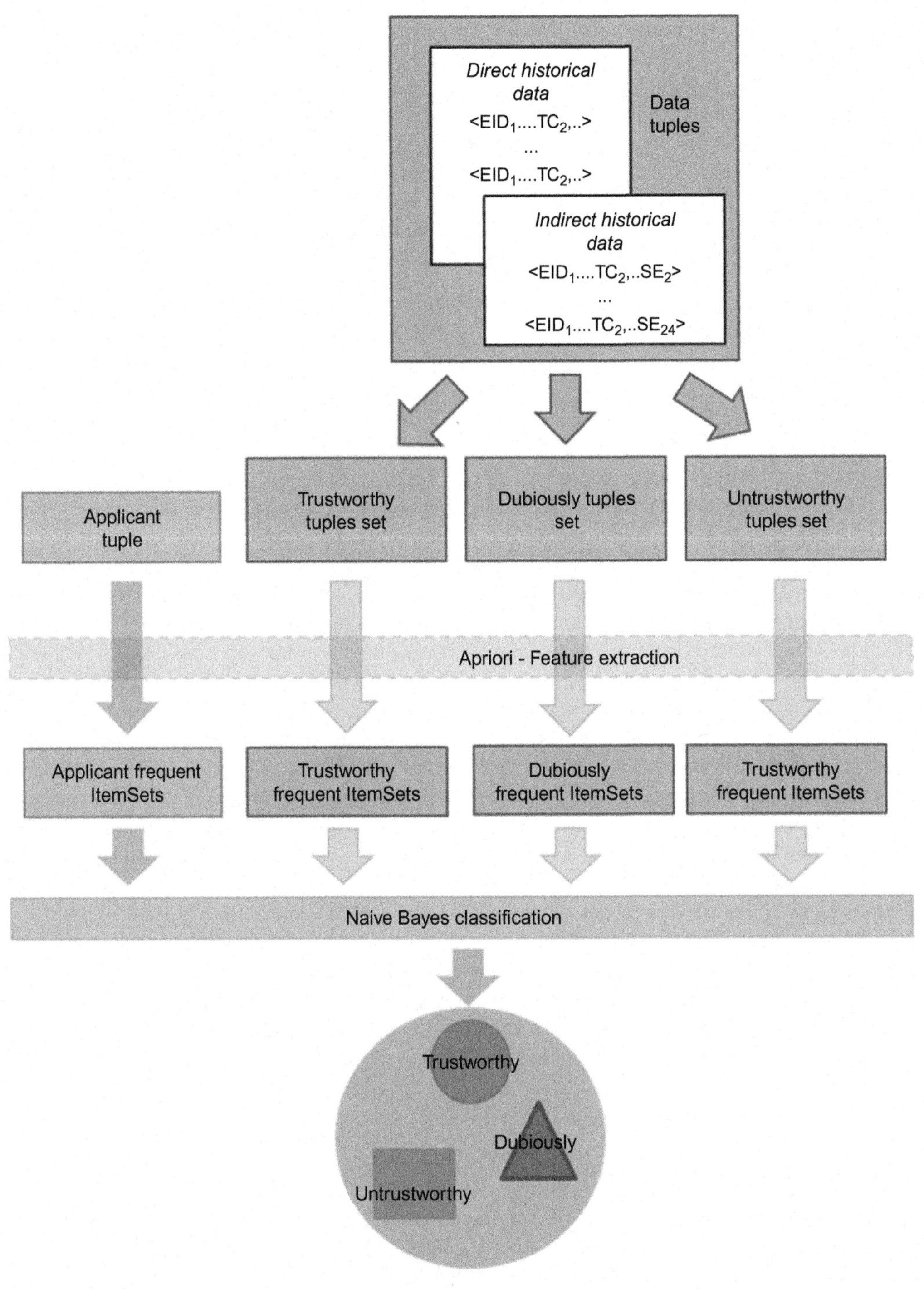


**FIG. 5**

The trust decision-making process.

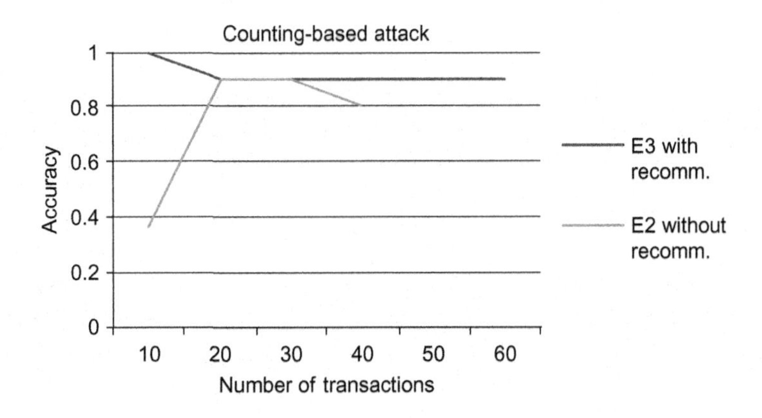


**FIG. 6**

Comparison of performance between two user receiving a counting-based attack. $E_3$ uses recommendations, while $E_2$ does not use recommendations.

appear, which would not be discovered by the conventional approach in which only the global score is used as trustworthiness measure.

The application of the distributed cognition theory through the use of recommenders is able to solve the problem of the trust evaluation at the first interaction. In fact, the collective knowledge allows to exchange the past third-party experience, and then it extends the skill of a single user to make decisions.

In Fig. 6, the comparison of performance between two user receiving a connection request is depicted.

The malicious user, which asks for a connection, gains a good reputation by acting honestly for a number of transactions and dishonestly for other (counting-based attack). The user $E_2$ performs the trust evaluation without recommendations. So, he is not able to make a decision immediately because he does not have enough data, and he is able to make the right decision only after some interactions. For against the user $E_3$ is able to evaluate the trustworthiness of the incoming connection at the first interaction because he extends his dataset with the historical data of the recommenders.

## 5 CONCLUSION

This chapter is intended as a work covering the concepts of the distributed cognition in the context of human-computer interaction with an attempt to investigate how machine intelligence changes the way humans solve their daily problems and tasks, and how it extends the cognition of an individual beyond his skill.

We have started from the traditional definition of the human cognition, which does not consider the external world as a part of the human cognitive process, and we have showed the consequences of the introduction of the machine intelligence in such a definition. We have demonstrated how the artificial intelligence has led to an increasing cooperation between humans and nonhuman agents. Consequently, the solution of a task is not only determined by the cognitive capability of any single individual, but it is the result of the interaction between individuals, and between individuals and a complex suite of tools.

Particularly, the distributed cognition theory has been examined in the Web framework. So as the nervous system of human brain, composed by many cells interconnected with other to implement brain functions, humans and nonhuman agents are interconnected through the Web to form a global brain. This new brain has its own cognition which offers performance many times higher than single individuals. It is like a living organism which self-regulates in order to adapt itself with the aim to achieve purposes.

The new potentialities offered by this new great brain have been showed and highlighted. In particular, the increasing use of the artificial intelligence, together with the Web, has led to a living environment in which the machines constantly help humans in their activity by adapting themselves to different contexts, situations, and even to user's humors. The machines, automatically and without the users' conscious knowledge, interact with the daily human tasks by offering contents, and contacts useful to solve the tasks in which humans are employed.

The outcoming result is the extension of human cognition toward a collective and distributed cognition.

As a consequence, it has been explored how the introduction of new technologies has changed the IT *security* scenario. So, in this chapter, we have described the implementation of an architecture for pervasive computer based on a distributed trust model. The interaction among users on the network been made trustworthy through the use of both the cognition of each single interacting entity and the collective cognition deriving from the whole pervasive environment. In the resulting system, any user, by observing the other ones while interacting on the network, dynamically learns the behavioral patterns, and is able to evaluate their trustworthiness in the network.

Eventually, we can conclude that the potentiality offered by the distributed cognition could have an impact on our life that should go beyond our imagination, opening new scenarios, and, consequently, new perspective in our life. For example, the natural consequence of what said, should be that Earth itself could be considered as a component of a higher level system having its own cognition. In such a view, humans need to be more responsible of how their actions affect the planet. As Elisabet Sahtouris argued: "*We are natural creatures which have evolved inside a great life system. Whatever we do that is not good for life, the rest of the system will try to undo or balance in any way it can.*"

## ACRONYMS

**HCI** human-computer interaction. It is an area of research that studies how people interact with computers.

**IT** information technology. It indicates the use of any computers, storage, networking and other physical devices, infrastructure and processes to create, process, store, secure, and exchange all forms of electronic data.

**NASA** National Aeronautics and Space Administration. It is a government agency responsible for the United States of America's space program and aerospace research.

**RFID** radio-frequency identification. It is a technology for identification and/or automatic storage of information through the use of a special electronic, called tag (or even transponder). Such a tag can reply to the special portable equipment, called reader, when this reader is close, and not necessarily in contact, to the tag.

## GLOSSARY

**Biota** The term is used to describe the set of plant and animal life that characterize a specific region or area.

**Bots** They are software algorithms that are designed to mimic the way the humans talk, offer news, assistance, and services.

**Chatbots** They are bots that simulate a conversation between robots and human, like a live chat.

**Cognition** It represents mental actions, such as memory, senses, reasoning, and attention.

**Data Mining** Data Mining is the set of techniques and methodologies used to extract knowledge from large amounts of data.

**Gaia** The definition includes the entire biosphere, that is everything living on the planet plus the atmosphere, the oceans, and the soil.

**Pervasive computing** In computer science, pervasive computing, also known as ubiquitous computing, refers to the environment in which the computers are present everywhere, and anytime.

**Recommenders** Third parties users able to give reputation information about another user.

**Slack** It is a real-time messaging platform.

**Socio-technical system** It is devoted to the effective blending of both the technical and social systems of an organization.

**Skype** It is a real-time messaging platform.

**Trust-model** It is an approach in which the access rights may change dynamically, depending on trust deserved by any user while interacting on the network.

**Value** It is the hidden information extracted from large databases through Data Mining techniques.

## REFERENCES


Aghaei, S., Nematbakhsh, M.A., Farsani, H.K., 2012. Evolution of the world wide web: from web 1.0 to web 4.0. Int. J. Web Semantic Technol. 3 (1), 1–10. doi:10.5121/ijwest.2012.3101.

Agrawal, R., Srikant, R., 1994. Fast algorithms for mining association rules in large databases. In: Proceedings of the 20th International Conference on Very Large Data Bases, VLDB '94. Morgan Kaufmann Publishers Inc., San Francisco, CA, USA, pp. 487–499. http://dl.acm.org/citation.cfm?id=645920.672836.

Agrawal, R., Imieliński, T., Swami, A., 1993. Mining association rules between sets of items in large databases. In: Proceedings of the 1993 ACM SIGMOD International Conference on Management of Data, SIGMOD '93. ACM, New York, NY, USA, pp. 207–216. doi:10.1145/170035.170072.

Akagi, T., 2006. Maintenance of environmental homeostasis by biota, selected nonlocally by circulation and fluctuation mechanisms. J. Artif. Life 12 (1), 135–151. doi:10.1162/106454606775186419.

Al-Aidaroos, K.M., Bakar, A.A., Othman, Z., 2010. Naïve Bayes variants in classification learning. In: International Conference on Information Retrieval Knowledge Management, (CAMP), pp. 276–281. doi: 10.1109/INFRKM.2010.5466902.

Argyris, C., Schon, D., 1978. Organizational Learning: A Theory of Action Perspective. Addison Wesley, Reading, MA.

Banavar, G.S., 2015. Watson and the era of cognitive computing. In: Proceedings of the International Conference on Pervasive Computing and Communications (PerCom). IEEE, New York, p. 95, doi:10.1109/PERCOM.2015.7146514.

Baxter, G., Sommerville, I., 2011. Socio-technical systems: from design methods to systems engineering. Interact. Comput. 23 (1), 4–17. doi:10.1016/j.intcom.2010.07.003.

Bernal, P.A., 2010. Web 2.5: the symbiotic web. Int. Rev. Law Comput. Technol. 24 (1), 25–37. doi:10.1080/13600860903570145.

Boden, M., 2006. Mind As Machine: A History of Cognitive Science. Oxford University Press, Oxford.
BotList, 2016. https://botlist.co/ (Accessed 21 November 2016).
Calvo, R.A., D'Mello, S., 2010. Affect detection: an interdisciplinary review of models, methods, and their applications. IEEE Trans. Affect. Comput. 1 (1), 18–37. doi:10.1109/T-AFFC.2010.1.
Celina, R., Hofkirchner, W., Fuchs, C., Schafranek, M., 2008. The web as techno-social system. The emergence of web 3.0. Cybern. Syst. 604–609. http://www.hofkirchner.uti.at/icts-wh-profile/pdf39.pdf.
da Rocha Gracioso, A.C.N., Suárez, C.C.B., Bachini, C., Fernández, F.J.R., 2013. Emotion recognition system using open web platform. In: 2013 47th International Carnahan Conference on Security Technology (ICCST), pp. 1–5. doi:10.1109/CCST.2013.6922065.
D'Angelo, G., Rampone, S., 2014. Towards a HPC-oriented parallel implementation of a learning algorithm for bioinformatics applications. BMC Bioinformatics 15 (5), 1–15. doi:10.1186/1471-2105-15-S5-S2.
D'Angelo, G., Rampone, S., 2016. Feature extraction and soft computing methods for aerospace structure defect classification. Measurement 85, 192–209. doi:10.1016/j.measurement.2016.02.027.
D'Angelo, G., Palmieri, F., Ficco, M., Rampone, S., 2015. An uncertainty-managing batch relevance-based approach to network anomaly detection. Appl. Soft Comput. 36, 408–418. doi:10.1016/j.asoc.2015.07.029.
DAngelo, G., Rampone, S., Palmieri, F., 2015. An artificial intelligence-based trust model for pervasive computing. In: Proceedings of the 10th International Conference on P2P, Parallel, Grid, Cloud and Internet Computing (3PGCIC). IEEE, New York, pp. 701–706, doi:10.1109/3PGCIC.2015.94.
D'Angelo, G., Rampone, S., Palmieri, F., 2016. Developing a trust model for pervasive computing based on Apriori association rules learning and Bayesian classification. Soft Comput. 1–19. doi:10.1007/s00500-016-2183-1.
Engestrom, Y., 2000. Activity theory as a framework for analyzing and redesigning work. Ergonomics 43 (7), 960–974. doi:10.1080/001401300409143.
Esposito, C., Palmieri, F., Castiglione, A., 2015. A knowledge-based platform for big data analytics based on publish/subscribe services and stream processing. Knowl.-Based Syst. 79, 3–117. doi:10.1016/j.knosys.2014.05.003.
Fermentas, I., 2004. Web2.0 conference. http://conferences.oreillynet.com/web2con/ (Accessed 21 November 2016).
Ferrara, E., Varol, O., Davis, C., Menczer, F., Flammini, A., 2016. The rise of social bots. Commun. ACM 59 (7), 96–104. doi:10.1145/2818717.
Ficco, M., D'Arienzo, M., D'Angelo, G., 2007. A Bluetooth infrastructure for automatic services access in ubiquitous or nomadic computing environment. In: Proceedings of the 5th ACM International Workshop on Mobility Management and Wireless Access. ACM, New York, NY, USA, pp. 17–24. doi:10.1145/1298091.1298095.
Fjeld, M., Lauche, K., Bichsel, M., Vo orhorst, F., Krueger, H., Rauterberg, M., 2002. Physical and virtual tools: Activity theory applied to the design of groupware. Comput. Supported Coop. Work 11 (1), 153–180. doi:10.1023/A:1015269228596.
Hakkani-Tür, D., Tur, G., Iyer, R., Heck, L., 2012. Translating natural language utterances to search queries for SLU domain detection using query click logs. In: International Conference on Acoustics, Speech and Signal Processing (ICASSP). IEEE, New York, pp. 4953–4956, https://doi.org/10.1109/ICASSP.2012.6289031.
Halfaker, A., Riedl, J., Los Alamitos, CA, USA, 2012. Bots and cyborgs: Wikipedia's immune system. Computer 45 (3), 79–82. doi:10.1109/MC.2012.82.
Harmelen, F.V., 2004. The semantic web: what why how and when. IEEE Distrib. Syst. Online 5 (3), 1–4. http://ieeexplore.ieee.org/stamp/stamp.jsp?arnumber=1285880.
Hollan, J., Hutchins, E., Kirsh, D., 2000. Distributed cognition: toward a new foundation for human-computer interaction research. ACM Trans. Hum. Comput. Interact. 7 (2), 174–196. doi:10.1145/353485.353487.
Hutchins, E., 1995. How a cockpit remembers its speed. Cogn. Sci. 19 (1), 265–288. doi:10.1207/s15516709cog1903_1.
Hutchins, E., 1995a. Cognition in the Wild. MIT Press, Cambridge, MA.
Hutchins, E., 1996. The integrated mode management interface (Tech. rep.). Final report for project NCC 92-578, NASA Ames Research Center, University of California at San Diego, La Jolla, CA.

Hutchins, E., Klausen, T., 1996. Distributed cognition in an airline cockpit. In: Engestrom, Y., Middleton, D. (Eds.), Cognition and Communication at Work. Cambridge University Press, Cambridge, pp. 15–34.

Kagal, L., Finin, T., Joshi, A., 2001. Trust-based security in pervasive computing environments. Computer 34 (12), 154–157. doi:10.1109/2.970591.

Karim, M.R., Hossain, M.A., Jeong, B.S., Choi, H.J., 2012. An intelligent and emotional web browsing agent. In: 2012 International Conference on Information Science and Applications, vol. 1, pp. 1–6. doi:10.1109/ICISA.2012.6220978.

Karkulahti, O., Kangasharju, J., 2012. Surveying Wikipedia activity: collaboration, commercialism, and culture. In: The International Conference on Information Networking (ICOIN) 2012, pp. 384–389. doi: 10.1109/ICOIN.2012.6164405.

Khoi-Nguyen Tran, P.C., 2015. Cross-language learning from bots and users to detect vandalism on Wikipedia. IEEE Trans. Knowl. Data Eng. 27 (3), 673–685. doi:10.1109/TKDE.2014.2339844.

Lew, M., Bakker, E.M., Sebe, N., Huang, T.S., 2007. Human-Computer Intelligent Interaction: A Survey. Springer Berlin Heidelberg, Berlin, Heidelberg, pp. 1–5. doi:10.1007/978-3-540-75773-3_1.

Liu, W., Zhang, J., Feng, S., 2015. An ergonomics evaluation to chatbot equipped with knowledge-rich mind. In: 2015 3rd International Symposium on Computational and Business Intelligence (ISCBI), pp. 95–99. doi:10.1109/ISCBI.2015.24.

Loghry, J.B., 2013. The Recreation of Consciousness: Artificial Intelligence and Human Individuation. Ph.D. thesis, AAI3605083.

Lovelock, J., 2000. Gaia: A New Look at Life on Earth. OUP Oxford, Oxford.

Mansour, O., 2009. Group intelligence: a distributed cognition perspective. In: Proceedings of the International Conference on Intelligent Networking and Collaborative Systems. IEEE, New York, pp. 247–250. doi: 10.1109/INCOS.2009.59.

Minsky, M., 1988. The Society of Mind. Simon & Schuster, New York.

Motion, I., 2016. Chatbots made easy. https://www.motion.ai/. (Accessed 21 November 2016)

Nardi, B.A., 1996. Chapter 4: Studying context: a comparison of activity theory, situated action models, and distributed cognition. In: Context and Consciousness: Activity Theory and Human-Computer Interaction. MIT Press, Cambridge, MA, pp. 69–102.

Norman, D.A., 1991. Cognitive artifacts. In: Carroll, J.M. (Ed.), Designing Interaction: Psychology at the Human-Computer Interface. Cambridge University Press, Cambridge, UK, pp. 17–38.

Norman, D.A., 1993. Things That Make Us Smart: Defending Human Attributes in the Age of the Machine. Addison-Wesley, New York.

Perkins, D., 1993. Person-plus: a distributed view of thinking and learning. In: Salomon, G. (Ed.), Distributed Cognitions: Psychological and Educational Considerations. Cambridge University Press, Cambridge, UK, pp. 88–110.

Raman, M., 2010. Wiki technology as a "free" collaborative tool within an organizational setting. EDPACS 42, 1–10. doi:10.1080/07366981.2010.531238.

Rogers, Y., Ellis, J., 1994. Distributed cognition: an alternative framework for analysing and explaining collaborative working. J. Inform. Technol. 9 (2), 119–128. doi:10.1057/jit.1994.12.

Russell, P., 2006. The Global Brain: The Awakening Earth for a New Millennium. Peter Russell.

Sahtouris, E., 1999. EARTHDANCE: Living Systems in Evolution. iUniverse.

Salomon, G., 1997. Distributed cognitions: Psychological and educational considerations. Cambridge University Press, Cambridge.

Smets, K., Goethals, B., Verdonk, B., 2008. Automatic vandalism detection in Wikipedia: towards a machine learning approach. In: Proceedings of the Association for the Advancement of Artificial Intelligence (AAAI) Workshop on Wikipedia and Artificial Intelligence: An Evolving Synergy (WikiAI '08), pp. 43–48, https://www.aaai.org/Papers/Workshops/2008/WS-08-15/WS08-15-008.pdf.

Smith, E., 2008. Social relationships and groups: new insights on embodied and distributed cognition. Cogn. Syst. Res. 9 (1–2), 24–32. doi:10.1016/j.cogsys.2007.06.011.

Stahl, G., 2006. Group Cognition: Computer Support for Building Collaborative Knowledge. MIT Press, Cambridge, MA.

Telegram, 2016. Telegram bot store. https://storebot.me/ (Accessed 21 November 2016).

Tur, G., Mori, R.D., 2011. Spoken Language Understanding: Systems for Extracting Semantic Information from Speech. John Wiley and Sons, New York, NY.

Turing, A.M., 1950. Computing machinery and intelligence. Mind 49 (236), 433–460. doi:10.1093/mind/LIX.236.433.

van Krevelen, D.W.F., Poelman, R., 2010. A survey of augmented reality technologies, applications and limitations. Int. J. Virtual Real. 9 (2), 1–20.

Wang, Y., 2014. From information revolution to intelligence revolution: big data science vs. intelligence science. In: Proceedings of the 13th International Conference on Cognitive Informatics & Cognitive Computing (ICCI*CC). IEEE, New York, pp. 3–5, doi:10.1109/ICCI-CC.2014.6921432.

Weiser, M., 1999. The computer for the 21st century. SIGMOBILE Mob. Comput. Commun. Rev. 3 (3), 3–11. doi:10.1145/329124.329126.

Weizenbaum, J., 1966. Eliza—a computer program for the study of natural language communication between man and machine. Commun. ACM 9 (1), 36–45. doi:10.1145/357980.357991.

Wright, P.C., Fields, R.E., Harrison, M.D., 2000. Analyzing human-computer interaction as distributed cognition: the resources model. Hum. Comput. Interact. 15 (1), 1–41. doi:10.1207/S15327051HCI1501_01.

CHAPTER

# A NOVEL CLOUD-BASED IoT ARCHITECTURE FOR SMART BUILDING AUTOMATION

# 10

**David Sembroiz*, Sergio Ricciardi*, Davide Careglio***
*Technical University of Catalonia (UPC) - BarcelonaTech, Barcelona, Spain**

## 1 INTRODUCTION TO THE INTERNET OF THINGS

The Internet of Things (IoT) is meant to be the future of the current Internet. It is commonly defined as a network of physical and virtual objects, devices, or things that are capable of collecting surrounding data and exchanging it between them or through the Internet. To enable data collection, devices are embedded with sensors, software, and electronics; the exchange capability is achieved by connecting them to local area networks or to the Internet.

The origins of the Internet of Things are diffuse. Even though the word was first coined in 1999 by Kevin Ashton, co-founder and executive director of the Auto-ID Center at MIT, for companies such as CISCO, the IoT was born in 2009, when more devices than people were connected to the Internet. At that time, the number of connected devices were 10 billion, but the expectations are generous. It is thought that by 2020, more than 50 billion devices will be connected to the Internet.

As it can be extracted from the numbers, during the last few years, the Internet of Things has seen an unexpected increase in popularity, mainly thanks to the following technology improvements:

- **Smaller, more durable, and powerful sensors.** Newly manufactured sensors are seeing their size substantially reduced, allowing their placement in small spaces and also in delicate and dangerous scenarios.
- **Increased efficiency.** One of the key aspects of the Internet of Things paradigm is the wireless interconnection between devices. Thus these devices must be equipped with autonomous power supplies that limit their lifespan. To cope with this problem, manufacturers are aiming for efficient processors and software engineers are specifically designing software and communication technologies for IoT in which lower energy consumption is the main requirement. To achieve this, sensors usually work in low-power-usage mode. Devices remain in sleep mode until a new sample message needs to be generated. Then, it wakes up, creates the message and transmits it by powering up the RF power amplifier. When the message is transmitted, both the RF power amplifier and the device are turned down until the next cycle.
- **Lower production cost.** The improvements in industry and the easiness in which mass production is currently achieved allow companies to lower the price of each component.

**Security and Resilience in Intelligent Data-Centric Systems and Communication Networks. https://doi.org/10.1016/B978-0-12-811373-8.00010-0**

Copyright © 2018 Elsevier Inc. All rights reserved.

The combination of these improvements created new market opportunities that companies have foreseen. Since the Internet of Things is in a very young state, the lack of coordination and the rapidness with which new gadgets are created are hampering the standardization of the future Internet.

The possibility to attach small hardware to any kind of electronic or mechanic device enables the possibility to monitor everything. That is why the Internet of Things is also defined as the interaction with *anything, anytime, anywhere*.

Smart buildings are gaining popularity and many companies are putting their efforts in this field since the number of potential clients is very high. Elements such as the smart fridge, smart thermostat, and smart lightning are appearing to ease daily life and to increase peoples comfort.

In the case of the health-care systems, companies aim for taking advantage of smart and small devices to monitor the condition of people in order to detect anomalies and immediately inform familiars or even hospitals.

## 2 MAIN ENABLING TECHNOLOGIES AND PROTOCOLS

Since the beginning of the current Internet, many groups have been created for helping in the standardization of protocols and technologies. The Internet Engineering Task Force (IETF), World Wide Web Consortium (W3C), Institute of Electrical and Electronics Engineers (IEEE) are some of the most important ones.

In the initial steps of the IoT, protocols such as RFID and NFC were the standard de facto mainly due to its low production cost. However, transmission limitations in terms of range coverage and the inability to communicate through the Internet hampers their usage in new IoT scenarios such as smart buildings or cities. In the industry sector, they are still widely used for packet tracking and object identification.

In recent years, groups have put their efforts in creating standards for protocols directly related to the Internet of Things. Even though it is possible to use the old communication protocols such as Bluetooth or Wi-Fi, their characteristics do not fit with IoT device requirements. Many IoT devices rely on the necessity of having external power supplies such as batteries to work, requiring reduced power consumption and cost, while maintaining similar communication range with respect to their analogous current Internet protocols. To cite some, Bluetooth Low Energy, Wi-Fi HaLow or LoRaWAN are IoT-focused protocols with the stated requirements. Even though the core of such protocols is very similar, depending on the kind of application being developed, one may fit best than another. Moreover, the usage of multiple protocols inside the same system is not forbidden. For instance, a general system combining many information sources, each of them using the specific sensor devices, can use the protocol that fits best by taking into account devices connectivity and location.

IoT enabling protocols can be divided into two major groups, named Infrastructure protocols and Application protocols. Infrastructure protocols refer to the ones that actuate inside the underlying infrastructure and create the communication between system layers. For instance, the connection between the perception layer and the network layer, or the one between the network layer and the cloud layer. Application protocols are responsible for interconnecting the infrastructure with the application.

In the following sections, the most relevant protocols of both groups are explained more in depth. To summarize, a table comparing the relevant features is also shown.

## 2.1 WIRELESS INFRASTRUCTURE PROTOCOLS

### 2.1.1 Bluetooth low energy (BLE)

Bluetooth Low Energy or Bluetooth Smart, as it has been branded, is an enhancement of the Bluetooth technology in which connectivity and power usage are smarter than its predecessor. However, devices with Bluetooth Smart technology attached are not compatible with previous versions. To cope with this problem, Bluetooth Special Interest Group completed the Bluetooth Core Specification version 4.0 to include compatibility between versions. Current devices include this new core protocol making them able to communicate with any Bluetooth device.

The shifting in the connection paradigm performed by the Internet of Things has forced new protocols to include new behavioral modes. Bluetooth Smart includes ultra-low peak, average, and idle modes. Once the pairing between two devices is performed, Bluetooth Smart focuses on sending small bits of data when needed and putting the connection in a low power consumption mode in order to drastically reduce energy usage.

According to the Bluetooth SIG specification, this protocol has been specifically designed for smart home, health, sports, and fitness sectors. These sectors can take advantage of the following Bluetooth Smart features (LitePoint, 2012):

- Low power requirements, allowing the devices to operate for months or even years.
- Small size and low cost.
- Compatibility with a large base of mobile phones, tablets, and computers, allowing the interoperability between such devices.

As for its technical details, Bluetooth Smart operates in the same spectrum range as its predecessor, the 2.4 GHz–2.4835 ISM band. However, the set of channels used vary significantly. Instead of the classic 79 1-MHz channels, Bluetooth Smart uses 40 2-MHz channels. Regarding bit rate and maximum transmission power, they are limited to 1 Mbit per second and 10 milliwatts respectively. Its range coverage is ten times that of the classic Bluetooth (10 m versus 100 m approximately). Latency wise, it is 16 times shorter (100 ms versus 6 ms) (Frank et al., 2014) (Table 1).

### 2.1.2 ZigBee

ZigBee is a standard based on the IEEE 802.15.4 specification specially targeted for long battery life devices in wireless mesh networks. This protocol has been evolving since its appearance in 1999, and

**Table 1 Comparison Between Bluetooth Classic and BLE (Frank et al., 2014)**

| | Bluetooth Classic | Bluetooth Smart |
|---|---|---|
| Spectrum Range | 2.4–2.4835 GHz | 2.4–2.4835 GHz |
| Channel Bandwidth | 1 MHz | 2 MHz |
| Number of channels | 79 | 40 |
| Max. Bit Rate | 3 Mbps | 1 Mbps |
| Max. Transmission Power | 100 mW | 10 mW |
| Avg. Range | 10 m | 100 m |
| Avg. Latency | 100 ms | 6 ms |

its last specification is called ZigBee PRO, from 2007. Even though it shares features with Bluetooth, ZigBee is intended to be simpler and less expensive.

Regarding operation bands, ZigBee uses the same as Bluetooth (Siekkinen et al., 2012), the 2.4 GHz band. In some locations, this band varies. For instance, China uses the 784 MHz band, Europe uses the 868 MHz band, while USA and Australia uses the 915 MHz one.

Its simplicity also limits some important aspects such as transmission rate and communication range. Unlike Bluetooth, data transmission is limited to a maximum of 250 Kbit per second, which may be enough depending on the scenario under use. Communication range varies between 10 and 20 m for indoor transmissions, depending on power output and environmental characteristics.

### 2.1.3 6LoWPAN

The IPv6 over Low power Wireless Personal Area Networks or 6LoWPAN was created by a concluded working group in the Internet area of the IETF to fulfill the necessity to allow any kind of device, even the smallest ones with limited power usage and processing capabilities, to participate in the Internet of Things.

6LoWPAN is a combination of IEEE 802.15.4 and IP in a simple, well understood way. The key features of this protocol are the encapsulation definition and header compression that allow the compatibility between local area networks and wide area networks with IEEE 802.15.4-based networks.

Since 6LoWPAN pertains to the network layer of the OSI model, it does not have a specific transmission specification. Instead, the underlying link layer protocol is responsible for providing them. As mentioned before, this protocol has been designed to work on top of IEEE 802.15.4 based networks which provides the transmission characteristics already explained in Section 2.1.1.

### 2.1.4 Wi-Fi HaLow

Wi-Fi HaLow is a very new technology presented in January 2016 in the Computer Electronic Show (CES) by the Wi-Fi Alliance (Wi-Fi, 2016). This new Wi-Fi specification is directly suited to meet the unique requirements of IoT environments such as Smart Homes, Smart Cities, and Industrial markets. It extends Wi-Fi, and specifically its 802.11ah specification, to operate in the 900 MHz band, enabling the low power connectivity necessary for applications including sensors and wearables which hardly rely on battery lifetime. Its range has been extended to almost twice that of current Wi-Fi, and will not only be capable of transmitting farther, but also providing a more robust connection in harsh environments thanks to its ability to penetrate walls or other barriers more easily.

Devices with HaLow support are expected to also support current 2.4 and 5 GHz Wi-Fi bands, allowing interoperability between current devices and new ones. They also support IP-based connectivity to natively connect to the Internet. Another important point worth mentioning is the ability to connect thousands of devices to a single access point to create dense device deployments. As for its transmission power, HaLow is expected to work between 150 Kbps and 18 Mbps depending on the requirements of the application. To support such transmission rates, different channel setups are required: 150 Kbps only requires a 1 MHz channel but the maximum transmission rate requires a 4 MHz-wide channel.

This new technology is the answer to Bluetooth, and Wi-Fi alliance expects to begin certifying HaLow products in 2018.

### 2.1.5 LoRaWAN

LoRaWAN (Sornin et al., 2015) is a Low Power Wide Area Network created by the LoRa Alliance as a solution for wireless battery operated devices. It specifically target the IoT main requirements such as secure communication, mobility, and localization services. In a typical LoRaWAN network, devices and gateways compose a star of stars topology in which only the gateways are connected to the Internet, whereas devices use single-hop wireless communication to transmit their data to single or multiple gateways simultaneously. The transmission between devices and gateways is bi-directional, but it also enables the possibility for multicast messaging for Over The Air software upgrade.

LoRaWAN supports a wide range of frequency channels and data rates. Moreover, transmission with different specifications to the same gateway do not interfere with each other. Every transmission is encapsulated in a separate *virtual* channel which increases the capacity of the gateway significantly. Data rates range between 0.25 Kbps and 50 Kbps.

LoRaWan defines three classes for end point devices to address the different needs reflected in the wide range of possible applications:

- Bi-directional end devices (Class A): asynchronous transmissions in which every uplink message is followed by 2 short downlink windows that the gateway can take advantage of to send messages to the end devices. After these windows have finished, the end devices is set to idle until the next uplink transmission. This class operates in the lowest power and is suitable for applications that only need end device to gateway communication.
- Bi-directional end device with scheduled receive slots (Class B): In addition to Class A random receive windows, end devices are told by means of a time synchronized Beacon from the gateway which time slot they must listen for any possible downlink communication.
- Bi-directional end devices with maximal receive slots (Class C): end devices are continuously listening for downlink messages and this window is only closed when transmitting to the gateway. This class is usually targeted for AC-powered devices because of its high power consumption.

Table 2 acts as a summary and comparison between the main infrastructure protocols presented. To clarify, communication range shows the distance to which these technologies can transmit depending if the transmitter and the receiver are in Line Of Sight (LOS) or not. Regarding spectrum usage, it can vary depending on the location since the legislation varies in every continent.

**Table 2 Comparison Between the Main Enabling Infrastructure IoT Protocols**

| | **BLE** | **ZigBee** | **6LoWPAN** |
|---|---|---|---|
| Spectrum Range | 2.4–2.4835 GHz | 2.4–2.4835 GHz | 868/915/2400 MHz |
| Bit Rate | 1 Mbps | 20–250 Kbps | 250 Kbps |
| Peak Consumption | <15 mA | 30–40 mA | <15 mA |
| Range | 10–100 m | 10–100 m | 10–200 m |
| | **Wi-Fi HaLow** | **LoRaWAN (EU)** | |
| Spectrum Range | 900 MHz | 868 MHz | |
| Bit Rate | 150 Kbps–18 Mbps | 0.25–50 Kbps | |
| Peak Consumption | ~ 50 mA | ~ 38 mA | |
| Range | 1 km | 2–22 km | |

## 2.2 APPLICATION LAYER PROTOCOLS

### *2.2.1 Hypertext transfer protocol (HTTP)*

The Hypertext Transfer Protocol (HTTP) is an application layer protocol designed for distributed, collaborative, hypermedia information systems. This protocol is the foundation of data communication for the World Wide Web.

HTTP was initiated in 1989 at the European Organization for Nuclear Research (CERN). However, the development of standards was coordinated by the IETF and the World Wide Web Consortium (W3C), culminating in the publication of a group of RFCs in 1997, with the definition of the HTTP/1.1 version in Fielding et al. (1997) firstly, and updated in Fielding et al. (1999). During many years this has been the standard *de facto*. In 2015, the successor HTTP/2 was standardized (Belshe et al., 2015).

HTTP works as a request-response protocol in the client-server computing model. The majority of the time, it uses TCP as a transport protocol for reliability. However, it can be adopted to use unreliable protocols such as UDP.

Even though its usage has been extended to the IoT world, it was not specifically designed for this purpose. If compared to other IoT-oriented protocols, HTTP may not be the best choice due to its protocol overheads and communication requirements. However, it has served as a strong base for newly developed protocols such as CoAP.

### *2.2.2 Constrained application protocol (CoAP)*

The Constrained Application Protocol (CoAP) is defined as a *specialized web transfer protocol for use with constrained nodes and constrained networks in the Internet of Things* (CoAP, 2014). As it can be extracted from the definition, this protocol is specifically tailored for the IoT and M2M applications. The major standardization of this protocol has been carried out by the IETF Constrained RESTful environments (CoRe) Working Group and the core is specified in Shelby et al. (2014). This application layer protocol can be seen as an enhancement of HTTP for low power devices. It is based on the successful REST model, in which resources are available under a URL and clients can access those resources using the GET, PUT, POST and DELETE methods. Additionally, CoAP also supports publish-subscribe thanks to the usage of an extended GET method.

Even though it shares similarities with HTTP, CoAP is specifically designed to run over UDP only. As UDP is inherently not reliable, CoAP defines two types of messages, namely *confirmable messages* and *nonconfirmable messages* to define its own reliability mechanism. The former requires an acknowledgment similar to the ACK used in TCP communications while the latter does not require any kind of acknowledgment.

### *2.2.3 Message queue telemetry transport (MQTT)*

Message Queue Telemetry Transport (MQTT) is a client-server publish-subscribe messaging transport protocol standardized under the ISO/IEC PRF 20922 (ISO, 2016). It is lightweight, simple, and very easy to implement. Its lightness characteristic make it ideal for environments in which communication capabilities are limited such as M2M or IoT scenarios. Unlike CoAP, MQTT has been designed to run over TCP/IP or other network protocols that provide ordered, lossless, and bi-directional communication. In this regard, MQTT is similar to HTTP. However, the former has been designed to have less protocol overhead.

**Table 3 Comparison Between the Main Enabling Application IoT Protocols**

| | HTTP | CoAP | MQTT |
|---|---|---|---|
| Main Transport Protocol | TCP | UDP | TCP |
| RESTful | ✓ | ✓ | ✗ |
| Publish/Subscribe | ✗ | ✓ | ✓ |
| Request/Response | ✓ | ✓ | ✗ |
| QoS | ✗ | ✓ | ✓ |

The reliability of messages in MQTT is taken care by three Quality of Service (QoS) levels:

- At most once: messages are delivered in a best effort manner and messages loss can occur. This QoS is tailored for scenarios in which the loss of a message is not relevant.
- At least once: messages are assured to arrive, but duplicates can occur.
- Exactly once: messages arrive exactly one time, without duplicates. This QoS is reserved for systems that must operate reliably all the time, such as banking systems.

Thanks to the publish/subscribe model, it also allows for one-to-many message distribution with application decoupling (Table 3). Table 3 summarizes the main characteristics of the protocols previously explained.

# 3 EVOLUTION OF IoT ARCHITECTURES

The rapid proliferation of gadgets and solutions for the everyday problems and situations is creating chaos inside the Internet of Things. The lack of agreement between companies for the usage of architectural and communication standards is making it impossible to create heterogeneous systems in which devices from different manufacturers interchange information smoothly. This is known as the vertical silos problem, in which every manufacturer creates his own close and private solution ranging from the sensors up to the end user application. This verticality does not allow the interoperability between company solutions, disabling the possibility to intercommunicate gadgets for different information sources inside the same scenario. Following sections present an overview of the evolution of IoT architectures since the appearance of such term up until the current ones making emphasis on how they coped with the stated *vertical silos* problem.

## 3.1 INITIAL MODELS

The lack of architectural standards and protocols during the initial stages of the IoT development hampered the creation of systems with the current minimum requirements such as scalability, interoperability, security, and reliability. During the first years, The Intranet of Things was a more accurate term to define the situation. Devices were only provided with physical wireless communication protocols such as Bluetooth or ZigBee, with no possibility to transmit through the Internet. Moreover, the connection between those devices and the application was directly performed without any intermediate layer to decouple the system. Fig. 1 shows an abstraction of such architecture.

**FIG. 1**

Initial IoT 2 layer architecture.

As can be seen, this architecture can only be divided into two separate layers, namely the perception and application layers. Although it was possible to use multiple devices inside the same system, they acted as individual elements, without communicating between them or helping each other during transmission. That is, a Network Layer combining them into a Wireless Sensor Network was not used. Instead, data being generated by the perception layer was directly sent to the application layer without any intermediate decoupling which made impossible the scalability or interoperability of such system.

Then, the architecture that can be seen as the birth of the Internet of Things appeared, which was comprised by three layers namely perception, network and application layer (Wu et al., 2010). The network layer grouped all the sensors and actuators of the system forming a Wireless Sensor Network, in which devices were aware of each other. Additionally, gateways were added to gather and forward all the raw messages generated by the perception layer devices. Even though initial systems continued using only physical wireless communication protocols for its devices, the insertion of gateways as a more powerful intermediate element allowed for Internet communication between the network and the application layers (Fig. 2).

The introduction of the network layer helped to slightly cope with the scalability problem by means of the placement of more gateways, if necessary, to handle all the device connections. However, this did not completely solved the problem. Regarding the interoperability and heterogeneity, the lack of standardization between companies for the usage of protocols and message structure hampered the combination of several devices into the same system.

At this point, researchers and manufacturers agreed in the necessity of having an abstraction layer to completely decouple the physical network from the application, in order to allow the creation of device-agnostic applications.

## 3.2 THE APPEARANCE OF A *MIDDLEWARE*

Many efforts have been combined to provide an abstraction and standardization layer from the WSN perspective. European Union projects such as SENSEI (Tsiatsis et al., 2010) and Internet of

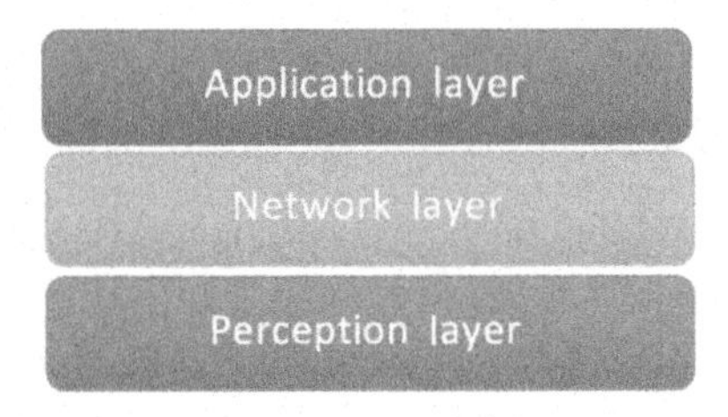


**FIG. 2**

IoT 3-layered architecture.

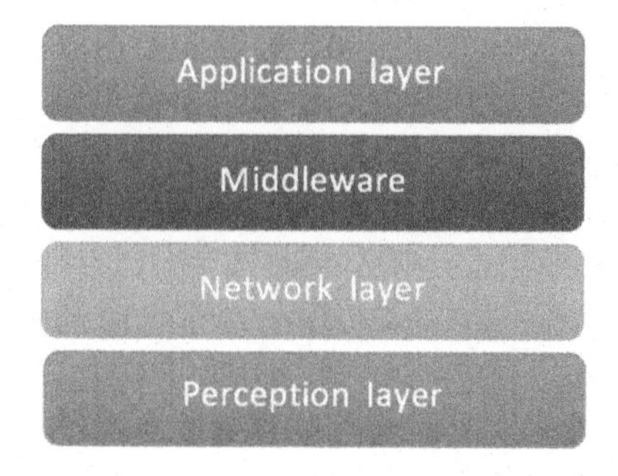


**FIG. 3**

IoT 4-layered architecture.

Things—Architecture (IoT-A) (Bauer et al., 2013) have been addressing this problem by means of creating and defining the architecture for different applications. However, there is still a lack of agreement when it comes to overall architectural standards in regard to upper layers.

A middleware generally abstracts the complexities of the system and hardware allowing the application developer to fully focus all his efforts on the task to be solved without the distraction of concerns regarding system or hardware level. A middleware provides a software layer between physical and application layers. As it has been seen before, the IoT interacts with many infrastructure and application technologies. Therefore a middleware must provide almost full compatibility (Fig. 3).

Even though of the agreement in the necessity of a middleware as an abstraction layer, during the last years diverse solutions have appeared in terms of their design approach, such as event-based, database-oriented, application-specific, or service-oriented (Al-Fuqaha et al., 2015; Milić and Jelenković, 2015). However, the usage of a single design approach might not be sufficient. Instead, successful middlewares have been built upon the combination of multiple designs.

Since all the generated data must traverse the middleware for abstraction, the inclusion of the database as an element of such layer seems the right decision. This is why many current middleware solutions include a database-oriented design. However, the connectivity to the database may vary depending whether data is directly exposed to the end users or it is privately stored and instead, events or services are offered. The former case follows a mere database-oriented design, while the later is a combination of database and either event or service-oriented design.

Middlewares based on events with database storage are gaining popularity due to the easiness of deployment and lightness of resource utilization. Since individual sensor messages can be seen as events, the storage is straightforward. Regarding event communication, this type of middlewares usually use the publish/subscribe pattern, in which a set of subscribers acquire events from a set of publishers. Protocols such as MQTT (Section 2.2.3) or CoAP (Section 2.2.2) are designed to this aim.

Service-oriented middlewares are based on Service-Oriented Architectures (SOA) that have been traditionally used in corporate IT systems. Characteristics such as service reusability, composability, or discoverability are also beneficial for IoT scenarios. However, large scale networks, constrained devices, and mobility make this approach challenging.

With this approaches, applications connected to the middleware benefit from the abstraction and are agnostic to the underlying hardware.

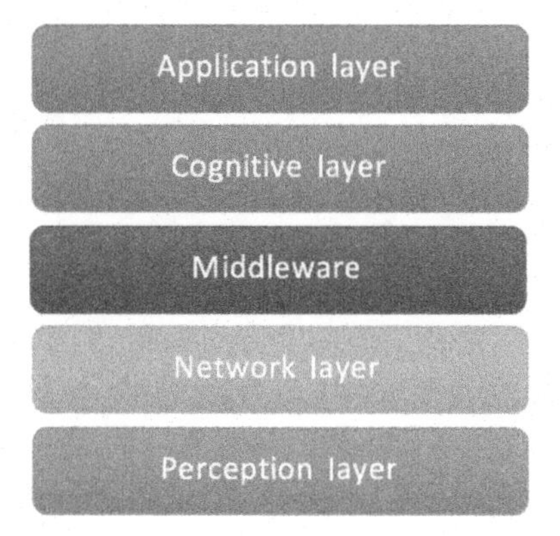


**FIG. 4**

IoT 5-layered architecture.

### 3.3 TOWARD INTELLIGENT IoT SYSTEMS

Up to this point, IoT applications started to exploit the benefits of new and well-designed architectures to solve daily problems or make lives easier. Moreover, many monitoring applications for different scenarios appeared, such as health monitoring systems, building energy monitoring, or city resource monitoring. However, the essence of such systems was merely informative.

New IoT or Future Internet is meant to go beyond that informative perspective. Instead, companies and researchers are aiming for the creation of intelligent and autonomous systems. To this aim, new elements are added to the previous defined architecture (Fig. 4).

Specifically, a new layer appears between the middleware and the application, commonly named knowledge-based layer, context awareness layer, or cognitive layer. It is responsible for requesting data and extracting valid information for acquiring new knowledge and act upon it.

Depending on the purpose of the application, many techniques can be used such as rule-based programming, machine learning, or predictive analysis. Rule-based applications are meant to modify the status of the scenario if certain events occur. Usually, rules are static. However, the combination of rules with machine learning techniques offers a richer system in which rules are modified depending on past actions. Predictive analysis is also being used to anticipate future actions and thus, increase comfort by adjusting the system to the desirable state beforehand.

## 4 CLOUD-BASED IoT ARCHITECTURE PRESENTATION

This section presents an architecture developed to try to cope with some of the problems stated in the previous sections. It offers interoperability in regard to the type of sensors and protocols that can be used. Moreover, reliability and data persistence is achieved by means of a cloud middleware capable of replicating services on demand. The cloud also allows for data exposure and possibly, utilize it as a service for third parties.

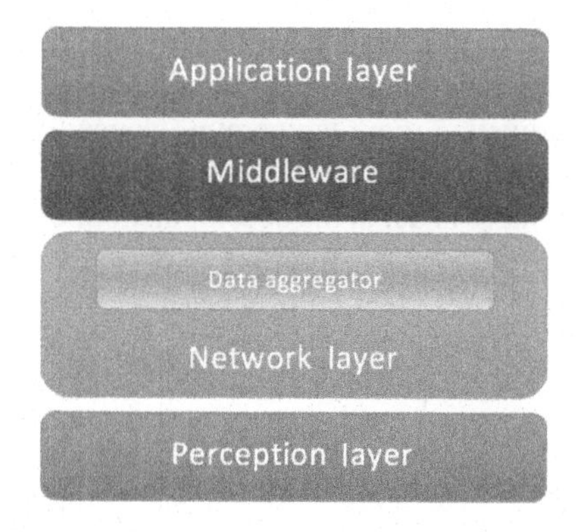


**FIG. 5**

Cloud-based IoT architecture abstraction.

Fig. 5 shows the architecture decomposed in the different layers. Starting from the bottom, the perception layer includes all the sensors and actuators of the network. It is responsible for sensing the environment and also for executing the necessary actions that are received from the above layer.

The network layer comprises and groups the gateways of the platform. Since these devices are resource-constrained in regard to the number of established connections, it is necessary to study the scenario under development in order to know how many of these devices need to be deployed. In terms of energy usage, gateways need a more powerful source than sensors. That is why these devices are usually placed inside buildings to maintain them fully operable. Moreover, received data might need to be uploaded to the Internet, which is another reason to locate them inside Internet-reachable buildings.

As previously mentioned, one of the main issues regarding the IoT and WSN is the heterogeneity at the physical level. The lack of agreement for communication and message structure make it necessary to endow the system with the possibility to upgrade gateway software to allow compatibility with new devices.

Even though the south gate (i.e., the communication between sensors and gateways) of the gateways may be heterogeneous, the north gate (i.e., the communication between gateways and the above layer) maintains its homogeneity by exclusively using one communication protocol.

The data aggregator and processing layer, as its name says, is responsible for receiving every sensing message in raw format. These messages are then processed in order to modify their structure to a standard one. Since JSON is the standard de facto inside the Big Data world and is also used by the middleware, raw sensing data is transformed into JSON-formatted files. This layer can be seen as a module of the network layer, and that is why it has been located inside it. This decision is explained later more in depth.

Regarding the middleware, it has been decided to use an external cloud platform called Villalba et al. (2015) that allows for data upload, storage, and retrieval using standard protocols such as HTTP and MQTT. Thanks to the usage of Big Data technologies, it also offers the possibility for server and database replication in case of necessity due to an increase in the number of connection requests.

Finally, the application layer contains the actual application. It is fed with standard data incoming from the middleware, which means the developer does not have to worry about the underlying hardware and communication protocols.

This architecture has been used for the development of a simulator for smart buildings that tries to reduce building energy consumption by avoiding unnecessary device states. For instance, switching off room lights when it is empty or adjusting room temperature depending on the environmental one and also on occupant desires.

The following sections explain more in detail the development of every layer and the communication between them.

## 4.1 PERCEPTION LAYER

The perception layer is formed by all the sensors and actuators of the system. The main task of this layer is gathering data from the elements of the scenario under monitoring. In the case of a Smart Building, sensors are usually placed to monitor environmental conditions such as temperature, humidity, luminosity, air quality, and also device state such as doors, windows, blinds, and computers. Moreover, actuators are deployed to allow state modification of those elements. For instance, if the system detects that a room has been left with lights on, it can send the signal to switch them off in order to avoid wasting energy unnecessarily. To achieve this feature, the communication between devices and the above layer is bi-directional.

The system can also take advantage of the communication bi-directionality to interact with the sensors. Since sensors can sometimes be located in hard-to-reach locations, this feature is needed to allow for software upgrade without having to manually access to them, commonly known as Over The Air (OTA) programming.

As it can be seen, there is a plethora of characteristics to monitor and actuate with, allowing for wide market opportunities for companies. The *vertical silos* problem previously stated starts in this layer. Companies usually specialize themselves in a single scenario or problem, without commonly agreeing on standards in design or communication. However, when a more general system is developed such as a Smart Building, it is necessary to combine multiple sensors to fulfill all the requirements stated above. Therefore it is needed a layer in which all this differences are solved by allowing the transmission and communication using different protocols.

## 4.2 NETWORK LAYER

The network layer groups and manages the gateways and it is responsible for creating a WSN between sensors, actuators, and gateways. Due to the heterogeneity of the perception layer, gateways must be rich in terms of protocol compatibility. To this aim, they must be endowed with multiple interfaces depending on the type of sensors they are managing. Due to this necessity, their requirement in terms of energy usage is higher and the usage of batteries is not sufficient. Instead, they are deployed inside buildings in order to be powered with an electrical current.

Up to this point, the communication between the devices of the system is locally performed. However, once messages are received by the gateway, the connectivity can vary. Local systems can opt to maintain a private network with no Internet connection in which gateways locally connect to the above layer to standardize and store messages. Another approach could be to endow the gateways with Wi-Fi interfaces to directly upload the data to the Internet.

Similarly to sensor software programming, gateways can also be enhanced with new compatibilities if needed. However, this can be more time and money consuming if the compatibility also needs to place new physical interfaces in every deployed gateway.

## 4.3 DATA AGGREGATOR LAYER

The data aggregator layer can be seen as the standardization message layer. It is responsible for receiving raw messages from every gateway and transforming them into a valid message format. It has been decided to use JSON as the data standard because of the compatibility with the above layer and the friendliness that it offers with big data technologies.

This layer can be deployed into multiple places inside the architecture. Specifically, these are valid locations for it:

- Multiple instances distributed across the gateways.
- Central server with replicability.
- Middleware module.

Depending on the power of the gateways, this layer can be deployed in each of them in order to avoid the necessity of a central server gathering the data from every gateway to later transform and upload it. However, this decision has some drawbacks. Firstly, it requires that all the gateways of the platform are capable of connecting to the Internet in order to upload the data. Secondly, processing power and storage for these gateways would need to be higher. Finally, in the case of needing to make a modification in the data aggregator to allow new data structures, it would be needed to completely flash all the gateways of the platform.

Another alternative is to develop a module for the middleware under usage in order to have a unique layer capable of standardizing the data and storing it. This is a good design approach but in our case it has not been followed in order to maintain the external middleware as it is.

The design approach finally followed has been to deploy this layer into a central server capable of creating multiple replicas if needed to cope with incoming connections. With this design, gateways do not need to have Internet connection and their processing power can be reduced to also reduce consumption, allowing a possible gateway deployment by means of battery power sources if strictly necessary.

## 4.4 MIDDLEWARE

As previously defined in Section 3.2, a middleware is an abstraction layer that hides the complexities of the system and hardware underneath.

Moreover, this middleware can have additional features depending on the requirements, which in our case are:

- Cloud storage with standard technologies.
- Big Data oriented with high scalability.
- Standard communication protocols for data upload and download.
- Public and private virtual objects for data scope control and sharing.

After reviewing the currently available IoT platforms, ServIoTicy was finally chosen since it covers all the requirements stated above. ServIoTicy is an online platform developed by the Barcelona Supercomputer Center (BSC) (2017) during the COMPOSE project (2015). It allows for fast and simple composition of IoT data streams, offering multitenant data architecture. As for its communication capabilities, both for data upload and download, it allows REST and publish/subscribe communication.

The transmitted content must be formatted into JSON data-objects. The extension and acceptance of this format as a standard allows to homogenize all the data independently of the transmitting platform, completely hiding the hardware from below.

### 4.5 APPLICATION LAYER

The application layer, as its name indicates, contains the application responsible for interacting with the user or showing the desired information. Inside the IoT world, current developed applications are first focused on monitoring the environment and acting as an information panel in which the user can read in real time, the values of the different sensors of the system, such as the inside temperature, power usage, and outside luminosity. There also exist interactive applications in which the user, apart from being able to see the sensor information, can also interact with the environment by sending actions to perform, such as close the door, lower the inside temperature, or switch elements ON or OFF.

One of the main advantages of the architecture presented is the freedom that the developer has when creating a specific application. By having a middleware with standard formats and transmission protocols, it allows the developer to fully focus their efforts into the use case without having to take into consideration hardware specifications.

The only coupling element between the middleware and the application is the message reception module. As it has been previously mentioned, messages can be requested via REST API or subscriptions thanks to the publish/subscribe protocol. The former allows for synchronous data requests, which can be necessary when a specific value needs to be obtained. However, the later is the standard widely used. The publish/subscribe protocol allows for asynchronous message reception by the application without the necessity to constantly query the middleware. Instead, when a new sensor message is stored inside the middleware, it is directly forwarded to the application by means of the subscription previously performed.

## 5 USE CASE: SMART BUILDING AUTOMATION

This section presents a Building Management System (BMS) for Smart Building automation using the architecture explained in Section 4. The main purpose of a BMS is to increase people's comfort by maintaining the building in the desired state every time and also reduce energy consumption by avoiding situations in which elements are being overused. For instance, by predicting the time of entrance of a person in a room, it is possible to adjust the temperature beforehand, or by detecting that a room is empty, lights can be turned *off* if they have been left *on* by mistake.

The first section explains how the data is generated and which elements are being monitored inside the building. Since the deployment of many sensors inside a building is costly, some of them are simulated in order to scale the system and create a more realistic scenario. Then, how this data is transformed into a standard format and later uploaded to the cloud platform is explained. After,

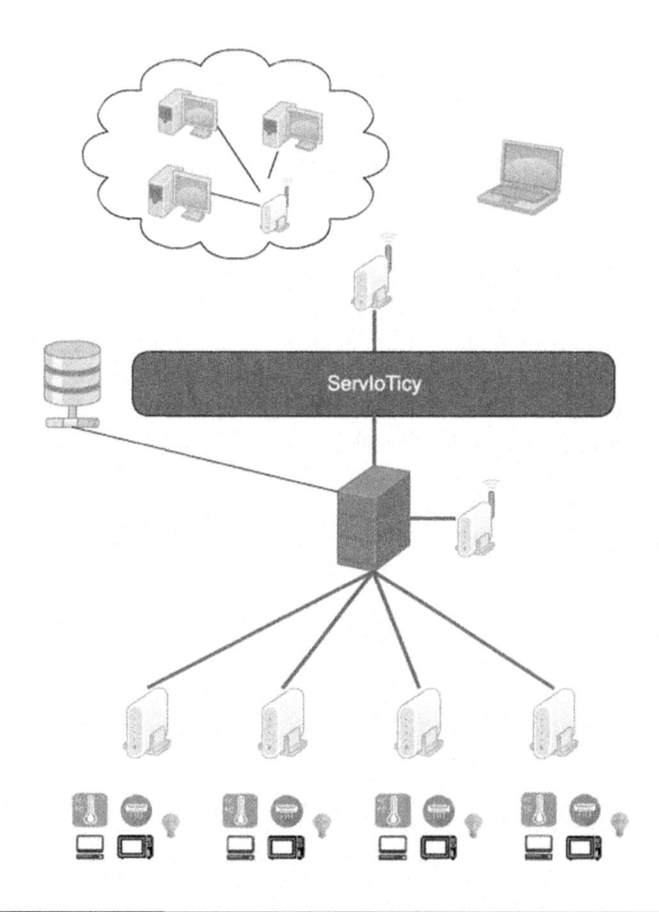


**FIG. 6**

Cloud-based IoT architecture.

the consumption of the data by means of the building management application is described. Finally, conclusions of the performance of the architecture and the benefits of this specific case are extracted.

Fig. 6 shows the elements composing the system divided into the different layers stated in Fig. 5

## 5.1 DATA GENERATION

The first step toward the enhancement of a building with smart features is the monitoring of all the necessary elements inside it. In the case of a building, important elements are lights, HVAC systems, computers, doors, and windows. Moreover, the environment needs to also be monitored to know whether we can take advantage of it. For instance, lights can be turned *off* if outdoor luminosity is high enough for indoor working.

Some of the aforementioned elements are endowed with small sensors capable of acquiring the necessary data to deduce their state. In the case of lights, HVAC systems and computers, potentiometers are used to read the amount of energy being consumed and thus, know their state. Alternatively, doors

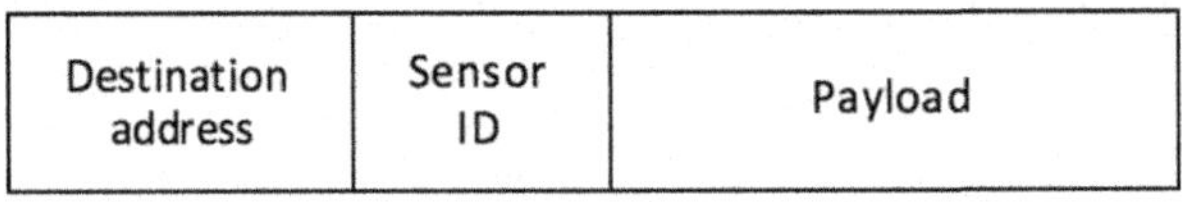

| Destination address | Sensor ID | Payload |
|---|---|---|

FIG. 7

Data packet abstraction.

and windows make use of electromagnetic sensors to know if they are opened or closed. In the case of environmental data, temperature, humidity and luminosity sensors allow us to exactly read the respective values. For mimicking the rest of the elements of the system, data is generated by means of a software capable of creating the exact same packets as the physical ones.

Even though the data generation may vary from one type of sensor to another, the packet containing the data and the corresponding headers for a transmission is equal. Fig. 7 shows the abstraction of the structure of such packets. As it can be seen, it is merely formed by the destination address, the sensor identifier and the payload containing sensor readings.

Data generation rate varies depending on the device under monitoring. In the case of environmental conditions and power usage, samples are taken every 5 minutes. Regarding doors and windows, the sample rate remains the same but additionally, if a change in their state is detected, a message is also generated.

Once the sensor data is read and encapsulated in a packet with the shape seen in Fig. 7, it is transmitted to the closest gateway in each case. Gateways receive the messages by means of different protocols such as ZigBee and Bluetooth Low Energy. Since gateways are enhanced with Internet connection, raw messages are directly forwarded to a central server responsible for the data aggregation and standardization.

As it can be seen, the architecture is capable of combining the usage of real sensors with software defined sensors, which allows for easy scalability and also fast adaptation testing on other possible scenarios.

## 5.2 DATA TRANSFORMATION AND STORAGE

When the messages reach the central server, it first reads the sensor identifier to know the type of message contained inside the payload. Once the type has been detected, the message is transformed into JSON standard format with a structure of *"type":"value"* for each value. Additionally, each message contains a time stamp to know when the value has been generated.

The standardized data is then pushed using the REST API to the cloud service explained in Section 4.4. In this cloud platform, each physical sensor corresponds to a virtual sensor. That means, each physical identifier is assigned to a virtual identifier. This relationship is privately stored inside the central server in order to be able to correctly push the messages to the corresponding virtual sensor. However, virtual identifiers along with sensor information such as model, type of sensor and location is publicly available thanks to an additional cloud database that stores this data. By doing so, external entities can take advantage of the platform and query specific sensors without the necessity of deploying their own ones.

As it has been mentioned before in Section 4.4, not all the sensors registered into the system are publicly available. The necessity to privatize some sensors is directly related to the security and privacy of the users under the monitored environment. For instance, if proximity and movement sensors are publicly available, third persons would be able to know whether the room containing the sensors is empty or not, and take advantage of such information for social hacking.

In order to avoid this, the only sensors that are shared correspond to environmental monitoring such as temperature, humidity and light. For the rest of the sensors, a password is needed to receive the updated values.

## 5.3 DATA CONSUMPTION

The application developed is composed by two differentiated elements. Firstly, the BMS is responsible for directly receiving sensor information via the subscriptions performed to the different sensors inside the building. By using a lightweight publish/subscribe client, once a new message is stored in the middleware database, it is also forwarded to the application, allowing the BMS to act accordingly if necessary. However, as it has been previously said, since the deployment and testing of such scenario in a real environment is too costly, simulation has been chosen as the second element for acquiring close-to-real results of the benefits of the building enhanced with smart capabilities.

The behavior of the whole system is as follows. The architecture developed feeds the BMS with both real and software generated sensor data. Once this data reaches the application, simulated elements modify their state in order to be synchronized with the corresponding sensor. For instance, if a sensor from a specific location tells that the lights are *off*, the simulated element must also be *off*. By using this pattern, the simulator maintains the synchronism between the real and virtual sensors with the building simulated elements. Consequently, if the simulator detects that an actuation must be performed, it automatically changes the state of the element and updates all the required software defined sensors in order to maintain the synchronism.

In addition to the simulation of the building elements, people inside it are also simulated to be able to repeat the tests multiple times. People are defined by a set of actions that can be performed inside the building along with the probability over time of this actions to actually be performed. For instance, if the building under simulation corresponds to an office, people are more susceptible to performing the action *enter* during the initial morning hours. The combination of such definitions is stored as the *profile* of the user, allowing different user profiles across the simulated people.

The last feature of the simulator corresponds to the smart capabilities of the BMS. That is, the system must be able to detect whether an action that increases comfort and possibly reduces energy consumption can be carried on by looking at the state of the building at each moment. The implementation is developed by means of a rule-based system that monitors conditions corresponding to every possible actuation to activate. For instance, to know whether the light of a room can be switched *on* or *off* directly depends on the presence of people inside the room, the current indoor light state, outdoor luminosity, and windows position.

By comparing the results of a building structure enhanced by the smart features previously explained it is possible to increase the comfort of the people. Even though there exists no metric capable of quantifying the comfort of a person, it is possible to deduce that by entering in a room with the desired temperature or by not having to interact with elements such as lights or HVACs, his comfort is increased. Moreover, energy consumption of the building can be reduced thanks to the activation of rules under wasteful scenarios.

## 6 CONCLUSIONS

The potential benefits of the designed architecture after analyzing how it behaves under a close-to-real simulated scenario are:

- Hardware abstraction at the lowest possible level. Network layer gateways completely hide the protocols underneath.
- Cloud middleware allows for rapid database and server replication under demand.
- Homogenization both in terms of file format and communication protocols.
- Possibility for sensor data sharing thanks to the public database and flexibility in the visibility of sensors.

Regarding the scalability of the system, the simulations show that under a standard office building, the number of sensors is no issue and does not affect on the performance of the system. Reliability is also another important factor to take into account. Even though the cloud middleware can be seen as a central attack point that could hamper the functioning of the system, the replicability offered substantially increases the reliability.

The testing of the platform by means of simulations shows that the platform is viable for a physical deployment. Moreover, the level of abstraction eases the development of the application by allowing the developers to fully focus their efforts in it, without worrying about heterogeneity of protocols and formats.

The future of the IoT will be directly related to the design of architectures aiming for a homogeneous Internet and the usage of abstraction middleware layers capable of interconnecting many solutions for the creation of more sophisticated systems. Additionally, the placement of such middlewares inside the Internet is gaining popularity due to the possibility to access data from everywhere and the integration of the *things* with the *Internet*, enabling the creating of multilocated systems.

## REFERENCES

Al-Fuqaha, A., Guizani, M., Mohammadi, M., Aledhari, M., Ayyash, M., 2015. Internet of Things: a survey on enabling technologies, protocols, and applications. IEEE Commun. Surv. Tutorials 17 (4), 2347–2376. doi: 10.1109/COMST.2015.2444095.

Bauer, M., Boussard, M., Bui, N., Carrez, F., 2013. Project Deliverable D1.5—Final Architectural Reference Model for IoT, 53–59. http://www.iot-a.eu.

Belshe, M., Peon, R., Thompson, M., 2015, May. Hypertext Transfer Protocol Version 2 (HTTP/2). RFC 7540. Internet Engineering Task Force (IETF). https://tools.ietf.org/html/rfc7540.

Barcelona Supercomputing Center, 2017. http://www.bsc.es.

CoAP, 2014. CoAP RFC 7252 Constrained Application Protocol. http://coap.technology.

COMPOSE, 2015. COMPOSE: Collaborative Open Market to Place Objects at your Service. http://www.compose-project.eu.

Fielding, R., Gettys, J., Mogul, J., Frystyk, H., Berners-Lee, T., 1997, January. Hypertext transfer protocol—http/1.1. RFC 7540. Network Working Group. https://tools.ietf.org/html/rfc2068.

Fielding, R., Gettys, J., Mogul, J., Frystyk, H., Leach, P., Berners-Lee, T., 1999, June. Hypertext transfer protocol—http/1.1. RFC 7540. Network Working Group. https://tools.ietf.org/html/rfc2616.

Frank, R., Bronzi, W., Castignani, G., Engel, T., 2014, April. Bluetooth low energy: an alternative technology for VANET applications. In: 2014 11th Annual Conference on Wireless On-demand Network Systems and Services (WONS), pp. 104–107.

ISO, 2016. ISO/IEC 20922:2016. Information technology: Message Queuing Telemetry Transport (MQTT) v3.1.1. International Organization for Standardization. https://www.iso.org/standard/69466.html.

LitePoint, 2012. Bluetooth Low Energy Whitepaper. Rev. 1.

Milić, L., Jelenković, L., 2015. A novel versatile architecture for Internet of Things. 2015 38th International Convention on Information and Communication Technology, Electronics and Microelectronics, MIPRO 2015—Proceedings, pp. 1026–1031. doi:10.1109/MIPRO.2015.7160426.

Shelby, Z., Hartke, K., Bormann, C., 2014, June. The Constrained Application Protocol (CoAP). RFC 7252. Internet Engineering Task Force (IETF). https://tools.ietf.org/html/rfc7252.

Siekkinen, M., Hiienkari, M., Nurminen, J.K., Nieminen, J., 2012, April. How low energy is Bluetooth low energy? Comparative measurements with zigbee/802.15.4. In: Wireless Communications and Networking Conference Workshops (WCNCW), 2012 IEEE, pp. 232–237.

Sornin, N., Luis, M., Eirich, T., Kramp, T., Hersent, O., 2015. LoRaWAN Specification. LoRa Alliance. Rev. 1.

Tsiatsis, V., Gluhak, A., Bauge, T., Montagut, F., Bernat, J., Bauer, M., Villalonga, C., Barnaghip, P., Krco, S., 2010. The SENSEI real world internet architecture. In: Towards the Future Internet: Emerging Trends from European Research, pp. 247–256. doi:10.3233/978-1-60750-539-6-247.

Villalba, A., Carrera, D., Pedrinaci, C., Panziera, L., 2015. ServIoTicy and iServe: A Scalable Platform for Mining the IoT. Proc. Comput. Sci. 52, 1022–1027. https://doi.org/10.1016/j.procs.2015.05.097.

Wi-fi, 2016. Wi-Fi alliance introduces low power, long range Wi-Fi HaLow. https://www.wi-fi.org/news-events/newsroom/wi-fi-alliance-introduces-low-power-long-range-wi-fi-halow.

Wu, M., Lu, T.J., Ling, F.Y., Sun, J., Du, H.Y., 2010. Research on the architecture of Internet of Things. ICACTE 2010, 2010 3rd International Conference on Advanced Computer Theory and Engineering, Proceedings, vol. 5, 484–487. doi:10.1109/ICACTE.2010.5579493.

CHAPTER

# MONITORING DATA SECURITY IN THE CLOUD: A SECURITY SLA-BASED APPROACH

11

**Valentina Casola***, **Alessandra De Benedictis***, **Massimiliano Rak**[†], **Umberto Villano**[‡]

*University of Naples Federico II, Naples, Italy** *University of Campania "Luigi Vanvitelli", Aversa, Italy*[†]
*University of Sannio, Benevento, Italy*[‡]

## 1 INTRODUCTION

Thanks to the wide diffusion of the *cloud computing* paradigm, more and more organizations and individual customers are relying upon cloud services to carry out their business. Unfortunately, Cloud Service Customers (CSCs) do not have full control over cloud infrastructures, and so a CSC has no possibility to manage and to react to intrusions and to traditional attacks targeting availability, confidentiality and integrity of cloud resources and services. This lack of control over leased resources, which implies that customers perceive the services as *less secure*, is considered one of the main inhibitors to a wider cloud paradigm adoption (Pearson and Benameur, 2010; Pearson, 2013; Sengupta et al., 2011).

Recently, Security Service Level Agreements (SLAs) have been proposed as a solution to the cloud security problem, as outlined by ENISA, and explored by research projects like SPECS (2013), SLA-Ready (2015), and SLALOM (2015). Security SLAs are contracts regulating the conditions under which target services are to be delivered to customers, and include security-related terms and guarantees that specify the level of security that services have to guarantee. According to such agreements, security requirements, like the ones regarding the protection from intrusions and the correct vulnerability assessment, may be granted by a Cloud Service Provider (CSP) and regulated through dedicated Service Level Objectives (SLOs) that have associated *security metrics* to be used for monitoring purposes.

One of the main issues to solve in order to implement such an approach is related to security (continuous) monitoring: cloud customers need security metrics to measure in a quantitative way the level of security and need effective tools to verify that SLOs are being respected. The security-related monitoring tools for cloud environments that have appeared recently are primarily intended to monitor a whole infrastructure, and not *per-user* or *per-service* defined agreements. In other words, SLA-based cloud security monitoring services and tools are not yet available to customers for monitoring their SLAs (Petcu, 2014). The design of such services and tools strictly depends on an accurate analysis of available security mechanisms and controls, aimed at identifying what can be actually negotiated with the customer and automatically enforced, and what are the related monitorable metrics and the policies to configure to detect SLA violations or to raise alerts.

In order to demonstrate that the use of Security SLAs is a viable solution to cloud data security, we will exploit the SPECS Framework, which enables the development of secure cloud services through

Security and Resilience in Intelligent Data-Centric Systems and Communication Networks. https://doi.org/10.1016/B978-0-12-811373-8.00011-2
Copyright © 2018 Elsevier Inc. All rights reserved.

an SLA-based approach. The SPECS project[1] was intended to improve the state-of-the-art in cloud computing security by creating, promoting, and exploiting a platform devoted to offering Security-as-a-Service using an SLA-based approach. The SPECS framework makes it possible to enhance the offerings of existing providers by means of the activation, through suitable components, of security mechanisms and controls that can be negotiated by customers, automatically enforced through an enriched supply chain,[2] and continuously monitored according to a signed Security SLA.

In this chapter, we will focus on two classes of security issues: denial of service detection, and mitigation and vulnerability assessment. They involve completely different technologies, but are among the most important user security requirements to reply to questions like: is my software and data protected if a malicious attacker aims at breaking it? Is my software protected against well-known vulnerabilities in the hardware/software stack I am using? Thanks to the SPECS framework, we propose a monitoring architecture that is automatically configured and activated, based on a signed Security SLA. Such monitoring architecture integrates different security-related monitoring tools (either developed *ad-hoc* or already available as open-source or commercial products) to collect measurements related to specific metrics associated with the set of security SLOs that have been specified in the Security SLA.

The remainder of this chapter is organized as follows. The next section briefly summarizes the state-of-the-art of security monitoring and IDS solutions, while Section 3 focuses on the problem of SLA-based monitoring. Section 4 introduces the SPECS framework, and Section 5 illustrates the techniques adopted to automatically enforce and monitor denial of service detection and mitigation and vulnerability assessment. Section 6 applies the above techniques and tools in order to build up a secure cloud service. The chapter ends with a section devoted to summarizing our conclusions.

## 2 CLOUD SECURITY MONITORING

Cloud monitoring typically involves dynamically tracking the Quality of Service (QoS) parameters related to virtualized resources (e.g., virtual machines, storage, network, and appliances), the physical resources they share, the applications running on them, and the hosted data. The continuous monitoring of the cloud and of its SLAs, mostly expressed in terms of performance-related guarantees, is of paramount importance for both cloud providers and customers. As for providers, in particular, they aim at preventing SLA violations to avoid penalties, and at ensuring an efficient resource utilization to reduce costly maintenance.

While several tools exist for performance and QoS monitoring in cloud environments, both open source and commercial security-related monitoring tools are less diffused. When talking about security monitoring, many questions and open issues should be addressed. In particular, any monitoring solution should cope with the following questions: (1) *What to monitor?* Physical resources? Physical infrastructures? Or even virtual machines and related software assets? (2) *Where are the monitoring agents?* Many options are configurable in the cloud (monitoring on-premises, monitoring on hosting IaaS, monitoring via SaaS or via other third parties), so which is the configuration that best fits the

[1] www.specs-project.eu.

[2] A supply chain consists of the set of components invoked to provide the negotiated service along with their configurations.

signed SLA? and (3) *What data should be monitored?* How to manage the huge amount of data? Last, but not least, (4) *Which security metrics should be used?*

Security monitoring of cloud infrastructures and services is based on the generation, collection, analysis, and reporting of security-relevant data. To this aim, the so-called *probes* can be installed on the infrastructures themselves and can be adopted to gather information on users, applications, and systems activities, as well as on the on-line exchanged traffic. Security monitoring is fundamental to enable threat detection and allow for the identification and enforcement of proper countermeasures to avoid system compromise, to verify if the set security controls are properly working, and if any bug or vulnerability is exposed, and to provide legal evidence in case of a security incident.

Cloud monitoring has been extensively studied in the recent literature, and several monitoring solutions are currently offered by big cloud providers (refer to Aceto et al. (2013) and Fatema et al. (2014) for a survey on cloud monitoring and monitoring tools). Most of the current cloud monitoring tools are focused on specific aspects of cloud operation, providing only a partial solution for the cloud monitoring problem. For example, the open source tool Nagios[3] offers complete monitoring and alerting for servers, switches, applications, and services, while Ganglia[4] is a scalable distributed monitoring system for high-performance computing systems such as clusters, and Grids, which collects dozens of system metrics related to CPU, memory, disk, network, and process data.

With regard to security monitoring, a number of commercial and open source products and of research prototypes exist that are mainly devoted to the detection and management of known software vulnerabilities, to the tracking of user activities and system changes, and to the detection of malicious behavior. Among the monitoring solutions proposed by big providers, Amazon provides the CloudWatch[5] service for AWS resource monitoring, which can be invoked to obtain information related to the status of the system, such as resource usage and application performance. In addition to this, Amazon has also set up a vulnerability reporting process to notify possible vulnerabilities found in its services, and releases periodic security bulletins to communicate security and privacy-related events to customers. Moreover, Amazon has established a policy for the execution of customers' penetration testing procedures. Google also performs security monitoring on its network infrastructure by exploiting both commercial and open-source tools in a complex integrated system. Specific analyses are carried out to detect suspicious activities perpetrated by employees, and the security management team continuously checks security bulletins to identify security incidents that can negatively impact Google services.

Many existing security monitoring tools perform an analysis of the vulnerabilities exposed by a system. Among these, Nessus,[6] a commercial product from Tenable Network Security Inc., promises a fast asset scanning of vulnerabilities. It supports the coverage and profiling of a broad set of network devices, virtualization resources, operating systems, databases, and Web applications, and is also available as an AWS AMI. The Open Vulnerability Scanner Assessment System (OpenVAS)[7] is an open source project derived from the Nessus product that provides a framework for the monitoring and management of vulnerabilities as well. It is characterized by a modular architecture and defines

[3]Nagios—The Industry Standard In IT Infrastructure Monitoring, http://www.nagios.org/.

[4]Ganglia Monitoring System, http://ganglia.sourceforge.net/.

[5]http://aws.amazon.com/cloudwatch/.

[6]http://www.tenable.com/products/nessus-vulnerability-scanner.

[7]http://www.openvas.org/.

protocols suitable for SSL-protected communications among internal components. NeXpose,[8] released by Rapid7 Inc. in several different versions including a free community edition, is a solution to manage vulnerabilities at several levels (from the single user to the organization). It supports automatic detection, scanning, and remediation in virtualized environments. Metasploit[9] is a popular open source penetration testing solution, also offered by Rapid7, that can be used to validate vulnerabilities discovered by tools like NeXpose and enable related remediation actions.

There exist other tools that provide Security Information and Event Management (SIEM) solutions. SIEM technologies provide real-time analysis of security alerts generated by network hardware and applications. SIEM products are able to gather, analyze, and present information from network and security devices, identity and access management applications, vulnerability management and policy compliance tools, operating systems, databases and application logs and external threat data. NetIQ Security Manager[10] is a SIEM solution that provides host-based security: it supports the monitoring of security-related activities, the collection of logs, the management of threats, the response to incidents and the detection of changes in the system.

There is a relatively large literature on the use of IDS in clouds to prevent DoS and DDoS attacks coming from the outside and targeting resources in the cloud (Roschke et al., 2009; Dhage et al., 2011; Arshad et al., 2011; Gul and Hussain, 2011; Modi et al., 2012). An interesting solution of IDS working at hypervisor level, which does not require additional software to be installed in virtual machines, is presented in Nikolai and Wang (2014). Good surveys on these topics are Mehmood et al. (2013) and Roy et al. (2015). There is also on-going, yet unpublished, work on the reverse problem, such as on the use of cloud resources to perform DoS attacks externally to the cloud.

Despite the number of tools currently available for cloud monitoring and security monitoring, to the best of our knowledge not so much work has been done in the area of managing security requirements specified through SLA documents. The listed tools and solutions, although allowing for the tuning of the monitoring policy based on user-selected metrics, are not directly related to Security SLAs nor to the security SLOs they specify. As pointed out by Petcu (2014), SLA-based cloud security monitoring services and tools are not yet available. Several research projects are recently addressing this issue, including the above-mentioned SPECS, and a few papers on the topic are currently available.

Ullah and Ahmed (2014) proposed a Security SLA management solution that automatically deploys services, represented by virtual machines (VMs). These services are provided with different security levels depending on the security requirements negotiated by the customer and included in a Security SLA. While launching the VM, a custom monitoring agent is also installed onto the VM for its monitoring. The different security levels are obtained by deploying different predefined system configurations at the physical and application level; the related monitoring systems and monitoring policies are predefined as well. Karjoth et al. (2006) introduce the concept of Service-Oriented Assurance (SOAS). SOAS adds security providing assurances (an assurance is a statement about the properties of a component or service) as part of the SLA negotiation process.

Smith et al. (2007) present a WS-Agreement (Andrieux et al., 2007) approach to a fine-grained security configuration mechanism that allows an optimization of application performance based on specific security requirements. They present an approach to optimize Grid application performance

[8]https://www.rapid7.com/products/nexpose/compare-downloads.jsp.
[9]https://www.rapid7.com/products/metasploit/index.jsp.
[10]https://www.netiq.com/products/sentinel/.

by tuning service and job security settings based on user supplied WS-Agreement specification. Brandic et al. (2008) present advanced QoS methods for metanegotiations and SLA-mappings in Grid workflows. the prerequisites to be satisfied for negotiation, the supported negotiation protocols and document languages for the specification of SLAs. In the prerequisites there is the element < *security* >, which specifies the authentication and authorization mechanisms that the party wants to apply before starting the negotiation.

Ficco et al. (2012a,b, 2013) proposed architectures to detect intrusions on cloud services with a focus on cloud-specific attacks, and considered the possibility to use SLAs in the IDS service offering. Examples of SLA-oriented monitoring tools are represented by CloudComPaaS,[11] LoM2HiS (Emeakaroha et al., 2010) and CASViD (Emeakaroha et al., 2012) or the solution proposed in mOSAIC (Rak et al., 2011). CloudComPaaS is an SLA-aware PaaS for managing a complete resource lifecycle, and features an extension of the WS-Agreement SLA specification for cloud computing. The monitor module performs the dynamic assessment of the QoS rules from active SLAs. The three basic operations of the monitor are: updating the SLA terms state, checking the guarantees state, and performing self-management operations. SLAs registered in the monitor are set to be continuously updated after a given period of time, commonly defined as monitoring cycle. The monitor evaluates the formulas of the guarantee terms and sets the value of the guarantees to either Fulfilled or Violated.

CASViD (cloud application SLA violation detection, Emeakaroha et al., 2012) aims at monitoring and detecting SLA violations at the application layer, and includes tools for resource allocation, scheduling, and deployment. It is an SNMP-based monitoring approach for SLA violation. Service requests are placed through a defined interface to the front-end node, acting as the management node. The VM configurator sets up the cloud environment by deploying preconfigured VM images. The request is received by the service interface and delivered to the SLA management framework for validation, then it is passed to the application deployer for resource allocation and deployment. CASViD monitors the application and sends information to the SLA management framework for detection of SLA violations.

## 3 SLA-BASED SECURITY MONITORING

The first challenge to face to enable SLA-based monitoring is to provide a mapping between the end-users' security requirements and the specific security SLOs reported in an SLA, related to measurable security metrics. For instance, let us consider the *availability* of high-level security requirement, related to a cloud application. The application is actually running on physical or virtual resources, which can be characterized by low-level metrics such as CPU, memory, uptime, and downtime, which are those actually measurable. Thus there is a gap between the low-level metrics and the high-level SLA parameters that can be negotiated by an end-user.

According to ENISA (Dekker and Hogben, 2011), the security parameters for a security monitoring framework can be classified as in Fig. 1. For each parameter, the monitoring and testing methodology have to be defined, as well as the related thresholds to trigger events (e.g., incident reports or response and remediation). In terms of security requirements, the monitoring tests are quite complex. One of the reasons is the restricted access to the monitoring data, represented in Fig. 2.

[11]GRyCAP CloudComPaaS, http://www.grycap.upv.es/compaas/about.html.

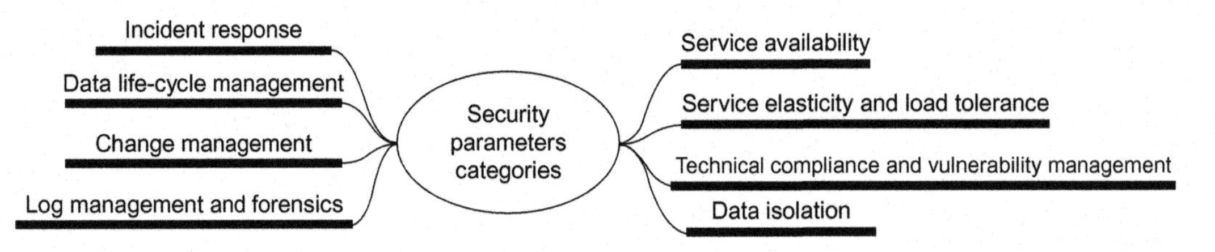


**FIG. 1**

ENISA security parameters.

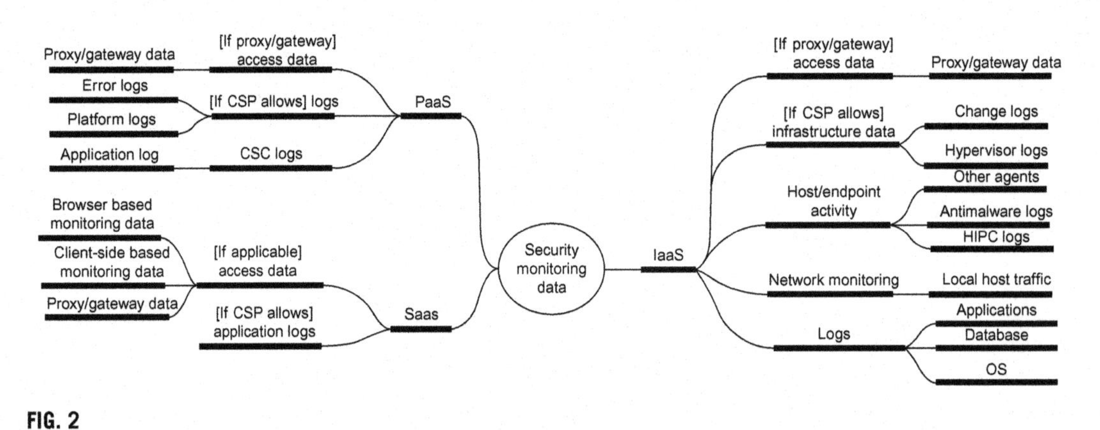


**FIG. 2**

Security monitoring data.

Different key issues arise when choosing and configuring a suitable SLA monitoring system. Indeed, once the security parameters to be monitored have been defined, it is necessary to determine appropriate monitoring agents and the data collection architectures, considering the balance between the early detection of possible SLA violations and the intrusiveness of the monitoring tools on the whole system. In addition to these aspects, the rate of acquiring information about the usage of the resources and the current resource availability status is also an important factor, influencing the overall performance of the system and the profit of the provider. On the one hand, monitoring at a high rate delivers fast updates about the resource status to the provider, but it can lead to high overhead, which possibly degrades the performance of the system. On the other hand, monitoring at a low rate may cause a miss of information: for instance, it may happen to miss an SLA violation detection, which implies the payment of penalties by the provider. Therefore to address this issue, techniques to determine the optimal measurement intervals to efficiently monitor to detect SLA violations are required.

Another key issue related to the selection of the parameters to monitor is the monitoring granularity. Three main options are possible: client-oriented monitoring, virtual system monitoring, and physical system monitoring. Finally, another related issue is the approach adopted to gather monitoring data. Again, three options are possible: use suitable APIs offered by the public cloud providers themselves to collect logs, install custom monitoring agents on the monitored infrastructure, or use third-party tools able to gather information on the services under monitoring from the outside.

The adoption of the best configuration of monitoring systems to activate should be automatically related to the security parameters included in the SLA, and it should affect both infrastructures that host many user virtual resources (multitenancy) and user-specific resources to protect. Indeed, the above mentioned SPECS project tried to cope with this problem by defining, during the SLA enforcement phase, the number and typology of monitoring services to activate based on the negotiated specific security features and controls.

In particular, as we will discuss in the next section, SPECS provides a flexible architecture that manages the whole SLA lifecycle and exploits cloud automation technologies to automatically deploy, configure, and activate a set of available security services and monitoring systems and agents based on the security SLOs specified in a Security SLA. To enable the automatic launch and configuration of such monitoring systems according to security SLOs, a clear mapping is needed between these SLOs and a set of measurable metrics. Producing such a mapping is critical and requires a deep analysis from a security expert. The mapping can be done with the adoption of Security SLA templates, which include all available security and monitoring systems. The SLA template is built according to a novel Security SLA model, proposed in Casola et al. (2016).

In this chapter, we assume the SLA as an input to our solution and focus on the automation of the monitoring task. In particular, we aim at designing the solution reported in Fig. 3. During a negotiation process, the user submits a set of desired SLOs related to specific security attributes, and a Security

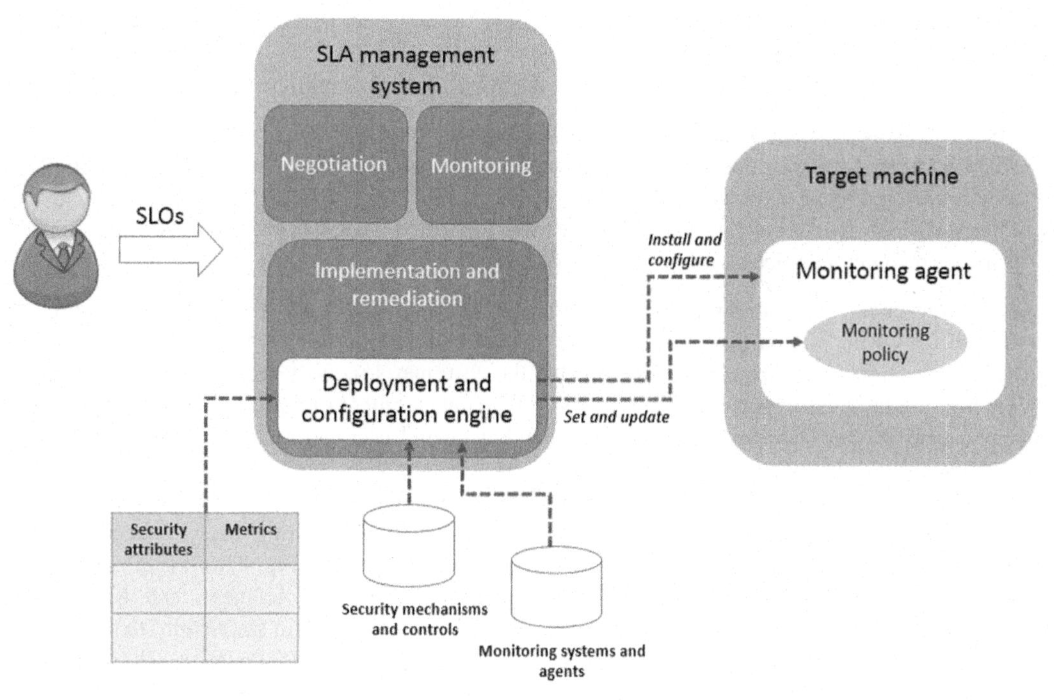


**FIG. 3**

The SLA-based security monitoring approach.

SLA is built and signed. The security requirements defined in the SLA are automatically enforced by deploying and activating a set of components implementing related security mechanisms and controls. Similarly, a set of monitoring systems able to monitor the metrics associated with the security attributes involved in the SLOs is automatically deployed and configured. In the next sections, we will present the architecture designed to achieve this goal and we will provide some details related to a case study.

## 4 THE SPECS FRAMEWORK AND THE SLA-BASED MONITORING ARCHITECTURE

The main goal of the SPECS project is the development of a framework for the management of the SLA lifecycle, intended to build applications (SPECS applications) offering services whose security features are stated in and granted by a Security SLA (Casola et al., 2014; Rak et al., 2013).

The SPECS framework addresses both CSPs' and CSCs' needs by providing techniques and tools for (1) enabling user-centric negotiation of security parameters in a cloud SLA, by providing a trade-off evaluation process among customers and CSPs, in order to compose cloud services guaranteeing a minimum security level; (2) monitoring in real-time the fulfillment of the SLA agreed upon, notifying both CSCs and CSPs when an SLA is not being fulfilled, and (3) enforcing the agreed SLAs in order to keep a sustained Quality of Security (QoSec) that fulfills the specified security parameters. The SPECS enforcement framework is also able to "react and adapt" in real-time to fluctuations in the QoSec, by advising and/or applying suitable countermeasures.

In the typical SPECS usage scenario there are three main involved parties:

- **SPECS Customer**: the end-user, such as the CSC of the cloud services covered by Security SLAs;
- **SPECS Owner**: the provider of the SPECS security services to provide Security SLAs to cloud services offered by an External CSP;
- **External CSP**: an independent (typically public) CSP, which is unaware of the SLAs, and provides resources/services without security guarantees.

As for the interactions among such parties, the SPECS Customer uses the cloud services offered by the SPECS Owner, which mainly acts as a broker by acquiring resources from External CSPs and by reconfiguring/enriching them in order to match the customer's security requirements. The SPECS Owner may possibly be supported by a *Developer* in the development of new SPECS applications and of new security mechanisms, which can be negotiated by an end-user and used to add security features to the delivered cloud services.

The SPECS framework makes it possible to easily enrich an existing cloud service with Security SLAs, by re-using a set of available security mechanisms and by exploiting a set of services (*Core services*) devoted to the management of the SLA lifecycle. As shown in Fig. 4, a SPECS application orchestrates the SPECS Core services dedicated to Negotiation, Enforcement, and Monitoring, respectively, to provide the desired service (referred to as "Target Service" in the picture) to the SPECS Customer (i.e., to the end-user). The Core services run on top of the SPECS Platform, which provides all the functionality related to the management of Security SLAs' lifecycle and needed to enable the communication among Core modules. In addition to this functionality, provided by the "SLA Platform services," the SPECS Platform also provides support for developing, deploying, running, and managing

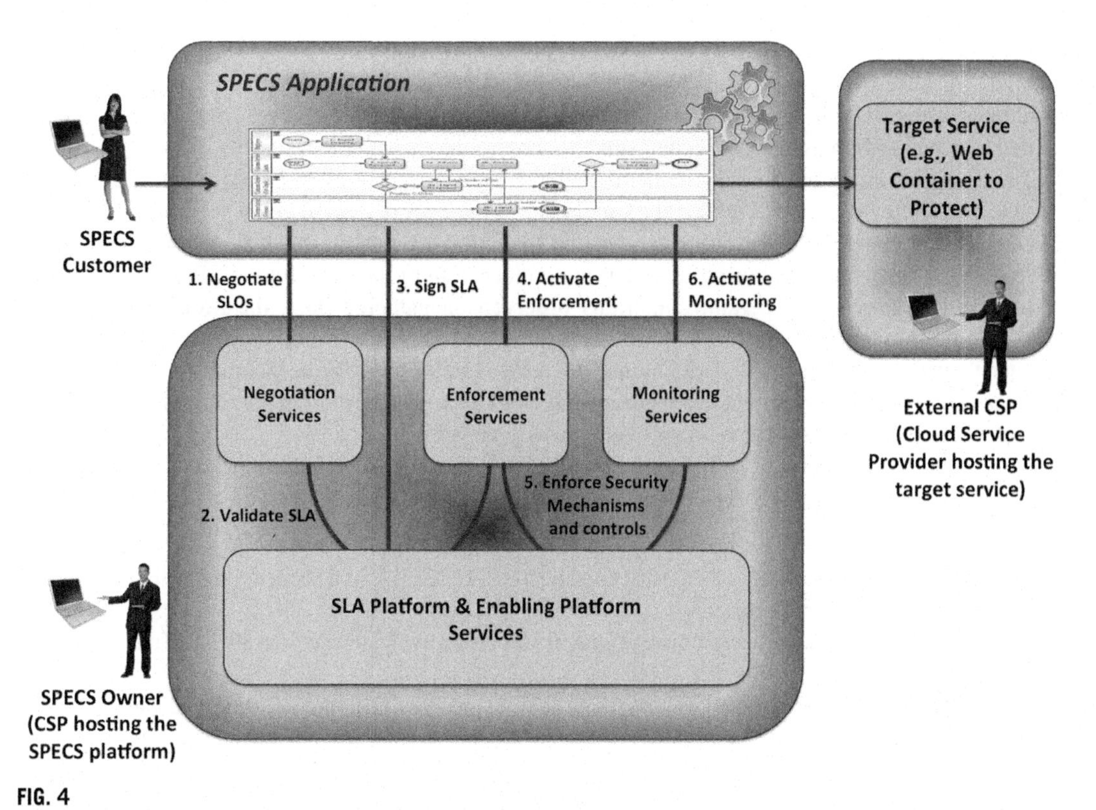


**FIG. 4**

Overview of SPECS.

all SPECS services and related components. In the figure, these services are referred to as "Enabling Platform services,"

Security-related SLOs are negotiated *(Step 1)* based on the SPECS Customer's requirements. A set of compliant offers, each representing a different supply chain to be implemented, is identified with the help of an interoperability layer (represented by the SPECS SLA Platform services), which is also responsible for their validation (e.g., for verifying their actual feasibility based on the current system configuration) *(Step 2)*. Of course, given a set of security requirements expressed by the SPECS Customer, multiple supply chains may be identified, each characterized by its own cost and associated security level. The resulting supply chains may be ranked to help the SPECS Customer choose the desired configuration. The agreed terms are included in a Security SLA that is signed by the SPECS Customer and the SPECS Owner *(Step 3)*. Afterward, the agreement is implemented through the Enforcement services, which acquire resources from external CSPs and activate suitable components that provide, in an *as-a-service* fashion, the security capabilities needed to fulfill the SLOs included in the signed Security SLA *(Steps 4 and 5)*. At the same time, suitable services and agents are activated for the monitoring of the specific parameters included in the Security SLA *(Step 6)*. Monitoring data are collected by the SPECS Monitoring module and analyzed based on a monitoring policy: if needed, they are forwarded to the Enforcement module, which performs a diagnosis to verify whether they reveal an incoming (or already occurred) violation of the signed SLA. As a consequence, countermeasures may be adopted, consisting in re-configuring the service being delivered, or applying remediation actions defined together with the security mechanisms.

The activation of monitoring components includes the start-up of proper services and agents that are able to monitor the specific parameters included in the Security SLA. Such services and agents, which may be represented by existing monitoring tools that are simply invoked or integrated within the framework, generate monitoring data that are collected and processed by the SPECS Monitoring module. In the presence of certain conditions observed on gathered data, the Monitoring module may generate monitoring events, which are further processed by the Enforcement module to verify whether they reveal a violation of the signed SLA or indicate a possible incoming violation. As a consequence, if present, proper countermeasures may be adopted consisting in reconfiguring the service being delivered, or the needed remediation actions may be performed.

The Security SLA format defined in the SPECS project (De Benedictis et al., 2015; Casola et al., 2016), based on the WS-Agreement standard, represents security features using the following concepts:

- **security capabilities**: the set of security controls (NIST, 2013) that a security mechanism is able to enforce over the target service;
- **security metrics**: the standard of measurement adopted to evaluate security levels of the services offered;
- **SLOs**: the conditions, expressed over security metrics, representing the security levels that must be respected according to the SLA.

Security capabilities declared in the SLA are enforced *as-a-service* thanks to the activation of related software security mechanisms. Similarly, reported SLOs are automatically put under observation thanks to the installation and configuration of suitable monitoring systems, also implemented in the form of security mechanisms. As will be made clear later, the automatic activation and configuration of security mechanisms is enabled by the Chef[12] cloud automation technology. The information needed to automate a security mechanism's deployment and configuration based on the content of a Security SLA is included in its *metadata*, defined by the mechanism developer. In light of the above, the development of SPECS security mechanisms mainly consists in the development of suitable Chef *cookbooks* (more details on this will be given later in this chapter), possibly obtained by adapting existing security software, and in the definition of the associated metadata.

As mentioned before, the goal of this chapter is to present a monitoring solution based on existing security-related monitoring tools, which can be automatically configured and activated according to a signed Security SLA to carry out a continuous monitoring of the security guarantees there defined. Such monitoring solution has been integrated within the SPECS framework, in order to offer a complete Security-as-a-Service solution by which cloud services can be delivered with specific security guarantees. In particular, all the guarantees depend on the Security SLA signed, and are used to configure in an automatic way the monitoring tools and agents and the monitoring policy, in order to continuously evaluate and measure the involved security metrics. In the next subsections, we will

[12]https://www.chef.io/chef/.

illustrate in detail the monitoring architecture and its automatic enforcement and configuration through the SPECS framework.

## 4.1 THE SPECS MONITORING ARCHITECTURE

Fig. 5 shows the high-level architecture of our monitoring solution, identifying the main involved components with their relationships in terms of provided and required interfaces, and specifying their deployment in the system.

Since we refer to the SPECS framework and monitoring services, we explicitly reported the set of components belonging to the SPECS framework that are involved in the monitoring task. In particular, we reported the components needed to carry out the automatic configuration and deployment tasks (SPECS Enforcement module), the processing of monitoring events and the diagnosis and remediation activities, in addition to the component of the SLA Platform module that is responsible for the management of SLAs.

For what regards the target machine to be monitored, we identified the components devoted to performing the actual monitoring, namely the monitoring agents, and those needed to properly activate and configure the monitoring agents.

In particular, in the SPECS framework we identified two main packages devoted to Enforcement and Monitoring respectively and the SLA Platform (cf. Section 4). The related components can be summarized as follows:

- the *Deployment and Configuration Manager*, belonging to the SPECS Enforcement module, is responsible for automating the installation, execution and configuration/reconfiguration, on target machines, of the security services and of the components devoted to automating the management of the deployed monitoring systems. This is done by processing the Security SLAs stored by the SLA Platform, from which the Enforcement module extracts the security metrics and the list of services to activate;
- the *Event Aggregator and Filter*, belonging to the SPECS Monitoring module, collects monitoring events generated by the deployed monitoring systems and performs aggregation and filtering operations before sending them to the Enforcement module for diagnosis purposes;
- the *Diagnosis and Remediation* component, belonging to the SPECS Enforcement module, is in charge of analyzing monitoring events generated by the Monitoring module to perform a diagnosis, in order to verify if they represent an alert or a violation related to an SLA and to identify proper countermeasures or remediation activities accordingly;
- the *SLA Manager* component, belonging to the SPECS SLA Platform, provides the basic functionalities in order to manage the entire lifecycle of SLAs negotiated and signed with End-users (SLA API);
- the *Service Manager* component, belonging to the SPECS SLA Platform, provides the Service API to query for registered services and to instantiate and to stop them.

On the target machine side, the *Deployment and Configuration Agent* is in charge of the deployment and configuration of local components according to the plans established by the *Deployment and Configuration Manager*. These include the components belonging to the Monitoring System package, namely the *Monitoring Manager* and the *Monitoring Agent*. The *Monitoring Manager* manages the *Monitoring Agent* installed on the target machine, which actually collects relevant data through its

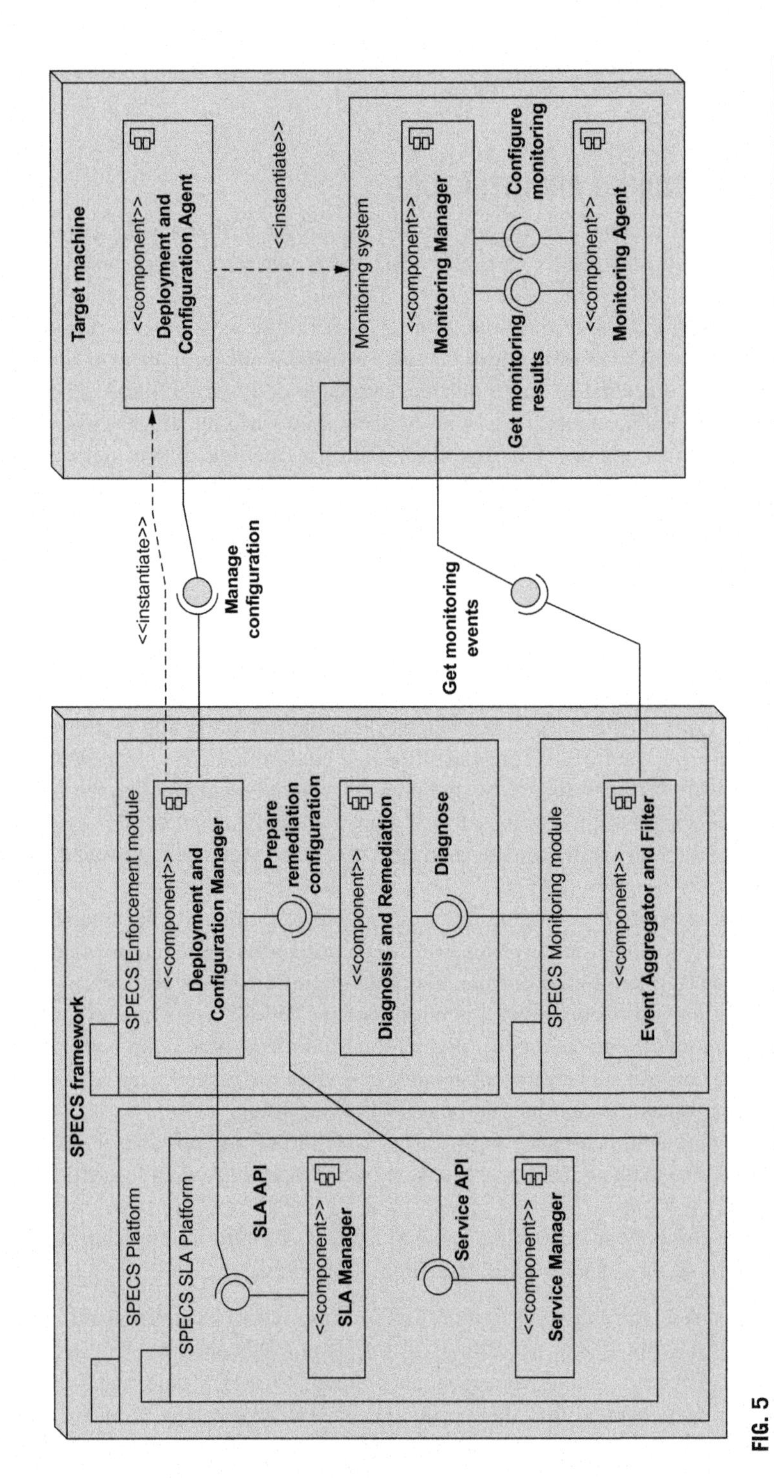


**FIG. 5**

The high-level monitoring architecture.

probes, by taking care of its configuration and tuning, and is also responsible for collecting the monitoring results and making them available to the SPECS Monitoring module for further processing, as discussed above.

It is worth mentioning that the discussed deployment is an optimized configuration of the system taking into account different security and resource constraints as discussed in Casola et al. (2016), but different deployment configurations are possible. Indeed, the *Monitoring Manager* may be installed on a separate machine rather than on the monitored target machine, to control remotely the *Monitoring Agent*. Clearly, in this case a specific *Deployment and Configuration Agent* must be installed and configured also on this machine. We will show such a solution in the next section, where we will specialize the presented architecture for a case study based on a vulnerability monitoring system. Finally, it should be noted that sometimes it may be not possible to install directly a probe on the target machine; in these cases, both the *Monitoring Manager* and the *Agent* may be deployed on an external machine.

#### 4.1.1 The enabling cloud automation technology: Chef

As previously mentioned, the proposed monitoring solution requires the automatic installation, configuration and execution on the target machines of specific software components devoted to the monitoring management. For this reason, we adopted a configuration management solution that enables us to integrate several monitoring tools within the SPECS framework by supporting all needed automatic configuration functionalities. Several configuration management solutions are currently available (i.e., CFEngine,[13] Puppet,[14] or Chef[15]). We chose Chef for its openness, high scalability and power.

Chef is a framework for the automation of cloud systems and infrastructures that simplifies the deployment of servers and applications on every physical, virtual, or cloud resource. The key concept of Chef is the "translation of infrastructure into code," which implies that the development environment is versionable, testable and repeatable, as the code of an application. Chef uses *recipes* to specify the infrastructure and the related configuration tasks. Recipes use building blocks called resources, representing pieces of an infrastructure, included in a *cookbook*, which is the fundamental unit for the configuration and deployment of content on a target machine, referred to as *node*.

Chef cookbooks, as well as other configuration data, are stored in the *Chef server*, which is the main component of the Chef architecture and may be either installed on local machines or invoked as a remote SaaS service. *Chef clients* are installed on the network nodes.[16] They periodically poll the *Chef server* for the latest recipes and check if the node is in compliance with the policy defined by these recipes. If the node is out of date, the Chef client runs them on the node to bring it up to date. Finally, the *Chef workstation* allows to communicate with the *Chef server* and to execute all operations needed to configure and execute Chef components. For example, it allows for writing new cookbooks and uploading them to the server. The communication between the *Chef workstation* and the *Chef server* is performed to the *knife* command line tool, which offers management functionalities for nodes, recipes and cookbooks, roles, and cloud resources. Different *recipes* are available in SPECS to provide different security controls and related monitoring systems.

[13]http://cfengine.com.
[14]https://puppetlabs.com/.
[15]https://www.chef.io/chef/.
[16]A node can be a physical server, a virtual server or a container instance.

# 5 A COMPLEX MONITORING SYSTEM FOR DoS DETECTION AND VULNERABILITY ASSESSMENT

As mentioned in Section 4, SPECS offers security capabilities that can be negotiated by customers and enforced automatically during the SLA implementation phase. Monitoring systems have been developed to monitor the metrics associated with these security capabilities: they are configured and deployed as illustrated in the previous section, and communicate with the SPECS Monitoring module to enable the detection of alerts and violations.

Some of these monitoring systems are offered as security capabilities, thus enabling the set-up of an SLA-based monitoring solution. In the following, we will present two monitoring capabilities that, as discussed in Section 2, have many applications in current cloud monitoring tools, namely *vulnerability scanning and management* and *denial of service detection and mitigation*. For both capabilities, we will provide the set-up details, including the identification of covered security controls and the selection of relevant security metrics used during negotiation. Moreover, we will discuss the technologies used to implement the monitoring mechanisms and we will describe the resulting monitoring architecture.

## 5.1 DoS DETECTION AND MITIGATION

In this section, we present the *DoS detection and mitigation* security capability, offered by SPECS to protect a generic Web server (hosted on a cloud infrastructure) against DoS attacks. In particular, the capability is intended to be applied whenever it is required to detect and thwart unauthorized and/or anomalous access attempts to a Web server hosting the End-user's Web applications.

As prescribed by the SPECS framework, the *DoS detection and mitigation* capability can be defined as a set of security controls. In particular, the following controls, belonging to the NIST Security Control Framework (NIST, 2013), have been identified:

- `SECURITY ASSESSMENT AND AUTHORIZATION | CONTINUOUS MONITORING`—(CA-7): The organization develops a continuous monitoring strategy and implements a continuous monitoring program, which includes the selection of proper metrics, the definition of a monitoring frequency, the establishment of response actions to address results of the analysis of security-related information and the reporting of the security status with a defined frequency.
- `INCIDENT RESPONSE | INCIDENT REPORTING`—(IR-6): The organization requires personnel to report suspected security incidents to the organizational incident response capability within organization-defined time period and reports security incident information to organization-defined authorities.
- `SYSTEM AND COMMUNICATION PROTECTION | DENIAL OF SERVICE PROTECTION` (SC-5): The information system protects against or limits the effects of denial of service attacks by employing proper security safeguards.
- `SYSTEM AND INFORMATION INTEGRITY | MALICIOUS CODE PROTECTION` (SI-3): The organization: a. Employs malicious code protection mechanisms at information system entry and exit points to detect and eradicate malicious code; b. Updates malicious code protection mechanisms whenever new releases are available in accordance with organizational configuration management policy and procedures; c. Configures malicious code protection mechanisms to perform periodic scans of the information system and real-time scans of files from external sources, to take required actions (block code, quarantine code, send alert) in response to malicious

code detection, and addresses the receipt of false positives during malicious code detection and eradication and the resulting potential impact on the availability of the information system.

- `SYSTEM AND INFORMATION INTEGRITY | INFORMATION SYSTEM MONITORING` (SI-4): The organization monitors the information system to detect attacks and indicators of potential attacks in accordance with a defined monitoring objectives and unauthorized local, network, and remote connections. It identifies unauthorized use of the information system and deploys proper monitoring tools to collect information about the system.
- `SYSTEM AND INFORMATION INTEGRITY | SOFTWARE, FIRMWARE AND INFORMATION INTEGRITY` (SI-7): The organization employs integrity verification tools to detect unauthorized changes to organization-defined software, firmware, and/or information.

The above controls represent the security features that are automatically provided by activating the *DoS detection and mitigation* capability. They can be requested by an end-user during negotiation and constitute the security declaration section of an SLA (Casola et al., 2016). However, in order to allow for the monitoring of such security features, and of the SLA in general, it is necessary to associate a set of measurable metrics to the above defined controls. The security metrics identified for the *DoS detection and mitigation* capability are the following:

- *Detection latency*: [covers IR-6] It represents the time interval between the first symptom of a (detected) attack and the generation of a message event.
- *False positives*: [covers CA-7, IR-6, SI-3, and SI-4] It reports the number of detected false positives in a predefined time interval.
- *Detected attacks*: [covers CA-7, IR-6, SC-5, SI-3, SI-4, and SI-7] It reports the number of attacks detected in a predefined time interval.
- *Attack report generation frequency*: [covers IR-6] It represents the frequency of attack report generation.

These metrics can be used to set SLOs, which define the security guarantees included in the SLA. Their values are sampled by suitable agents belonging to the security/monitoring mechanism implemented for the capability, named *DoS Protection* mechanism. As discussed in Section 2, there are multiple implementation options of this capability in a cloud. The one chosen here relies on the integration, configuration, and activation of external IDS tools. In particular, the popular OSSEC[17] tool was exploited, an open-source host-based IDS that performs log analysis, file integrity check, policy monitoring, rootkit detection, real-time alerting and active response. Some technical background on this tool and on its architecture is given in the following subsection.

### 5.1.1 OSSEC

The main goal of OSSEC is to find anomalies in the behavior of systems, which can be originated by security issues. OSSEC is able to detect anomalous activities in a system by performing log analysis: it collects, analyses, and correlates all logs recording the activity of running processes, generating alerts in case of suspicious behavior.

From an architectural point of view, OSSEC is based on a client-server paradigm. A *Manager* (the server) stores all configuration options, the databases used for integrity checking, the logs and the

[17]http://www.ossec.net/.

auditing system entries, and executes the main analysis logic. Small *Agents*, typically residing on the systems to be monitored (the clients), collect all relevant information and forward them to the Manager. However, it is possible to configure the Manager in order to act in an agent-less mode, for those systems that do not allow the installation of Agents on acquired resources.

Both the Manager and the Agents execute a set of processes in background. On every host running an Agent, a *LogCollector* collects generated logs, and an *Agentd* process compresses and encrypts them before forwarding them to the server. Here, an *Analysisd* process sends them to a filtering chain: in a predecoding phase, static variables (e.g., hostname, program name, timestamp, ...) are first extracted. In the actual decoding phase the key variables (e.g., the IP addresses that try to contact the host) are processed. The filtering is performed on the basis of specific rules or patterns: if an alarm is generated, OSSEC can provide either a passive or an active response. In the former case, an e-mail is sent to the system administrator, while in the latter it is possible to invoke specific actions specified in a script (e.g., isolation of a malicious IP).

For each rule match that occurs during the filtering process, the OSSEC Manager triggers an alert, and generates a message in the format reported in Table 1.

In the next subsection, we will illustrate how OSSEC has been used to build the *DoS Protection* mechanism, which has been integrated with the SPECS Enforcement services and which can be configured and activated based on an SLA.

**Table 1 OSSEC Alerts**

| Field | Description |
|---|---|
| crit | the level of severity associated with the alert |
| id | identifies the rule that was violated |
| component | the location of the log file that triggered the alert |
| classification | identifies the group or the groups that the violated rule belongs to |
| description | a textual description of the violated rule |
| message | the log string or group of strings that triggered the alert |
| acct | a user identifier of the machine that was found responsible of the alert (e.g., the username of the attacking machine) |
| src_ip | source IP address that was found responsible of the alert |
| src_port | source port used for perpetrating the attack |
| dst_ip | destination IP address of the machine were the alert was generated |
| dst_port | port of the agent where the alert was generated |
| file | path to the file where a change was detected |
| md5_old | MD5 hash before file modification |
| md5_new | MD5 hash after file modification |
| sha1_old | SHA1 hash before file modification |
| sha1_new | SHA1 hash after file modification |
| src_city | geographical location of the machine that caused the alert |
| dst_city | geographical location of the machine where the alert was generated |

### *5.1.2 The DoS protection mechanism*

In order to make OSSEC available on-demand as an automatically-enforceable security mechanism, we developed an OSSEC Adapter. This is a REST interface that allows to invoke and to control remotely both the server and the client components. The OSSEC Adapter acts as a gateway that interfaces with the OSSEC Manager and receives all the events that the OSSEC Manager generates. In this way, it is possible to automatically create and configure several Agents and to bind them to the Manager.

OSSEC is configured in order to monitor the metrics defined above, which can be selected and defined by a customer during negotiation. It should be noted that the listed metrics have different meanings and scopes. *False positives* and *Detected attacks* are parameters that can be derived from the logs generated by OSSEC and that give information on how much the system is being subject to attacks and on how well the detection is going on. This information can only be used to observe the system behavior, while no SLOs can be set on such metrics. For what regards the *Detection latency*, a customer may wish to set a clear SLO on such metric, but in this case the fulfillment of this SLO would not simply depend on how the monitoring system is configured, since the nature of each attack is also relevant. Finally, the *Attack report generation frequency* constitutes not only a measurable attribute (it is possible to check the timestamps of generated reports), but it can be also used to configure the *DoS Protection* mechanism.

For each rule match that occurs during the filtering process, the OSSEC Manager triggers an alert, and generates a message in the format reported in Table 1. Such a message is translated in order to be compliant with the format adopted by the *Event Hub* component, residing in the SPECS Monitoring module and devoted to collecting all event notifications coming from the deployed monitoring systems. This component is responsible for forwarding such event notifications to *Filtering* and *Aggregator* components for the actual detection and classification of the event in terms of SLA alerts or violations. The SPECS monitoring event format is shown in Table 2.

All OSSEC modules, along with the modules for communication and for the management of the alerts, have been packed to enable automatic configuration and deployment. In particular, two different packages are available, for the client and the server side, respectively. The *DoS Protection* mechanism is publicly available on the SPECS Team Repository on BitBucket (https://bitbucket.org/specs-team/specs-mechanism-monitoring-ossec), which contains the Chef cookbook for its installation and configuration. In particular, the cookbook includes the recipes for the installation of the OSSEC

**Table 2 Event Hub Event Format**

| Field | Description |
|---|---|
| component | identifies the specific OSSEC Agent that originates the message. |
| object | identifies the specific log that generated the alert. |
| labels | labels the event to enable aggregation and filtering (e.g., ossec-switch-user to identify that the event is related to an attempt to change user). The Adapter will add standard OSSEC labels to every message received from the OSSEC Manager. |
| type | identifies the type of the payload represented by the "data" field, in order to enable its parsing (e.g., syslog or JSON for OSSEC logs). |
| data | contains the alert message payload, as generated by OSSEC. |
| timestamp | contains the message timestamp. The Adapter will report the time at which the event is generated. |

Agents and of the OSSEC Server. By executing these recipes, it is possible to configure the agents on the target machines and to register them with the OSSEC Server, which typically resides on a separate machine.

## 5.2 VULNERABILITY SCANNING AND MANAGEMENT

As done for the *DoS detection and mitigation* capability, we now list the security controls identified for the *Vulnerability scanning and management* capability:

- VULNERABILITY SCANNING—(RA-5): The organization scans for vulnerabilities in the information system and hosted applications with a defined frequency and when new vulnerabilities potentially affecting the system/applications are identified and reported by adopting vulnerability scanning tools that are able to enumerate platforms, software flaws and improper configurations. The organization analyzes vulnerability scan reports and results and is able to perform a remediation in a given time period.
- VULNERABILITY SCANNING | UPDATE BY FREQUENCY—(RA-5(1)): The organization updates the information system vulnerabilities scanned with a defined frequency.
- PENETRATION TESTING (CA-8): The organization conducts penetration testing with a defined frequency on the system.

The metrics that we found relevant to these controls are the following:

1. *Vulnerability Report Max Age*: [covers RA-5 and CA-8] It is the frequency of report generation (e.g., 7*24h requires that report needs to be generated at least once per week).
2. *Vulnerability List Max Age*: [covers RA-5(1)] It is the frequency of vulnerability list updates (e.g., 24h means that list of known vulnerabilities needs to be updated at least once per day).
3. *Number of vulnerabilities (for each family) in a period*: [covers RA-5] It represents the number of vulnerabilities detected for each family in the period (the period depends on the metric n.1).
4. *Number of vulnerabilities (for gravity) in a period*: [covers RA-5] It represents the number of vulnerabilities detected for each gravity in the period (the period depends on the metric n.1).
5. *Number of vulnerability tests executed in a period*: [covers RA-5] It represents the number of vulnerability tests executed in the period (the period depends on the metric n.1).
6. *Number of available vulnerability tests*: [covers RA-5] It represents the number of the available and executable vulnerability tests.

The *Vulnerability scanning and management* capability has been implemented through the *Vulnerability Scanning* mechanism, based on the OpenVAS scanner tool.[18] Some technical background on this tool is given in the following subsection.

### 5.2.1 OpenVAS

OpenVAS is the open source spin-off of the popular security scanner Nessus, and provides a modular architecture composed of a scanning agent and a manager with a powerful interface. It is highly configurable and all internal communications are carried out according to well-defined protocols and

[18]http://www.openvas.org/.

are provided with SSL support. In addition to its adaptability features, OpenVAS enables the scanning of a target both from the outside and from the inside by installing the agent directly on it. Moreover, it provides weekly updates of vulnerabilities so that the vulnerability database is always kept up-to-date. In order to perform vulnerability detection, OpenVAS carries out specific tests that exploit the vulnerabilities themselves to verify whether they affect or not a system (in a penetration testing-like fashion). Such tests can be configured and tuned, which is very useful to manage different monitoring policies.

OpenVAS is a framework of several tools and services. The architecture of OpenVAS consists of three main components:

- *OpenVAS Scanner*: It performs, on the target machines, the latest network vulnerability tests (NVTs) that are downloaded with daily updates through the OpenVAS NVT feed or another feed service. The tests, which are performed in an extremely efficient way, can be launched on many targets simultaneously. The OpenVAS Scanner adopts an ad-hoc communication protocol, namely the OpenVAS Transfer Protocol (OTP), which is provided with SSL support and allows for the control and management of the scanning executions.
- *OpenVAS Manager*: It manages the *OpenVAS Scanner* through the OTP protocol and offers itself an interface accessible through the XML-based OpenVAS Management Protocol (OMP). The *OpenVAS Manager* includes all the management logic, while lightweight clients can be developed to accomplish simple tasks such as the filtering or ordering of scanning results. The *Manager* also controls a sqlite-based database that centrally stores all configuration data and all scanning results. Finally, the *OpenVAS Manager* takes care of the access control and management of users.
- *OpenVAS Client*: There are two different OMP clients, namely the Greenbone Security Assistant (GSA) and the OpenVAS CLI. The GSA is a Web service that offers a lightweight user interface for Web browsers. It uses an XSLT (Extensible Stylesheet Language Transformation) to convert OMP responses, which are XML-based, to HTML. The OpenVAS CLI provides a command line tool that allows to create batch processes and to operate the OpenVAS Manager.

### 5.2.2 The vulnerability scanning mechanism

In order to make OpenVAS available on-demand as an automatically-enforceable security mechanism, we developed an OpenVAS Adapter (custom-openvas-adapter) able to monitor the metrics previously discussed and the cookbook needed to install and configure respectively the OpenVAS client and the OpenVAS manager. The cookbook, publicly available at https://bitbucket.org/specs-team/specs-monitoring-openvas, includes a recipe to install and configure the OpenVAS Manager and OpenVAS Scanner on the target machine (the machine that has to be monitored), and a recipe to install the OpenVAS CLI and the custom-openvas-adapter on a provisioned VM. Both are provided as tar.gz files that are hosted within the cookbook. The latter recipe activates also the adapter component, starting a scan on each agent that has been previously installed.

## 5.3 THE REFINED MONITORING ARCHITECTURE: INTEGRATING THE SECURITY MECHANISMS

Fig. 6 shows a refined version of the architecture shown in Fig. 5, in which are shown the components belonging to the *DoS Protection* and *Vulnerability Scanning* mechanisms and to the Chef tool. As shown in Fig. 6, the *Deployment and Configuration Manager* is implemented by the *Chef Server* and the *Chef*

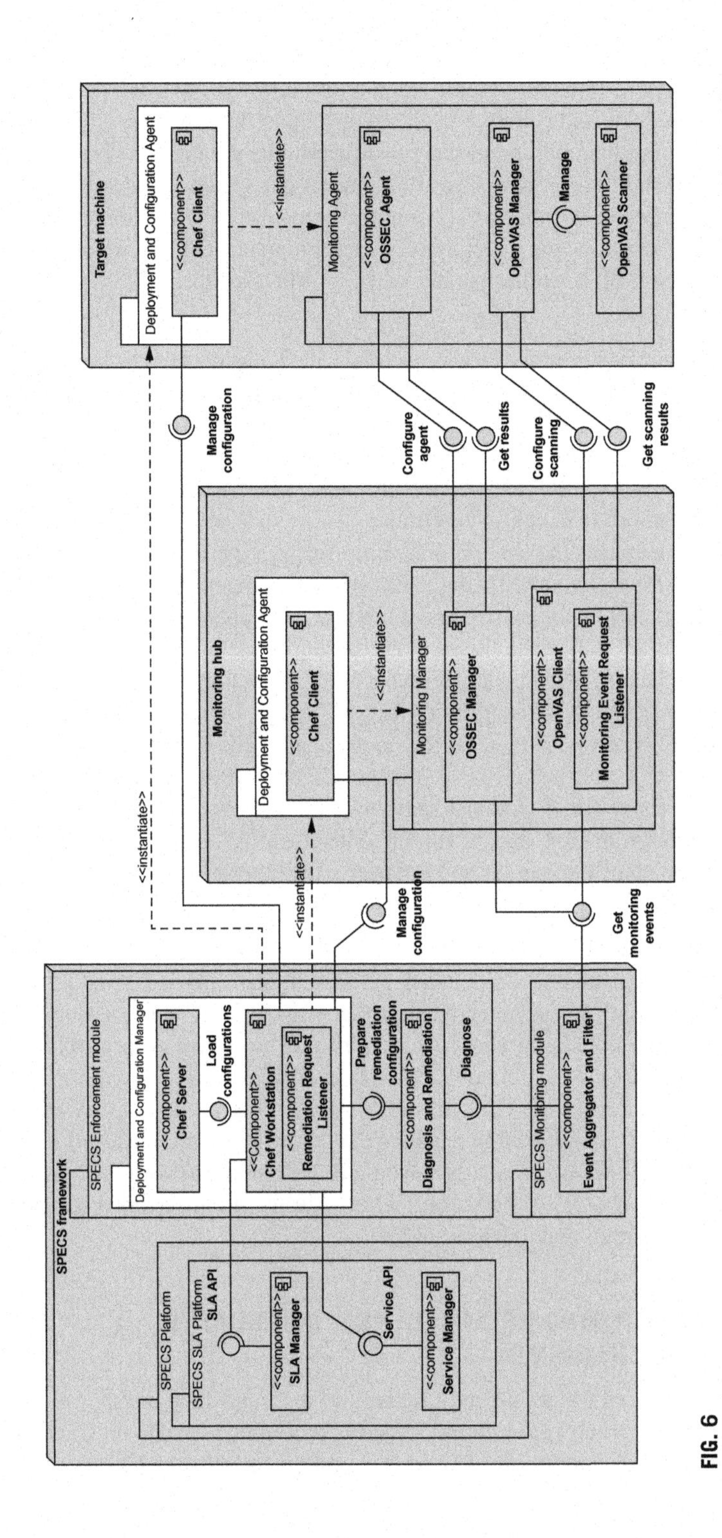


**FIG. 6**

The refined monitoring architecture.

*Workstation* components. In particular, the *Chef server* is devoted to storing the information related to target nodes and the description (as recipes and cookbooks) of the tasks that the *Deployment and Configuration Agents*, represented by the *Chef Clients*, must execute on them. The *Chef Workstation* loads the Chef recipes (depending on the security metrics extracted from the SLA), onto the *Chef Server* through the Knife interface, and manages operations such as the installation and execution of the agents (*Chef clients*) on the target nodes and the assignment of tasks. This component includes the *Remediation Request Listener*, a Java component that is invoked by the *Diagnosis and Remediation* component when a remediation action must be performed, so that the *Chef Workstation* can prepare a remediation configuration for the *Chef Clients* deployed on the target machines.

The Monitoring System depicted in Fig. 6 is split into two parts deployed on different machines. A machine that we called Monitoring hub hosts the *Monitoring Manager*, which runs an *OpenVAS Client* and the *OSSEC Manager*. It includes the *Monitoring Event Request Listener* Java component, which is invoked by the SPECS Monitoring module to retrieve monitoring events and process them. Similarly, the SPECS Monitoring module retrieves the events generated by the *OSSEC Manager*. The *Monitoring Manager*, along with the above-discussed components, is automatically installed and executed on the Monitoring hub by a *Chef Client* suitably installed by the *Chef Workstation* running in the Enforcement module.

The *Chef Workstation* also installs and executes a *Chef Client* on the target machine, which is responsible for the installation and execution of the *Monitoring Agent*. In this case, the *Monitoring Agent* includes both an *OSSEC Agent* and an *OpenVAS Manager*. The former is configured by the *OSSEC Manager* running on the Monitoring hub, while the latter is configured by the *OpenVAS Client* and is devoted to managing the *OpenVAS Scanner*.

## 6 CASE STUDY

In this section, we discuss a case study related to a Web developer (the end-user) that needs to acquire a Web hosting service to host his Web applications. The end-user has limited security skills, but he is aware of the main security threats to Web applications and is willing to harden the hosting Web server with denial of service protection and continuous scanning for vulnerabilities.

Let us first consider the scenario *without SPECS*. At the state-of-the-art, a Web developer that wants to deploy a Web container on resources leased from a public CSP is responsible for applying suitable security configurations. There are indeed existing appliances that offer predefined services (e.g., a preconfigured Web server), but there is no standard way to check for the security features that are *possibly* provided by CSPs. Hence, the Web developer has to (1) manually check each CSP's offers, (2) evaluate one by one the offers and compare them to his own security requirements, and (3) apply suitable configurations by means of external tools if the required security level is not natively supported (and this is almost always the case). But the hardest task is to monitor continuously the running service, in order to spot security issues. This is unsupported by CSPs, and is completely up to the Web developer.

This is the point where SPECS comes profitably into play. Providing the service through a SPECS application gives a number of significant advantages. As a matter of fact, the SPECS application (1) offers a single interface to select among different offers of multiple providers, (2) enables the Web developer to specify explicitly the needed security capabilities on the target Web container by negotiating and signing an SLA with the CSP, (3) automatically acquires and configures the resources (i.e., virtual machines) to enforce the security controls requested, (4) enables continuous monitoring of the security metrics negotiated, and (5) automatically finds and applies remedies to (some of) alerts in the case of SLA violations.

In this section, we propose a brief description of the Web container SPECS application behavior when the *DoS detection and mitigation* and *Vulnerability scanning and management* capabilities are enforced on the top of a Web server provided by a CSP. In particular, we focus on the DoS protection capability and show an example of operation in case of an attack. The Monitoring Hub depicted in Fig. 6, called Event Hub in SPECS, is a standard component devoted to sending alert notifications to the SPECS monitoring module. It receives published events from the monitoring adapters and processes them by operating a translation from the SPECS internal format to the format manageable by the *Heka Router*. Heka[19] is an open source stream processing software system developed by Mozilla, useful for loading and parsing log files and for performing real time analysis, graphing, and anomaly detection on any data flow.

Following up the negotiation process of the SPECS application, we acquired a pool of three virtual machines on Amazon AWS (t2.micro instances), preconfigured with the OpenSUSE operating system distribution, hosting respectively:

- the OSSEC Manager package;
- the OSSEC Agent package together with an Apache v2.2.2 instance (the client);
- the OSSEC Adapter;
- the Event Hub component.

Moreover, we acquired an additional instance of VM with a different user to run the attacker software. As for the type of attack to be performed, we searched for exploits of Apache's vulnerabilities enabling the execution of unauthorized code, administration privilege upscaling and also leading to the unavailability of the service. We identified several exploit databases, classified according to the vulnerabilities present in the CVE dictionary.[20] From the Exploit Database[21] in particular, we chose the well-known *Slowloris* attack.[22] This attack tries to open many connections to the target Web server, and to hold them open as long as possible. It accomplishes this by opening connections to the target Web server and sending partial requests, so that affected servers will keep these connections open, filling their maximum concurrent connection pool, eventually denying additional connection attempts from clients.

We prepared the attacker machine in order to run the Slowloris attack and, on the defense side, we prepared a script for counteracting the attack (by closing all active ports) when the Manager requires an active response to a detected alert. It is worth pointing out that the Manager already has several built-in responses that can be applicable for this attack, but they are not satisfactory to stop a running attack. Therefore we built our own solution. The script was included in the client package so as to be deployed together with the Agent on the Web server machine. It is activated when a specific rule is matched on the Manager (too many `400 error` codes returned) and closes all the active ports on 80/TCP by

[19] https://hekad.readthedocs.org/en/v0.9.2/.
[20] https://cve.mitre.org.
[21] http://www.exploit-db.com/.
[22] http://ha.ckers.org/slowloris/.

means of the command `fuser -k 80/TCP`. Then it re-activates the Apache Web server. This way, the malicious connections are shut down and the normal operation is re-established.

In our experiments, we run the attack on the attacker machine by means of the command:

```
./slowloris.pl -dns www.example.com -port 80 -timeout 2 -num 500 -tcpto 5
```

The command parameters identify the address and port to attack (`dns` and `port` respectively), the re-transmission time period (`timeout`), the number of sockets that are opened to send packets (`num`), and timeout window of TCP (`tcpto`). The result of the attack without the *DoS Detection and Mitigation* capability would be the unavailability of the server, as the tool opens several sockets and periodically sends requests with high frequency. This would be registered in the Apache access_log with several entries showing multiple requests from the same IP with response code `400-Bad Request`.

With the capability activated, the high number of `400-Bad Request` responses is instead notified to the Manager, which finds a match with one of its rules and activates, on the Agent, the script to stop the attack. The Manager also activates the OSSEC Adapter, which notifies the detected attack to the SPECS EventHub. In its turn, this collects the events and counts the number of detected attacks. The actual attack detection latency, which is measured and sent as an event to the EventHuB by the OSSEC Adapter, depends on the request timeout of Apache, which has to expire before a `400-Bad Request` is generated. The SPECS application enables the Customer to monitor the Web container, reporting the actual value of the security metrics agreed in a dedicated webpage, thus allowing the continuous monitoring of the agreed SLA.

## 7 CONCLUSIONS

In this chapter, we have dealt with security monitoring in the cloud, and in particular with SLA-based monitoring. We have discussed the approach followed in a recent EU project, SPECS, focused on the management of the whole Security SLA lifecycle and on the provisioning of secure cloud services through SLAs. With respect to the SPECS framework, we have presented an SLA-based monitoring architecture enabling the automatic deployment and set-up of monitoring capabilities obtained by integrating available/custom cloud monitoring tools. We have discussed two monitoring capabilities developed within SPECS, related respectively to denial of service detection and mitigation and to vulnerability scanning. Such capabilities are offered as-a-service and tuned based on the content of a negotiated SLA. For both capabilities, we have identified the declared security features and the measurable attributes, and we have illustrated the software mechanisms that implement them. Finally, we have provided a case study that shows the effectiveness of the DoS detection and mitigation capability against a popular DoS attack.

## REFERENCES

Aceto, G., Botta, A., de Donato, W., Pescapé, A., 2013. Cloud monitoring: a survey. Comput. Netw. 57 (9), 2093–2115. doi:10.1016/j.comnet.2013.04.001.

Andrieux, A., Czajkowski, K., Dan, A., Keahey, K., Ludwig, H., Nakata, T., Pruyne, J., Rofrano, J., Tuecke, S., Xu, M., 2007. Web Services Agreement Specification (WS-Agreement).

Arshad, J., Townend, P., Xu, J., 2011. An automatic intrusion diagnosis approach for clouds. Int. J. Autom. Comput. 8 (3), 286–296. doi:10.1007/s11633-011-0584-2.

Brandic, I., Music, D., Dustdar, S., Venugopal, S., Buyya, R., 2008. Advanced QoS methods for grid workflows based on meta-negotiations and SLA-mappings. In: 2008 Third Workshop on Workflows in Support of Large-Scale Science.

Casola, V., De Benedictis, A., Rak, M., Villano, U., 2014. Preliminary design of a platform-as-a-service to provide security in cloud. In: CLOSER 2014—Proceedings of the 4th International Conference on Cloud Computing and Services Science, Barcelona, Spain, April 3–5, 2014, pp. 752–757.

Casola, V., De Benedictis, A., Erascu, M., Modic, J., Rak, M., 2016. Automatically enforcing security SLAs in the cloud. IEEE Trans. Serv. Comput. 1.

De Benedictis, A., Rak, M., Turtur, M., Villano, U., 2015. REST-based SLA management for cloud applications. In: Proc. 2015 IEEE 24th International Conference on Enabling Technologies: Infrastructures for Collaborative Enterprises (WETICE 2015), pp. 93–98.

Dekker, M., Hogben, G., 2011. Survey and Analysis of Security Parameters in Cloud SLAs Across the European Public Sector. ENISA.

Dhage, S.N., Meshram, B.B., Rawat, R., Padawe, S., Paingaokar, M., Misra, A., 2011. Intrusion detection system in cloud computing environment. In: Proceedings of the International Conference & Workshop on Emerging Trends in Technology, ICWET '11. ACM, New York, NY, USA, pp. 235–239.

Emeakaroha, V.C., Brandic, I., Maurer, M., Dustdar, S., 2010. Low level metrics to high level SLAs—LoM2HiS framework: bridging the gap between monitored metrics and SLA parameters in cloud environments. In: International Conference on High Performance Computing and Simulation (HPCS), 2010, pp. 48–54. doi: 10.1109/HPCS.2010.5547150.

Emeakaroha, V.C., Ferreto, T.C., Netto, M.A.S., Brandic, I., De Rose, C.A.F., 2012. CASVID: application level monitoring for SLA violation detection in clouds. In: Computer Software and Applications Conference (COMPSAC), 2012 IEEE 36th Annual, pp. 499–508. doi:10.1109/COMPSAC.2012.68.

Fatema, K., Emeakaroha, V.C., Healy, P.D., Morrison, J.P., Lynn, T., 2014. A survey of cloud monitoring tools: taxonomy, capabilities and objectives. J. Parallel Distrib. Comput. 74 (10), 2918–2933. doi:10.1016/j.jpdc.2014.06.007.

Ficco, M., Rak, M., Di Martino, B., 2012a, An intrusion detection framework for supporting SLA assessment in cloud computing. In: Fourth International Conference on Computational Aspects of Social Networks (CASoN), 2012 , pp. 244–249.

Ficco, M., Venticinque, S., Di Martino, B., 2012b, Mosaic-based intrusion detection framework for cloud computing. In: On the Move to Meaningful Internet Systems: OTM 2012. Springer, Berlin, Heidelberg, pp. 628–644.

Ficco, M., Tasquier, L., Aversa, R., 2013. Intrusion detection in cloud computing. In: Eighth International Conference on P2P, Parallel, Grid, Cloud and Internet Computing (3PGCIC), 2013, pp. 276–283.

Gul, I., Hussain, M., 2011. Distributed cloud intrusion detection model. Int. J. Adv. Sci. Technol. 34, 71–82.

Karjoth, G., Pfitzmann, B., Schunter, M., Waidner, M., 2006. Service-oriented assurance, comprehensive security by explicit assurances. In: Gollmann, D., Massacci, F., Yautsiukhin, A. (Eds.), Quality of Protection, Advances in Information Security, vol. 23. Springer US, New York, pp. 13–24, doi:10.1007/978-0-387-36584-8_2.

Mehmood, Y., Habiba, U., Shibli, M.A., Masood, R., 2013. Intrusion detection system in cloud computing: challenges and opportunities. In: 2nd National Conference on Information Assurance (NCIA), 2013, pp. 59–66.

Modi, C., Patel, D., Borisanya, B., Patel, A., Rajarajan, M., 2012. A novel framework for intrusion detection in cloud. In: Proceedings of the Fifth International Conference on Security of Information and Networks, SIN '12. ACM, New York, NY, USA, pp. 67–74.

Nikolai, J., Wang, Y., 2014. Hypervisor-based cloud intrusion detection system. In: International Conference on Computing, Networking and Communications (ICNC), 2014, pp. 989–993.

NIST, 2013. NIST Special Publication 800-53 Revision 4: Security and Privacy Controls for Federal Information Systems and Organizations.

Pearson, S., 2013. Privacy, Security and Trust in Cloud Computing. Springer London, London, 3–42. doi:10.1007/978-1-4471-4189-1_1.

Pearson, S., Benameur, A., 2010. Privacy, security and trust issues arising from cloud computing. In: IEEE Second International Conference on Cloud Computing Technology and Science (CloudCom), 2010, pp. 693–702. doi: 10.1109/CloudCom.2010.66.

Petcu, D., 2014. A taxonomy for SLA-based monitoring of cloud security. In: Computer Software and Applications Conference (COMPSAC), 2014 IEEE 38th Annual, pp. 640–641. doi:10.1109/COMPSAC.2014.50.

Rak, M., Venticinque, S., Mahr, T., Echevarria, G., Esnal, G., 2011. Cloud application monitoring: the mosaic approach. In: IEEE Third International Conference on Cloud Computing Technology and Science (CloudCom), 2011, pp. 758–763. doi:10.1109/CloudCom.2011.117.

Rak, M., Suri, N., Luna, J., Petcu, D., Casola, V., Villano, U., 2013. Security as a service using an SLA-based approach via SPECS. In: IEEE 5th International Conference on Cloud Computing Technology and Science (CloudCom), 2013, vol. 2, pp. 1–6.

Roschke, S., Cheng, F., Meinel, C., 2009. Intrusion detection in the cloud. In: Eighth IEEE International Conference on Dependable, Autonomic and Secure Computing, 2009. DASC '09, pp. 729–734.

Roy, A., Sarkar, S., Ganesan, R., Goel, G., 2015. Secure the cloud: from the perspective of a service-oriented organization. ACM Comput. Surv. 47 (3), 41:1–41:30.

Sengupta, S., Kaulgud, V., Sharma, V.S., 2011. Cloud computing security-trends and research directions. In: 2011 IEEE World Congress on Services, pp. 524–531. doi:10.1109/SERVICES.2011.20.

SLA-Ready, 2015. Making cloud SLAs readily usable in the EU private sector. http://sla-ready.eu/.

SLALOM, 2015. Legal and open terms for cloud SLA and contracts. http://slalom-project.eu/.

Smith, M., Schmidt, M., Fallenbeck, N., Schridde, C., Freisleben, B., 2007. Optimising security configurations with service level agreements. In: Proceedings of the 7th International Conference on Optimization: Techniques and Applications (ICOTA 2007). IEEE Press, New York, pp. 367–381.

SPECS, 2013. Secure Provisioning of Cloud Services Based on SLA Management. http://www.specs-project.eu.

Ullah, K.W., Ahmed, A.S., 2014. Demo paper: automatic provisioning, deploy and monitoring of virtual machines based on security service level agreement in the cloud. In: 14th IEEE/ACM International Symposium on Cluster, Cloud and Grid Computing (CCGrid), 2014, pp. 536–537.

CHAPTER

# HARDENING iOS DEVICES AGAINST REMOTE FORENSIC INVESTIGATION

12

**Luis Gómez-Miralles***, **Joan Arnedo-Moreno***

*Open University of Catalonia, Internet Interdisciplinary Institute (IN3), Barcelona, Spain**

## CHAPTER POINTS

- Recent reports strongly suggest that traditional forensic mechanisms are being used for remotely attacking iOS devices.
- We present an overview of the iOS security and trust model, and introduce a proof of concept tool that applies several antiforensic techniques to mitigate the risk of such attacks.
- Moreover, the possible anti-antiforensic measures are discussed.

## 1 INTRODUCTION

Smartphones have rapidly become ubiquitous in our life. In barely one decade these small devices have managed to enter our pockets and nowadays accompany us at all times, storing a vast amount of personal information—often without the user's knowledge: phonecall logs, emails and SMS messages, calendars, address books, to-do lists, history of visited places, photographs, voice memos, etc., as well as 3rd-party application data such as chat logs from apps like WhatsApp and Telegram. Moreover, vendors have already started to produce wearable devices which hold an even closer relation with their users, gathering and quantifying diverse data about their life habits—a tendency that will only grow in the future years with the Apple Watch and other similar devices.

The rise of mobile technologies has introduced great changes in the information security landscape. Blackberry, the platform that dominated every corporate environment for years thanks to its security features, failed to keep up with its competitors and by Q2 2014 its market share was below 1% (International Data Corporation, 2014). In contrast, 84.7% of the devices sold in that period were Android devices, and 11.7% were iOS devices. When it comes to business environments, 67% of new devices activated in a corporate context during the same period were iOS ones (Good Technology, 2014).

As tends to happen with every software product, the iOS operating system and the core applications shipped with it have suffered from a number of vulnerabilities in the past, with different degrees of criticality. The most serious ones, for instance, made it possible for remote websites to gain full control over a device browsing them with MobileSafari, the integrated web browser (Allegra and Freeman, 2011). Fortunately, many of the vulnerabilities uncovered by researchers have been duly patched by

**Security and Resilience in Intelligent Data-Centric Systems and Communication Networks. https://doi.org/10.1016/B978-0-12-811373-8.00012-4**

Copyright © 2018 Elsevier Inc. All rights reserved.

Apple in subsequent iOS versions. However, it is possible that, at a certain point in the past, the same particular vulnerabilities were used by malicious actors to attack devices running specific iOS versions.

During 2013, Edward Snowden's revelations about the capabilities of the United States National Security Agency (NSA) showed that this body has more robust and persistent surveillance mechanisms over mobile devices (Rosenbach et al., 2013) than could be reasonably expected based on the general security posture of those platforms. As noted by Zdziarski (2014b) this can be achieved by abusing a series of background services available in all iOS devices. Under certain conditions, these services can leak all kinds of personal data stored in the device, bypassing the optional backup encryption password, and showing no indication at all to the user. These mechanisms can be used by forensic software solutions in order to collect information from the device. However, the same mechanisms are likely being exploited to gain unauthorized access to users' personal and corporate devices. As an example, it has been documented that the NSA habitually targets the personal resources of innocent people working as system administrators, with the purpose of getting into the networks of the companies from whom they work (Gallagher and Mass, 2014).

In this chapter, we present an analysis of several mitigation techniques that can be used to reduce the attack surface exposed by these services. As a result of this analysis, we introduce *Lockup*, an accompanying software tool that we have created to implement those measures, some of which are novel and, to our knowledge, have not been implemented before. This tool also serves as a proof of concept that such measures can be deployed in an iOS device.

This chapter is structured as follows. Section 2 provides an overview of the iOS security architecture, and presents the problem of potentially dangerous services that can be abused to extract an enormous amount of user data from the device. Section 3 discusses a number of possible mitigation strategies that can be applied to enhance the device security. Section 4 presents *Lockup*, the software tool that we have developed in order to implement those mitigations. In Section 5, a number of important questions are discussed, including the consequences of the jailbreak process, the antiforensic implications, and the anti-antiforensic measures that can be used to bypass our tool. Concluding the chapter, Section 6 summarizes the chapter contributions and outlines future work.

## 2 SECURITY AND TRUST IN THE iOS ENVIRONMENT

In this section, we present an overview of the main components of the iOS security architecture, its trust model, and the existing privacy threats, as well as the different approaches that exist for the forensic collection of data. This will allow us to show that some of the risks originating from certain weaknesses in the iOS trust model have an impact much higher than expected because of a number of iOS background services, which have no known legitimate purpose.

Over time, the basic iOS system has incorporated a number of security protections, including application sandboxing, mandatory code signing, DEP (data execution prevention) and ASLR (address space layout randomization). These measures are aimed at reducing the attack surface, thus complicating both bug exploitation and the subsequent privilege escalation.

A thorough analysis of each of these layers can be found in Miller et al. (2012).

In addition, every major iOS version has incorporated an increasing number of enterprise features (Apple Computer, Inc., 2014a,b), especially since the release of iOS 4 and the iPad in 2010—from

Exchange and Mobile Device Management (MDM) support to biometric authentication. Some of these features require the device to be remotely manageable somehow. The main method to achieve this would be generating a trust relationship with an external device, which, from then on, will be able remotely access a set of special services made available by the iOS operating system.

However, as we will explore later, this might be a double-edged sword, as those same capabilities can be used by malicious actors to read the data stored in the device or to surreptitiously install applications capable of tasks as dangerous as recording audio and capturing network data.

## 2.1 REMOTE ACCESS VIA DEVICE TRUST RELATIONSHIP

When it comes to sharing information with an external device (be it a desktop computer, an alarm clock that can play music, or a car audio system) the iOS security model works as follows. Whenever the iOS device is connected via cable to a previously unknown computer (or another external device), it presents a dialog on screen prompting the user whether the computer should be trusted, as seen in Fig. 1. Upon receiving the user's consent, both devices create and interchange a series of certificates, which from that moment will be used to authenticate each other and initiate a secure, encrypted connection. A pairing record consisting of these certificates is stored in well-known filesystem paths in both the computer and the iOS device.

As exposed by Lau et al. (2013) and Zdziarski (2014b), a computer that has successfully paired with an iOS device can initiate a connection to it and invoke a number of services exposed via the *lockdown* daemon, even wirelessly and without the user receiving any visual indication. The same can be done from any other computer or device, as long as the pairing record is extracted from the trusted computer.

There is no way for the user of an iOS device to review the list of external devices he has chosen to trust, or to revoke that trust other than to reinstall the device completely.

Unfortunately, a number of the *lockdown* services are designed in such a manner that they may leak significant amounts of personal information, even bypassing the user's backup encryption password. Given that any trusted device (alarm clock and car stereo) gets a pairing record that gives access to all the services, this can be exploited by either placing malicious devices in common areas, such as airports and coffee shops (Lau et al., 2013) or compromising trusted devices to steal the pairing record stored in it. Then, those pairing records can be used to establish connections to the iOS device, even over the air through either Wi-Fi or cellular connection, in order to perform surreptitious actions such as deploying malicious software or extracting information from the device.

## 2.2 SENSITIVE iOS DEVICE SERVICES

A computer that connects to an iOS device (either via USB cable or through the network in TCP port 62078) can invoke a series of services which, from the iOS side, are offered through the *lockdown* daemon (Zdziarski, 2014b; Lau et al., 2013). These services have diverse roles, such as allowing iTunes syncing or remote management for MDM purposes, while others have no known purpose and seem to be the perfect backdoors to be exploited by intelligence agencies, forensic products, and malicious actors all alike.

The complete list of services can be explored by checking the file `/System/Library/Lockdown/Services.plist`. A fresh installation of iOS 7.1.2 on an iPhone 5 exposes a total of 32 services via

**FIG. 1**

An iPhone prompting the user whether it should trust the connected computer.

*lockdown*, of which Zdziarski (2014b) identified the following ones as being valuable from a forensic standpoint:

- **com.apple.file_relay.** There is no known legitimate use for this service. It is designed to obtain huge amounts of information: the user's complete address book, calendar, SMS database, call history, voicemail, notes, photos, list of known Wi-Fi networks, GPS positioning logs, list of email accounts configured in the device, a list of all files existing in the device, with their metadata (size, creation and modification dates), even a list of every single word typed in the device, together with its word count; as well as and a number of additional system logs. No indication is shown to the user when this is done. Note that, although the user can set a backup encryption password through iTunes, data sent through this service is not encrypted in any manner.

- **com.apple.pcapd.** A network sniffer for which, again, no known legitimate purpose is known. It can be activated remotely and leaves no trace to the user. Apart from seeing the network traffic of other devices near the victim, and possibly performing man-in-the-middle attacks, this could also have other interesting uses for attackers: they could obtain information about the networks available at a certain location (possibly a restricted facility where the victim has access) in order to prepare more advanced attacks in which they could impersonate those networks; another possible use would be confirming or discarding the presence of other people at that location, by examining unique identifiers of the nearby devices, such as Bluetooth MAC addresses.
- **com.apple.mobile.MCInstall.** Installs managed configurations, such as the ones used in MDM deployments. This makes sense in corporate environments, where the company may need to enforce security restrictions on the device, preload applications, or provision encryption certificates; but is very rarely required in the case domestic users, and is a possible entry point for an attacker willing to deploy hidden applications to the victim's device, for instance with the purpose of recording background audio.
- **com.apple.mobile.diagnostics_relay.** Provides diagnostics information such as hardware state and battery level.
- **com.apple.syslog_relay.** Exposes various system logs.
- **com.apple.iosdiagnostics.relay.** Presents per-application network usage statistics.
- **com.apple.mobile.installation_proxy.** Used by iTunes to install applications.
- **com.apple.mobile.house_arrest.** Used by iTunes to transfer documents in and out of applications.
- **com.apple.mobilebackup2.** Used by iTunes to backup the device. If the owner has set a backup encryption password through iTunes, the data sent through this service will be encrypted with that password—something that does not happen in the case of the *com.apple.file_relay* service.
- **com.apple.mobilesync.** Used by iCloud and iTunes to sync 3rd-party application data as well as data belonging to core iOS applications: Safari bookmarks, notes, address book entries, etc.
- **com.apple.afc.** Exposes the complete *Media* folder—audio, photographies and videos.
- **com.apple.mobile.heartbeat.** Used to maintain the connection to other services being accessed.

The two first services (*file_relay* and *pcapd*) are the most dangerous ones. When these were first identified by Zdziarski, Apple responded to news site iMore (Ritchie, 2014) stating: "*We have designed iOS so that its diagnostic functions do not compromise user privacy and security, but still provides needed information to enterprise IT departments, developers and Apple for troubleshooting technical issues.*"

However, we agree with Zdziarski (2014a) that there seems to be no realistic scenario in which opening the door to such an enormous amount of user data can be justified for "*diagnostic*" and "*troubleshooting*" purposes.

## 2.3 FORENSIC ACQUISITION APPROACHES

A very basic approach to acquiring user data in the iOS platform is the so called *logical acquisition*: connecting the device via the standard USB cable to a computer running iTunes, Apple's multimedia player which is in charge of synchronizing content to the device. Using its AFC protocol (*Apple File*

*Connect*), iTunes syncs existing information (contacts, calendar, email accounts, and apps) and can even retrieve a complete backup of the device; however, there are two important caveats in this process:

1. The device needs to be correctly paired with the iTunes software in order to sync. If the device is protected by a passcode (which is the most probable case in all devices introduced since 2013 with the TouchID biometric technology), the investigator cannot unlock the device to authorize trusting (and start syncing data to) this new iTunes installation. A workaround for this is presented in Zdziarski (2014b): impersonating a device known (and trusted) by the iOS device, such as the owner's computer, by retrieving from it a set of files known as *escrow keybags*.
2. A dump obtained this way will miss logs and system files that could be of interest in certain scenarios, as well as all the unallocated space, from which deleted files could be recovered.

This logical acquisition is the process followed by most iOS forensic tools such as Lantern or Oxygen, since their first versions.

After the first iOS jailbreak was available, Zdziarski (2008) proposed a basic method for obtaining a forensic image of the iPhone with a *physical acquisition* approach, by jailbreaking the device and using SSH access and the `dd` and `netcat` standard UNIX tools, which by that time had already been ported as a part of the growing iPhone jailbreak community (Similar methods are explored by Rabaiotti and Hargreaves (2010) against a Microsoft Xbox device); the data transfer process was done through the device's Wi-Fi interface. In this kind of physical acquisition, the whole storage area is dumped; this includes the unallocated space, from which deleted files could be recovered.

One particular vendor, iXAM (Forensic Telecommunications Services Ltd., 2010), developed a "zero-footprint" solution that relied in the same bugs and exploits used by jailbreak tools. Instead of completing the jailbreak and installing the *Cydia* package manager as usual, their software uploads a tiny, small-footprint software agent which takes control of the system, dumps the solid state storage, and then reboot the device back into its normal state. The problem with these methods is the need for continuous support and upgrades as new iOS versions become available; in fact, according to iXAM website their product works only in the iPhone 4 (introduced in 2010) and older devices.

A similar process was described by Iqbal et al. (2012), although the publication of that paper was supposed to be accompanied by the release of a tool that never saw the light. And other authors have explored the use of similar techniques in other platforms such as Android (Vidas et al., 2011) or Windows Mobile (Grispos et al., 2011). Another paper in 2013 presented the design and implementation of an iOS forensic tool (Chen et al., 2013) aimed at simplifying the forensic acquisition of devices running iOS 6, which had been released one year before. However, it seems that the tool itself was not released.

With the release of iOS 4 in 2010, Apple introduced hardware-based encryption, branded as *iOS data protection*). Bédrune and Sigwald analyzed (Bedrune and Sigwald, 2011a) the underlying technology and released (Bedrune and Sigwald, 2011b) a set of open source tools capable of decrypting disk images and even undeleting certain file types; and in fact, we used their tools for a previous publication (Gómez-Miralles and Arnedo-Moreno, 2015a).

Over time Apple has improved iOS' *data protection* (encryption) implementation at both the hardware and the software levels. At the hardware level, it is important to note that recent devices (introduced 2011 onwards) are shipped with a new bootrom that fixes the bugs exploited by tools such as Bedrune and Sigwald (2011b) in jailbroken iOS devices to decrypt and undelete files.

And unfortunately, a similar bug has not been found in modern iOS devices. Or at least, it has not been publicly announced. As a consequence, so far it is not possible to recover deleted files from modern iOS devices, nor to perform physical acquisition.

Having lost one of the main benefits of jailbreak, the ability to defeat iOS data protection mechanisms and decrypt files, even undelete them, commercial tools have returned to the *logical acquisition* method (Chang et al., 2015) which does not require the jailbreak of the device. The tools themselves can behave and be seen by the iOS device as the iTunes software, and can only acquire whatever information the device is willing to expose or sync to iTunes. These tools can also operate on iTunes backups extracted from a computer, without access to the original device. On the other hand, well-resourced attackers probably count on more advanced tools even capable of exploiting bugs that are not publicly known in order to surreptitiously retrieve data from the device.

# 3 MITIGATION STRATEGIES

There are different mitigation measures that can be applied to cope with the weaknesses introduced by the most sensitive iOS services. We try to summarize the most relevant ones.

## 3.1 DELETE EXISTING PAIRING RECORDS

One way to mitigate the problem would be to control the number of trust certificates in the iOS device. This is the approach adopted by the *unTrust* tool (Stroz Friedberg, 2014): it runs in a computer connected to an iOS device connected via USB and removes all pairing records existing in the device except the one for the computer being used to execute the tool.

One drawback of this approach is that the iOS device still keeps trusting one computer, hence, there is still the risk that the pairing record is stolen from the computer and used to connect to the device services. In addition, if the user decides or needs to temporarily trust an external device, away from that computer (such as an audio system), there is no way to revoke that trust or purge the list of trusted devices until the user can get access to the trusted computer and execute *unTrust* again.

## 3.2 LIMIT SENSITIVE SERVICES TO USB (DISABLE OVER WIRELESS)

Another approach would be to limit the sensitive services to run only over USB, minimizing the risk for over-the-air attacks. The *lockdown* daemon, responsible for all the sensitive services described in this chapter, implements an option (*USBOnlyService*) to limit certain services to USB connections only, disabling the connection to those services over wireless networks. However, none of the sensitive services in iOS 7 requires this option. Starting in iOS 8, the option is applied by default to *com.apple.pcapd* (the network sniffer).

## 3.3 DISABLE SOME SERVICES

Finally, it would be ideal to disable the most sensitive services—something that has not been done so far. This is the approach chosen for our tool, *Lockup*. Given the access level required, it runs only on jailbroken devices, although it would be trivial for Apple to implement these changes in stock iOS versions.

## 3.4 LOCK PAIRING WITH NEW DEVICES

Another option worth mentioning is to block pairing with new devices, as implemented by Zdziarski in *pairlock*. This was useful up to iOS 6, given that in those versions external devices would be trusted blindly, without the iOS device presenting any prompt to the user. Since iOS 7 addressed this concern by asking for user permission before trusting new devices, *pairlock* has not been updated to work in iOS 7. Its approach leaves some doors open, as it does not allow the user to revoke existing trust relationships, nor does it address the risk of a pairing record being stolen from a computer or other trusted device.

# 4 *LOCKUP*: iOS HARDENING AND ANTIFORENSICS

As a proof of concept, in this chapter we present *Lockup*, a software tool that can be installed in devices running iOS versions 7 and 8. *Lockup* hardens the security of the device by addressing the issue of sensitive services using three different approaches:

1. Reducing the attack surface by disabling the most sensitive services: *com.apple.file_relay* (the service that retrieves lots of data bypassing the backup encryption password) and *com.apple.pcapd* (the network sniffer), both with no known purpose and vaguely defended by Apple (Ritchie, 2014), are disabled right away. In addition, the user is offered several *profiles*, allowing him to tailor which services are published, and enabling only those needed for the intended use of the device. The rest are eliminated. For instance, most users are not enrolled in corporate *Mobile Device Management* systems; hence they do not need to allow remote installation of software and configuration profiles, which are indeed very dangerous attacks vectors.
2. Limiting exploitation opportunities by restricting the rest of services to USB only, eliminating over-the-air threats. This is automatically done in most of the *profiles* mentioned previously.
3. Limiting trust relationships by automatically purging all pairing records after a configurable period of time. This constitutes an additional line of defense against attackers capable of stealing a trusted certificate from sources such as the user's computer.

When analyzing the trust model by which an accessory such as a car audio system can have access to all personal data stored in the device, one comes to the conclusion that Apple decided to sacrifice a certain degree of user security in order to simplify the user experience.

To a certain degree, the same problem affected application security up to iOS version 7. Citing Miller et al. (2012) about the iOS sandbox security feature: "*One thing to notice about the iOS sandbox is that every third party app from the App Store has the same sandbox rules. That means that if Apple thinks one app should have a certain capability, all apps must have that capability. This differs, for example, from the Android sandbox where every app can have different capabilities assigned to it based on its needs. One of the weaknesses of the iOS model is that it may be too permissive.*"

Starting in iOS 8, Apple implemented a more granular permissions system for applications (not so for peripherals). A custom prompt is presented to the user the first time that a particular application requests access to certain items (addressbook, camera roll, camera and microphone, location, or health

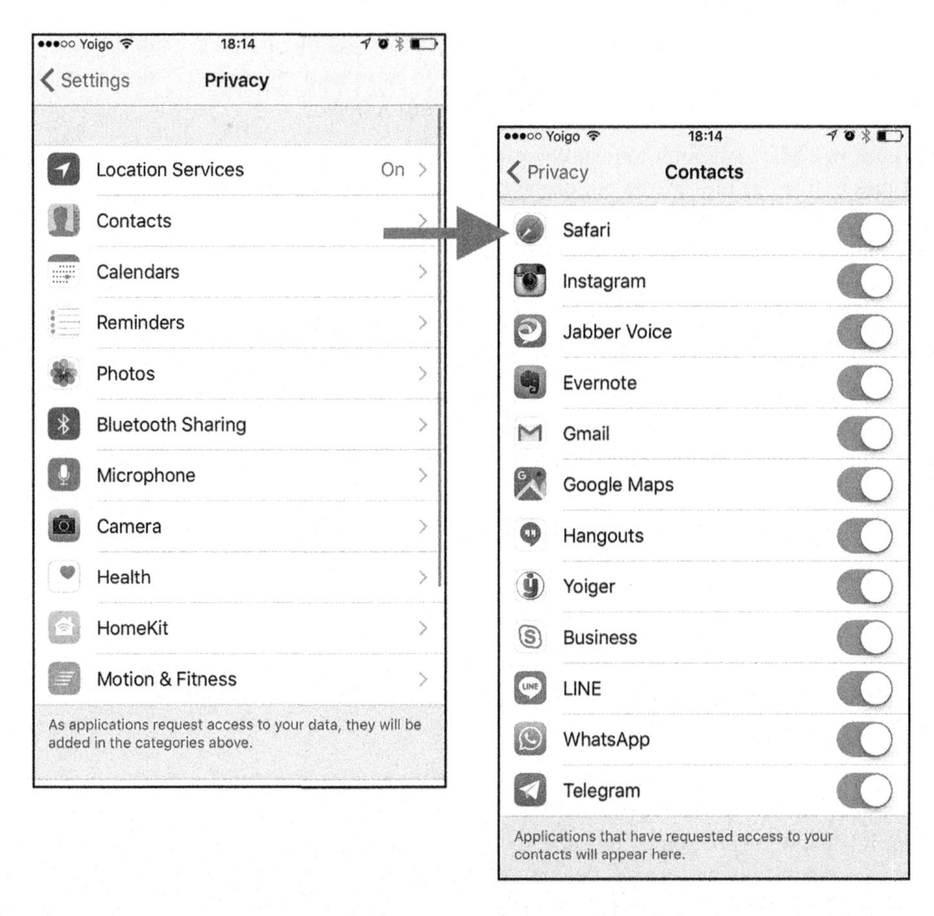


**FIG. 2**

New privacy options in iOS 8 and higher.

data) and the user can authorize or deny it. Furthermore, a new "Privacy" menu in the Preferences app allows the user to review which applications have access to each restricted set of data, and revoke it if desired, as shown in Fig. 2.

It would be desirable to be presented a similar prompt whenever external devices require access to at least the most sensitive *lockdown* services—those that expose personal user data. In our case, it did not seem feasible to implement those changes in *lockdown*, as it would only be possible via reverse engineering, and even then, such modifications would be complicated to maintain across future iOS updates. Hence, we have resorted to a different strategy. Instead of setting individual permissions for each external device, *Lockup* allows the user to choose between a series of *profiles*, each one increasingly restrictive, depending on what the user needs to do with the device at any given time. We believe it may be worth having the choice to sacrifice a fraction of the simplicity to improve the security of iOS devices.

In order to define the various *profiles*, we tried a number of configurations, enabling and disabling each service selectively, and attempting various common actions to make an iOS device interact with other external devices. In particular, we tried the following actions:

- Use iTunes in a Mac computer to install applications in the iOS device.
- Use iTunes to transfer files in and out of the applications installed in the iOS device.
- Use iTunes to perform a backup of the data stored in the device.
- Use a Bluetooth hands-free device to access the address book of the iOS device and place calls through it.
- Use iPhoto in a Mac computer to import the device's camera roll.
- Use a stereo system to play the audio coming out from the iOS device.

This list illustrates the problems of granting excessive privileges to external devices that access the iOS device's *lockdown* services. If a user does not regularly backup to iTunes, why should those services be exposed when the device is connected to, say, an alarm clock? With *Lockup*, the user can adjust the behavior of the device as needed.

## 4.1 TOOL CAPABILITIES

The main capabilities of *Lockup* can be summarized as follows.

- Controlling the device's trust relationships, by periodically purging the stored pairing records; and
- Disabling certain *lockdown* services and prevent others from being invoked over Wi-Fi connections, in order to prevent over-the-air attacks.

One common concern about the potential abuse of iOS services is that iOS lacks a way to see which other devices or computers have been paired with in the past, or to revoke those trust relationships. If the user just hits the wrong button by mistake, the connected device will be trusted forever (unless a full restoration of the device is performed, deleting all user data). This poses a significant risk, especially considering the possibility of an attacker stealing the pairing record from inside the trusted device and using it to establish remote connections to the iOS device.

In our solution we opted for including a background task that will wipe all trust relationships from the iOS device after a configurable period of time, applying the mitigation strategy discussed in Section 3.1. Once that happens, connecting to that device will require the user to confirm the trust relationship from the iOS screen. We observed that even with values as low as 1 minute standard features such as iTunes syncing still work normally (once the user authorizes the pairing by pressing "Trust" in the device screen). Note that these features keep working even when the pairing record is deleted in the middle of a session, as long as both devices are connected, the iOS device will not need to re-trust the external device.

In addition, as we have introduced earlier in this chapter, there are sensitive services potentially very dangerous and with no known purpose (such as *com.apple.mobile.file_relay*, that can be used to extract all kinds of personal information bypassing the backup encryption protection, or *com.apple.pcapd*, which can be used to turn the device into a sniffer that will capture the network traffic it can receive), and it seems obvious to us that these offending services should be removed from every device.

There are also other services that, despite having a legitimate purpose, can also be exploited to leak significant amounts of personal information or inject malicious software into the device.

Examples include *com.apple.mobile. installation_proxy* (used by iTunes to install applications in the device), *com.apple.house_arrest* (used by iTunes to copy application files from or to the device) and *com.apple.mobilebackup2* (used by iTunes to backup the data stored in the device).

We propose to define different service levels and keep the device in the most restrictive level that is suitable depending on the user needs, a measure that has not been implemented before, to the authors' knowledge. For instance, it is not necessary to keep all the iTunes-related services enabled unless the user wants to connect the device to iTunes, and even then, it is not necessary to expose those services over-the-air if the user prefers to sync using a USB cable. Similarly, a lot of users will prefer to disable the MDM-related services, which can be exploited to install software into their devices. This approach applies the mitigation strategies explained in Sections 3.2 and 3.3.

## 4.2 SERVICE PROFILES

Next, we describe the different profiles that we have implemented in *Lockup*, with each profile being increasingly restrictive and consequently more secure. In order to decide which services should be disabled in each profile, we have followed two different criteria.

On one hand, the services that we disable first are those that pose a higher privacy risk to the user. These are, for instance: the services that make it possible to bypass the backup encryption password, to capture network traffic, to deploy configuration profiles and applications to the device, etc.

At the same time, the first services that we disable are the ones likely to be needed by a reduced number of users. We first disable the totally unneeded services, afterwards we disable MDM, and then we disable other features that users may need at particular moments (such app installation via iTunes) while we still allow iTunes to obtain backups of the device data.

### 4.2.1 Level 1: Suitable for MDM

In the first security level that we propose, we disable only those services that have no known purpose and can leak significant amounts of information. In particular, this level disables the following services:

- *com.apple.file_relay.*
- *com.apple.pcapd.*

Devices configured in this mode will still be fully functional, even for remote management in MDM environments. Although many of these environments enforce policies in which no jailbroken devices are allowed, which makes sense from the security point of view given that, as we have already mentioned, jailbreaking a device disables a number of security mechanisms. In any case, this should be the bare minimum level used by any user with a jailbroken device.

### 4.2.2 Level 2: Suitable for syncing applications

In addition to the restrictions defined in the previous level, this level disables the remote installation of configuration profiles and a number of diagnostics services:

- *com.apple.mobile.MCInstall.*
- *com.apple.mobile.diagnostics_relay.*
- *com.apple.syslog_relay.*
- *com.apple.iosdiagnostics.relay.*

Moreover, the rest of the sensitive services (the ones subsequently disabled in the levels described below) are set to USB only, meaning that it is no longer possible for a malicious party to attack them over-the-air, using either Wi-Fi or cellular. Apart from this, the most important implication is that we disable the installation of management configurations, which can be abused by malicious parties in order to track a device via GPS, wipe its data remotely, or deploy additional certificates to the device with effects such as facilitating man-in-the-middle attacks against HTTPS connections. This change may affect the device's ability to enroll in certain MDM environments.

The vast majority of domestic users should be able to use this profile without noticing any adverse effect. An iOS device configured at this level can still sync applications and media with iTunes, and consequently, still exposes services that could be abused to install additional applications, or to retrieve the user content stored in any application installed.

### 4.2.3 Level 3: Suitable for backup

In addition to the restrictions previously defined, this level disables the following services:

- *com.apple.mobile.installation_proxy.*
- *com.apple.mobile.house_arrest.*

This change disables the remote installation of applications to the device. Automatic application syncing with other iOS devices will still work if enabled (from Settings—iTunes Store and App Store—Automatic Downloads—Applications).

It also keeps iTunes from transferring files in and out of the applications installed in the iOS device. Note that if a particular application offers its own mechanism for uploading files to the device (for instance many applications can activate an integrated web server for this), this will still work properly.

The use of iTunes to manage the device applications and data is becoming less and less common as modern iOS versions, except maybe at the time of acquiring a new device and populating it with a backup of the previous one. Thus this profile would still be appropriate for most domestic users.

A device configured in this mode will still allow iTunes to perform backups of the user data stored in the device. This service could be abused by an attacker to obtain the information stored in the device; however, if the user has set a backup encryption password, files transferred through this service will be encrypted with that password.

### 4.2.4 Level 4: Suitable for syncing media files

In addition to the restrictions already described, this level disables the following services:

- *com.apple.mobilebackup2.*
- *com.apple.mobilebackup.*

This level disables iTunes capability to backup the data stored in the device.

A device configured at this level will still be able to sync media files with iTunes, and consequently can expose the contents of the *Media* folder (audio, pictures and video), which could potentially be reached by an attacker abusing a trust relationship. However, all the content belonging to 3rd-party apps should be safe—or at least not reachable via *lockdown* services.

If bookmarks, address book, and calendar data are not being synchronized through iCloud, it may also be possible to obtain these items from a device configured at this level through the *com.apple.mobilesync* service.

### 4.2.5 Level 5: Suitable for media sharing

Apart from the restrictions previously defined, this level disables:

- *com.apple.mobilesync.*

With this change, iTunes syncing capabilities stop working completely, and the only piece of sensitive information exposed through *lockdown* services is the *Media* folder, containing the pictures and videos stored in the device, as well as voice memos, music and podcasts. This may be necessary for some programs or peripherals to access the media files stored in the device.

This profile is suitable for users as long as they do not rely on iTunes for the backup of their device data, something increasingly common as iOS now allows users to store their backups directly on their iCloud storage.

### 4.2.6 Level 6: No sensitive services

In addition to the changes performed in the previous levels, this level disables the following service:

- *com.apple.afc.*

This level breaks compatibility with programs such as iFunBox, which allow the user to browse files stored in the device. In addition, some peripherals may also rely on this service for accessing the files stored in the device, and consequently will stop working when this profile is applied.

Still, the profile should be appropriate for most users that don't use iTunes to manage nor backup their device.

### 4.2.7 Level 7: No lockdown services at all

This level completely removes every *lockdown* service, including *com.apple.mobile.heartbeat.*

It is worth mentioning that, even in this mode, the device is still capable of interacting with external devices through other mechanisms that do not rely on *lockdown*. In particular, we verified that the following actions work properly even when the device is configured in this mode:

- Connect the iOS device to a Parrot Minikit Smart hands-free device via Bluetooth, being able to import the address book and place calls.
- Connect the iOS device to a Mac computer via USB cable and using the iPhoto software in the computer to import the photographies and videos stored in the iOS device.
- Connect the iOS device to a Denon RCD-M39 audio system via USB cable so that the audio played in the iOS device sounds through the Denon audio system.

### 4.2.8 Additional considerations

It is worth mentioning that our solution will disable any additional services that may have been installed by the user, with or without his knowledge, at the moment of jailbreaking the device.

One case worth mentioning is *com.apple.afc2*, a service which is installed by many jailbreak tools and can also be installed separately by users through Cydia. This service exposes the whole filesystem of the iOS device through *lockdown*, making it possible for a trusted external device (or an attacker who has stolen the pairing record from it) to read and write any file in the device either through a USB connection or over the air. Given the privacy risks it represents, this service is always disabled in all profiles. If a user needs remote filesystem access, there are better alternatives from a security standpoint, such as installing OpenSSH and using a companion desktop applications such as *FileZilla* or *PuTTY*—if this is done, it is important to change the default password of the *root* and *mobile* users.

## 4.3 IMPLEMENTATION DETAILS

*Lockup* is designed to run in jailbroken versions of iOS 7 and 8. It can easily be ported to new major iOS versions as soon as a jailbreak is available for them, which typically happens a few weeks after the official iOS release. In the worst case so far, iOS 7.0 took 95 days until a public jailbreak was available for iOS 7; in contrast, iOS 8 was jailbroken 35 days after its official release. It is also remarkable that many users of jailbreak applications usually stick to an older iOS version until a jailbreak for the new one is available.

The different service profiles are defined by creating multiple copies of the */System/Library/Lockdown/Services.plist* file. In each profile, we disable an increasing number of services. In addition, in most of the profiles, the flag *USBOnlyService* is applied to sensitive services, so that these cannot be abused over the air, either via a Wi-Fi connection, or through the user's cellular connection.

In order to set a profile, the user executes the command *lockup-profile*. This can be done either using a terminal application such as *MobileTerminal* or accessing the device via SSH, if it has been installed. When the command is invoked, the user is presented with a menu as shown in Fig. 3. After the user picks

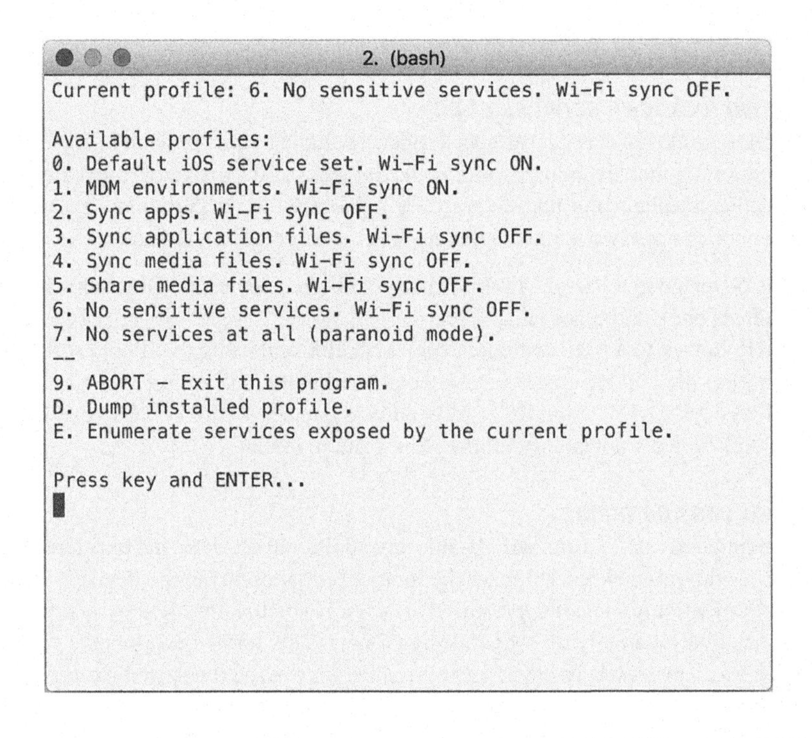

2. (bash)
Current profile: 6. No sensitive services. Wi-Fi sync OFF.
Available profiles:
0. Default iOS service set. Wi-Fi sync ON.
1. MDM environments. Wi-Fi sync ON.
2. Sync apps. Wi-Fi sync OFF.
3. Sync application files. Wi-Fi sync OFF.
4. Sync media files. Wi-Fi sync OFF.
5. Share media files. Wi-Fi sync OFF.
6. No sensitive services. Wi-Fi sync OFF.
7. No services at all (paranoid mode).
--
9. ABORT - Exit this program.
D. Dump installed profile.
E. Enumerate services exposed by the current profile.
Press key and ENTER...


**FIG. 3**

Menu presented by *lockup-profile*.

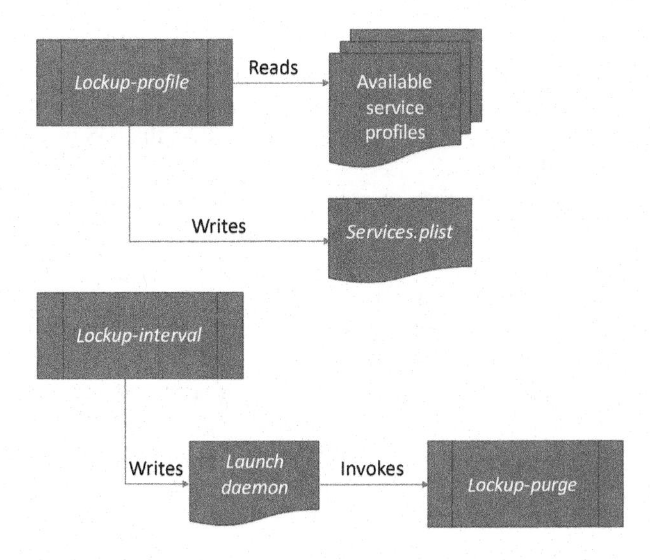


**FIG. 4**

*Lockup* components and main interactions.

a profile, the corresponding service list file is copied over */System/Library/Lockdown/Services.plist*. For the changes to take immediate effect, a *SIGTERM* signal is sent to the *lockdown* daemon with the *kill* command, which makes it restart and read its new configuration file. Additional options allow the user to enumerate the services exposed by the present profile and dump the whole contents of the *Services.plist* file, which may be specially useful to detect and investigate additional services that may have been installed inadvertently.

For the periodic purging of pairing records, *Lockup* uses various files. First, a shell script in charge of deleting the pairing records is installed. Secondly, a *launch daemon*, which will run the previous script periodically, is loaded through */System/Library/ LaunchDaemons/es.pope.lockup-purge.plist*. An additional script, *lockup-interval*, can be used to change the interval at which pairing records are deleted (one hour by default). Fig. 4 summarizes the main components and the interactions between them.

# 5 DISCUSSION

This section analyzes the security and forensics implications of using a tool like *Lockup*.

## 5.1 THE JAILBREAK PROCESS

One of iOS' main features is application sandboxing, meaning that an iOS application does not have full system access due to the security barriers put in place by the operating system. This, in turn, keeps us from developing and running custom software. In order to overcome this restriction we resort to the *jailbreak* technique, described here.

When the first iPhone model was released in 2007, the OS did not incorporate most of its current security features: there was no sandboxing, no code-signing enforcement, no DEP, and no ASLR. It took barely a week for George Hotz to figure out how to escape the OS limitations and gain root access to the device (Hotz, 2007): a process that is referred to as *jailbreak* in the iOS platform, or *rooting* in the Android platform.

A number of tools have been published over time which allow to jailbreak certain iOS versions; examples of these tools are *redsn0w*[1] (iPhone Dev Team, 2011), *greenpois0n* (Chronic Dev Team, 2010), and *evasi0n* (Wang et al., 2013). Whenever one such tool is released, it is analyzed by Apple to uncover the iOS bugs being exploited, which are patched in subsequent iOS releases, meaning that jailbreaking each iOS version requires finding new bugs that can be exploited, in what Steve Jobs himself once described as *a cat-and-mouse game* (Soghoian, 2007).

The use of jailbroken devices is very popular among developers and researchers, as it gives them much more control over the device's internals (Miller et al., 2012). Although it is hard to find global data about the number of jailbroken devices, a recent report focused in China found that over 30% of iOS devices being used in that country were jailbroken in January 2013. By December of the same year, this number had gone down (Umeng, 2014) to 13%. This fluctuation can be attributed to the fact that iOS 7, released in September, did not get a jailbreak until late December. In any case, the numbers show that a significant fraction of iOS devices are jailbroken.

Jailbreak has become increasingly popular among users and there are thousands of applications, both free and paid, that can be installed in jailbroken devices. These are applications that would never make their way into the official distribution channels, given that they infringe the App Store's rules in one way or another. Examples include software emulators and all kinds of system-wide tweaks that change the device's global aspect (Freeman, 2014b), alter global elements such as the Control Center or the Notification Center (Lisiansky, 2014), or inject code into other existing applications to change their behavior (Freeman, 2014a).

It is worth mentioning that a number of important features currently present in official iOS releases existed before as tweaks in the jailbreak community, in some cases influencing Apple's final implementation. Some examples are the Notification Center, the way the copy and paste feature works, and the "cards" interface for switching between applications.

In our case, we have leveraged the jailbreak technique to disable specific *lockdown* services and test different connectivity scenarios, and to develop and test the *Lockup* tool. Other users and researchers can install it and benefit from its features, provided that they are using an iOS version for which a jailbreak is available. And, of course, vendors could implement this kind of tool in future OS versions.

## 5.2 EFFECTS OF JAILBREAK ON THE SECURITY MODEL

It is often argued, and partly true, that jailbreaking a device introduces a number of security weaknesses (Apvrille, 2014; Porras et al., 2010) by disabling important security controls such as application sandboxing and mandatory code signing. Consequently, the use of this technique must always be weighed against the new risks introduced.

[1]Mind the spelling: that character in the middle of all those names is a *zero*, not a capital '*o*'.

In the case of the tool presented in this chapter, we consider that the security tradeoffs of jailbreaking the device make sense in at least three possible scenarios:

1. With research purposes, such as in order to test the tool and evaluate the adoption of similar approaches in other tools or as an iOS core feature.
2. In the case of obsolete devices which are stuck forever in old iOS versions with known vulnerabilities. For instance, dad's old iPhone which he used for two years, then handed over to mom who after another two years passed it to their teen child. This is the case of the 1st-gen iPad with iOS 5.1.1, every iPod Touch up to the 4th generation, and every iPhone up to the iPhone 4; the iPhone 4S and several iPad models are likely to remain stuck at iOS 9 when the next version is released in September 2016. Given that these iOS versions contain known vulnerabilities (exploited by the corresponding jailbreak tools), owners of these devices may prefer to jailbreak the devices themselves in order to install this and/or other protections.
3. Even in modern devices and iOS versions, as part of a more complex hardening strategy involving additional tools, possibly developed by power-users themselves.

When jailbreaking a device, users should always change the default passwords for users *mobile* and *root* (which is *alpine* in both cases), and install only the required software packages, always from trusted sources.

## 5.3 ANTIFORENSIC IMPLICATIONS

We did not have antiforensics in mind when we started developing this tool. However, as exposed by Zdziarski (2014b), most if not all of the forensic tools in the market extract information from iOS devices through the same sensitive services that we have discussed in this chapter. Consequently, using *Lockup* to restrict these services would result in those tools not being able to extract data from the device.

Given that our tool offers different service levels or *profiles*, the effectiveness of forensic tools against a particular iOS device with *Lockup* installed will depend on the profile being used at any particular moment. Although we have not been able to pit our tool against commercial forensic software tools, something we would like to do if we get resources, by definition (according to the services being disabled), when the forensic tools attempts to retrieve data from the device itself (i.e., excluding other analysis sources such as iTunes backups stored in a desktop computer) these should be the antiforensic effects of the different profiles available:

- *Level 1*. In this level we have disabled *com.apple.file relay*, preventing any tool that leverages this service from dumping nearly all data on the device and bypassing the backup encryption password. Although it is not known whether any publicly available forensic tool leverages this service, something we would like to address in future research, we can assume that it is trivial for an adversary to develop such tool.
- *Level 2*. Starting at this level, all services are forced to work over USB cable, disabling its invocation over Wi-Fi or cellular networks. This should not have any impact on forensic tools given that generally these tools recommend to retrieve the data over USB cable.
- *Level 3*. Having disabled *com.apple.mobile.house arrest*, forensic tools' ability to retrieve 3rd-party app data (in particular, user data stored within apps) may be somehow impacted.

Although most of those data is still reachable through *com.apple.mobilesync* and possibly also in the form of an iTunes backup (although in this case it may be protected with a backup encryption password, if set).

- *Level 4*. In this level the device will not provide data in the form of iTunes backup, because the services *com.apple.mobilebackup* and *com.apple.mobilebackup2* have been disabled.
- *Level 5*. This level disables *com.apple.mobilesync*, which should prevent forensic tools from retrieving most application data from the device; this would apply to both 3rd-party applications, as well as stock iOS apps (messages, phone logs, calendar events, etc.).
- *Level 6*. By disabling *com.apple.afc*, this level prevents access to the latest bit of personal information that was still being exposed: the *Media* folder containing the camera roll (pictures and videos), voice memos, music and podcasts.
- *Level 7*. This level definitely disables all *Lockdown* services. There is little difference between this, and disabling any profitable service as we have already done by level 6.

In summary, levels 4 and higher should keep most user data away from forensic tools, with the exception of photos and videos (as sharing multimedia materials is still enabled in levels 4 and 5, for those users who prefer it). For absolute isolation, levels 6 or 7 should be used.

It may be argued that using an iOS version for which a jailbreak is available would make it easier for forensic tools to access the data in the device. However, after jailbreaking the device it is possible to patch some or all of the bugs used by the jailbreak process, thus achieving a higher grade of protection.

With additional resources, we would like to conduct a more detailed study on how the existing commercial tools interact with an iOS device protected by *Lockup*.

## 5.4 COUNTERMEASURES: ANTI-ANTIFORENSICS

If *Lockup*, or a similar solution, gets popular, the forensic tools could be adapted to cope with it. Given that *Lockup* can only be installed in iOS versions for which a jailbreak is available, forensic tools could be modified to exploit the flaws that permit the jailbreak in each particular iOS version, in order to regain control over the device and reactivate the services needed to extract information. This, however, would most probably require: physical access to the device; knowing the device passcode (a password set by the user); and manual interaction in order to be able to powercycle the device and possibly press certain buttons during boot in order to enter a special boot mode known as DFU. This means that attacks via malicious devices such as the one described in Lau et al. (2013) are much less likely to success when *Lockup* is installed and a restrictive profile is applied.

As for attackers trying to abuse the *lockdown* services, their methods would stop working after applying a restrictive profile with *Lockup*. It would still be possible to retrieve user information by finding and exploiting other vulnerabilities in the particular iOS version running in the device, or through other means such as iCloud (Oestreicher, 2014; Ruan et al., 2013).

Assuming an attacker gets to execute code on the iOS device, she could modify the *Services.plist* file at will and re-enable any service to exploit it at a later time. The device owner could notice this by running *Lockup*, which in this case shows a warning informing the user that the installed *Service.plist* file does not correspond to any of the known profiles, as shown in Fig. 5. Future versions of *Lockup*

```
2. (bash)
Current profile: WARNING!! Unknown or corrupt Services.plist file.

Available profiles:
0. Default iOS service set. Wi-Fi sync ON.
1. MDM environments. Wi-Fi sync ON.
2. Sync apps. Wi-Fi sync OFF.
3. Sync application files. Wi-Fi sync OFF.
4. Sync media files. Wi-Fi sync OFF.
5. Share media files. Wi-Fi sync OFF.
6. No sensitive services. Wi-Fi sync OFF.
7. No services at all (paranoid mode).
--
9. ABORT - Exit this program.
D. Dump installed profile.
E. Enumerate services exposed by the current profile.

Press key and ENTER...
e
Enumerating services...

com.apple.crashreportmover
com.apple.mobile.notification_proxy
com.apple.mobile.heartbeat
com.apple.preboardservice
com.apple.misagent
com.apple.mobile.insecure_notification_proxy
com.apple.atc
com.apple.thermalmonitor.thermtgraphrelay
com.apple.mobile.MDMService
com.apple.rasd
com.apple.purpletestr
com.apple.mobile.mobile_image_mounter
com.apple.webinspector
com.apple.afc
com.apple.radios.wirelesstester.root
com.apple.mobile.debug_image_mount
com.apple.mobile.assertion_agent
com.apple.springboardservices
com.apple.crashreportcopymobile
com.apple.radios.wirelesstester.mobile
com.apple.hpd.mobile

bash-3.2$ ▌
```

**FIG. 5**

*Lockup* warning the user about the currently installed *Services.plist* file, and enumerating the services currently enabled.

could include periodic checks for this and warnings for the user—something that, again, could be targeted by an adversary.

## 6 CONCLUSIONS AND FUTURE WORK

In this chapter we have reviewed the security and privacy risks presented by certain background services that exist in the iOS operating system. We have presented a number of mitigation measures that can be used to reduce those risks. The main contribution of this chapter is *Lockup*, a software tool that hardens the security of iOS devices by defining a number of *profiles* which reduce the number of exposed services. In addition, we have discussed the antiforensic implications of our solution, and the anti-antiforensics countermeasures that could be used to bypass it. Given the huge amount of personal information that can be extracted by abusing these sensitive services, we believe it is worth exploring this kind of solutions. The expected rise of wearable devices will only increase the need for solutions that enhance the devices' security and privacy levels.

As it is usually said, a chain is only as strong as its weakest link. If all it takes is compromising a trusted device such as a desktop computer in order to put an iOS user under severe surveillance, that makes iOS not safer than the average desktop computer—with its unpatched OS, its obsolete Java, and it's always-vulnerable Flash Player. This level of security is acceptable to keep away casual attackers, but it will not stand a high-profile targeted attack against specific users.

*Lockup* has been released as free software (Gómez-Miralles and Arnedo-Moreno, 2015b) so that other researchers or developers can adapt it as they find convenient. We have the intention of continuing working in *Lockup*, maintaining it, and adding new features, such as: monitoring and logging connection attempts to *lockdown* services, alerting the user in real-time; adding a graphical interface to the software; monitoring the set of available services and alerting the user if new services are added. It would also be possible to integrate it with other solutions such as *activator* (Petrich, 2014). However, from a security standpoint, it would be preferable to keep the software as simple as possible, both in terms of size and in terms of dependencies.

The purpose of the proof-of-concept tool presented in this chapter is to fight the security risks presented by a number of iOS unwanted services. It must be kept in mind, however, that our solution will only work in jailbroken devices, and the process of jailbreaking itself implies circumventing and disabling a number of native iOS security mechanisms.

An interesting future line of research would be creating custom jailbreak tools that after deploying this software return the device to its original state to the best possible extent. This would keep most of the benefits and security features of stock Apple devices, while avoiding exposure through unwanted services. Another interesting point would be to create a survey of services used by common commercial accessories; in addition, we would like to obtain licenses for commercial tools, mainly those used by Law Enforcement, in order to test them against our software.

## ACKNOWLEDGMENT

This work was partly funded by the Spanish Government through project TIN2011-27076-C03-02 CO-PRIVACY.

## GLOSSARY

**Address Space Layout Randomization (ASLR)** A measure aimed at difficulting the exploitation of buffer overflow vulnerabilities by randomizing at run-time the position in memory of the data structures used by the application.

**Cydia** The de facto unofficial App Store for jailbroken devices.

**Data Execution Prevention (DEP)** A technique to complicate buffer overflow exploits, by marking the data pages in memory as not executable. This makes it no longer possible for an attacker to store exploit code in the memory space of a variable.

**Device Firmware Upgrade (DFU)** A special mode in which iOS devices can be put by entering a specific combination of button presses. A device in DFU mode will receive a firmware image through USB cable, will write it to the device internal storage and will try to boot the image.

**iCloud** Apple's cloud storage platform, capable of syncing documents and data across iOS devices and Mac computers.

**iOS** The operating system used by the iPhone, iPad, and iPod Touch. Before the introduction of the iPad in 2010 it was simply called *iPhone OS*, even when it was used in the iPod Touch as well.

**iPhoto** Apple's photo management software for desktop computers; normally it is also used to import pictures from external devices such as digital cameras and iOS devices.

**iTunes** Apple's multimedia software for desktop computers, which is also used to manage and backup the iPhone, iPad, and iPod.

**Jailbreak** A technique to suppress iOS' code execution restrictions, thus being able to run custom software in the iOS device. It is one of the many measures typically used by researchers for examining the iOS internals; a similar technique exists in the Android platform, called *rooting*. The process is also popular among users because it allows them to run all sorts of tweaks that would never be approved by Apple to be published in the official App Store.

**Lockdown** An iOS system process that presents a number of network services, which can be accessed from a host computer either through USB cable or wirelessly. Its behavior is somehow comparable to that of the `inetd` daemon present in many UNIX systems.

**MobileSafari** The stock web browser embedded in every iOS version.

**Mobile Device Management (MDM)** A set of software products designed to control mobile devices. These are typically used in corporate environments, where the company can enforce certain security settings, preload corporate applications, deploy certificates, and configuration profiles.

**Network sniffer** A software program designed to capture network data (wireless data in this case) regardless of whether the device is the intended destination of such traffic. Simply put, a wireless network sniffer can be used to intercept the traffic of other nearby devices.

## REFERENCES

Allegra, N., Freeman, J., 2011. JailbreakMe 3.0. http://jailbreakme.com.

Apple Computer, Inc., 2014a. iPhone in business. https://www.apple.com/iphone/business/ios.

Apple Computer, Inc., 2014b. Resources for IT and enterprise developers. https://developer.apple.com/enterprise/.

Apvrille, A., 2014. Inside the iOS/AdThief malware. https://www.virusbtn.com/pdf/magazine/2014/vb201408-AdThief.pdf.

Bedrune, J.B., Sigwald, J., 2011a. iPhone data protection in depth. Hack In The Box Conference.

Bedrune, J.B., Sigwald, J., 2011b. iPhone data protection tools. http://code.google.com/p/iphone-dataprotection/.

Chang, Y.T., Teng, K.C., Tso, Y.C., Wang, S.J., 2015. Jailbroken iPhone forensics for the investigations and controversy to digital evidence. J. Comput. 11, 2.

Chen, C.N., Tso, R., Yang, C.H., 2013. Design and implementation of digital forensic software for iPhone. In: Proceedings of the 8th Asia Joint Conference on Information Security, AsiaJCIS 2013.

Chronic Dev Team, 2010. greenpois0n. https://github.com/Chronic-Dev/greenpois0n.

Forensic Telecommunications Services Ltd., 2010. iXAM—Advanced iPhone Forensics Imaging Software. http://www.ixam-forensics.com/.

Freeman, J., 2014a. Cydia Substrate. http://www.cydiasubstrate.com.

Freeman, J., 2014b. WinterBoard. http://cydia.saurik.com/package/winterboard/.

Gallagher, R., Mass, P., 2014. Inside the NSA's secret efforts to hunt and hack system administrators. https://firstlook.org/theintercept/2014/03/20/inside-nsa-secret-efforts-hunt-hack-system-administrators/.

Gómez-Miralles, L., Arnedo-Moreno, J., 2015a. Airprint forensics: recovering the contents and metadata of printed documents from iOS devices. Mobile Inform. Syst. 2015 (2015). Article ID 916262.

Gómez-Miralles, L., Arnedo-Moreno, J., 2015b. Lockup. http://www.pope.es/lockup.

Good Technology, 2014. Good Technology mobility index report Q2. http://media.www1.good.com/documents/rpt-mobility-index-q2-2014.pdf.

Grispos, G., Storer, T., Glisson, W.B., 2011. A comparison of forensic evidence recovery techniques for a windows mobile smart phone. Digit. Invest. 8, 23–36.

Hotz, G., 2007. iPhone serial hacked, full interactive shell. http://www.hackint0sh.org/f127/1408.htm.

International Data Corporation, 2014. Worldwide Quarterly Mobile Phone Tracker Q2 2014. International Data Group.

iPhone Dev Team, 2011. redsn0w. http://redsn0w.com.

Iqbal, B., Iqbal, A., Obaidli, H.A., 2012. A novel method of iDevice (iPhone, iPad, iPod) forensics without jailbreaking. In: Proceedings of the 8th International Conference on Innovations in Information Technology.

Lau, B., Jang, Y., Song, C., Esser, S., Wang, T., ho Chung, P., Royal, P., 2013. Mactans: injecting malware into iOS devices via malicious chargers. https://media.blackhat.com/us-13/US-13-Lau-Mactans-Injecting-Malware-into-iOS-Devices-via-Malicious-Chargers-WP.pdf.

Lisiansky, D., 2014. CCControls. http://cydia.saurik.com/package/com.danyl.cccontrols/.

Miller, C., Blazakis, D., Dai Zovi, D., Esser, S., Iozzo, V., Weinmann, R., 2012. iOS hacker's handbook. Wiley, New York.

Oestreicher, K., 2014. A forensically robust method for acquisition of iCloud data. Digital Invest. 11 (suppl. 2), s106–s113.

Petrich, R., 2014. Activator. https://rpetri.ch/cydia/activator/.

Porras, P., Saïdi, H., Yegneswaran, V., 2010. An analysis of the iKee.B iPhone Botnet. In: Security and Privacy in Mobile Information and Communication Systems, Lecture Notes of the Institute for Computer Sciences, Social Informatics and Telecommunications Engineering. Springer Berlin Heidelberg, Berlin, Heidelberg.

Rabaiotti, J.R., Hargreaves, C.J., 2010. Using a software exploit to image RAM on an embedded system. Digital Invest. 66 (3–4), 95–103.

Ritchie, R., 2014. Apple reaffirms it has never worked with any government agency to create a backdoor in any product or service. http://www.imore.com/apple-reaffirms-never-worked-any-government-agency-backdoor-product-service.

Rosenbach, M., Poitras, L., Stark, H., 2013. iSpy: How the NSA accesses smartphone data. http://www.spiegel.de/international/world/how-the-nsa-spies-on-smartphones-including-the-blackberry-a-921161.html.

Ruan, K., Carthy, J., Kechadi, T., Baggili, I., 2013. Cloud forensics definitions and critical criteria for cloud forensic capability: an overview of survey results. Digital Invest. 10, 34–43.

Soghoian, C., 2007. A game of cat and mouse: the iPhone, Steve Jobs and an army of blind hackers. http://www.cnet.com/news/a-game-of-cat-and-mouse-the-iphone-steve-jobs-and-an-army-of-blind-hackers/.

Stroz Friedberg, 2014. unTRUST. https://github.com/strozfriedberg/unTRUST.

Umeng, 2014. Insight Report: China Mobile Internet 2013.

Vidas, T., Zhang, G., Christin, N., 2011. Toward a general collection methodology for android devices. Digital Invest. 8 (special issue), s14–s24.
Wang, Y.D., Bassen, N., et al., 2013. evasi0n. http://evasi0n.com.
Zdziarski, J., 2008. iPhone Forensics: Recovering Evidence, Personal Data, and Corporate Assets. O'Reilly
Zdziarski, J., 2014a. Apple responds, contributes little. http://www.zdziarski.com/blog/?p=3447.
Zdziarski, J., 2014b. Identifying back doors, attack points, and surveillance mechanisms in iOS devices. Digital Invest. 11, 3–19.

CHAPTER

# PATH LOSS ALGORITHMS FOR DATA RESILIENCE IN WIRELESS BODY AREA NETWORKS FOR HEALTHCARE FRAMEWORK

# 13

**Arif Sari*, Ahmed Alzubi***

*Girne American University, Kyrenia, Cyprus**

## 1 INTRODUCTION

In recent years, scholarly work on wireless body area networks (WBAN) has gained substantial attention both from practitioners and scholars. This is primarily due to the rapid growth in the field of information communications and technology, in particular protocols, an important aspect of WBAN internet communication. WBAN were developed solely for medical applications because most medical applications rely on sensors to collect data from sensitive areas in human body, such as the heart, vein, and the brain. As such the sensor nodes should be energy-efficiency to increase battery life time, moreover, patient data transfer are often subjected to path-loss and/or unauthorized use. Therefore the need for sensors nodes to be energy efficient and to restrict unauthorized access to patient data by intruder exists.

## 2 OVERVIEW OF WBAN FRAMEWORK

In a globalized world where our daily life becomes increasingly competitive, where pollution and infections becomes part of our life, and where the access and offer of medical services increases, the importance of advanced healthcare system is inevitable. People are prone to various kinds of diseases both natural and man-made; hence, there is a need for an exhaustive HealthCare framework. A framework that would satisfy mankind desires "to move around" and "to be free." One of those technologies is the wireless body area network (WBAN) system. Ragesh and Baskaran (2012) stated that WBAN has gained global attention for its remarkable medical services. This subsumes gadgets such as modest sensors that can be attached in or be embedded around the body in close vicinity to monitor those who require medical attention. This development has eliminated the need for real-time inspection of elderly and attention-needy patients, that is medical experts can monitor such people from a remote location, as noted by Preneel (2003).

A new WBAN communication protocol both for medical and nonmedical applications developed with a range of at least three meters (3m) was defined (Ryckaert et al., 2004). Kim and Kim (2014) added that "the protocol was developed to support a low complexity, low cost, ultra-low power, and highly reliable wireless communication for use in close proximity to, or inside, a human body (but not

**Security and Resilience in Intelligent Data-Centric Systems and Communication Networks. https://doi.org/10.1016/B978-0-12-811373-8.00013-6**

Copyright © 2018 Elsevier Inc. All rights reserved.

limited to humans) to satisfy an evolutionary set of entertainment and healthcare products and services." Therefore it is expected that WBAN would function as the fundamental element for electronic health service.

In health services, WBAN monitors sick person(s) suffering from diseases that requires consistent medical attention such as cardiac arrest, diabetes, and others. It was reported that medical service expenditure would reach 20% of the Gross Domestic product (GDP) in the next 10 years. The figure is huge and suggests that the world economy will be affected; as such there is a need to shift from real time treatment to remote form of medical services in order to free some resource. Wearable systems for periodic medical monitoring are important technology that can help the medical industry and sector transit to a more proactive and affordable health system (Kim and Kim, 2014).

Portable sensors are inserted into or attached to human body for monitoring potential changes. There are various ways information can be transmitted and exchanged through WBAN's and this depends on the type and characteristic of the device namely: Bluetooth, ZigBee, MICS, and UWB (Esat, 2004). At the point when information gathered by the sensors and gadgets are exchanged through remote medium to remote destination, then "path loss can happen. Path loss can be In-Body and On-Body, moreover, it relies on recurrence of operations and separation among transmitter and recipient. Avi (2016) proposed a basic path loss model for WBAN in his study. Recently utilizing MATLAB, Bellovin and Rescorla (2006) proposed mimic In-Body path loss model. Generally speaking, there are four path loss models: (1) profound tissue inserts to embed; (2) close surface inserts to embed; (3) profound insert to embed, and (4) close surface insert to embed.

## 2.1 CHARACTERISTICS OF WIRELESS CHANNEL IN WBANS

A WBAN meant to monitor medical environments is comprised of various sensor nodes that can checking the vital data and report the patients' health state and events. These sensor nodes are placed on the human body, the accurate area, and connection of the sensor nodes in patient body rely on upon the sensor sort, size, and weight. Mankind can use sensors in the form of stand-alone gadgets or be incorporated with gems, connected as small fixes on the skin, covered up in the client's garments or shoes, or even embedded in the client's body (Sana et al., 2009). WBAN devices can be placed on the surface of a patient body or inside the patient's body, these devices can work with varying range of frequency bands that is contingent upon the placed position and connection. "The allowable connection scenarios are implant to implant, implant to body surface, implant to an external point, and body surface to body surface (line-of-sight and non-line-of-sight)" (see Fig. 1).

Tables 1 and 2 provide examples of the two types of WBAN.

The accurate number and kind of the sensors utilized as a part of a WBAN for healthcare monitoring rely upon the end-client application and may incorporate a varieties of the supporting sensors (Bh et al., 2010; Ugent et al., 2011).

- An ECG sensor for observing heart movement
- An EMG sensor for observing muscle movement
- An EEG sensor for observing cerebrum electrical movement
- A SpO2 sensor for observing blood oxygen immersion
- A sleeve based weight sensor for observing circulatory strain
- A resistive or piezoelectric mid-section belt sensor for observing breath
- A blood glucose level sensor temperature sensor for observing body temperature

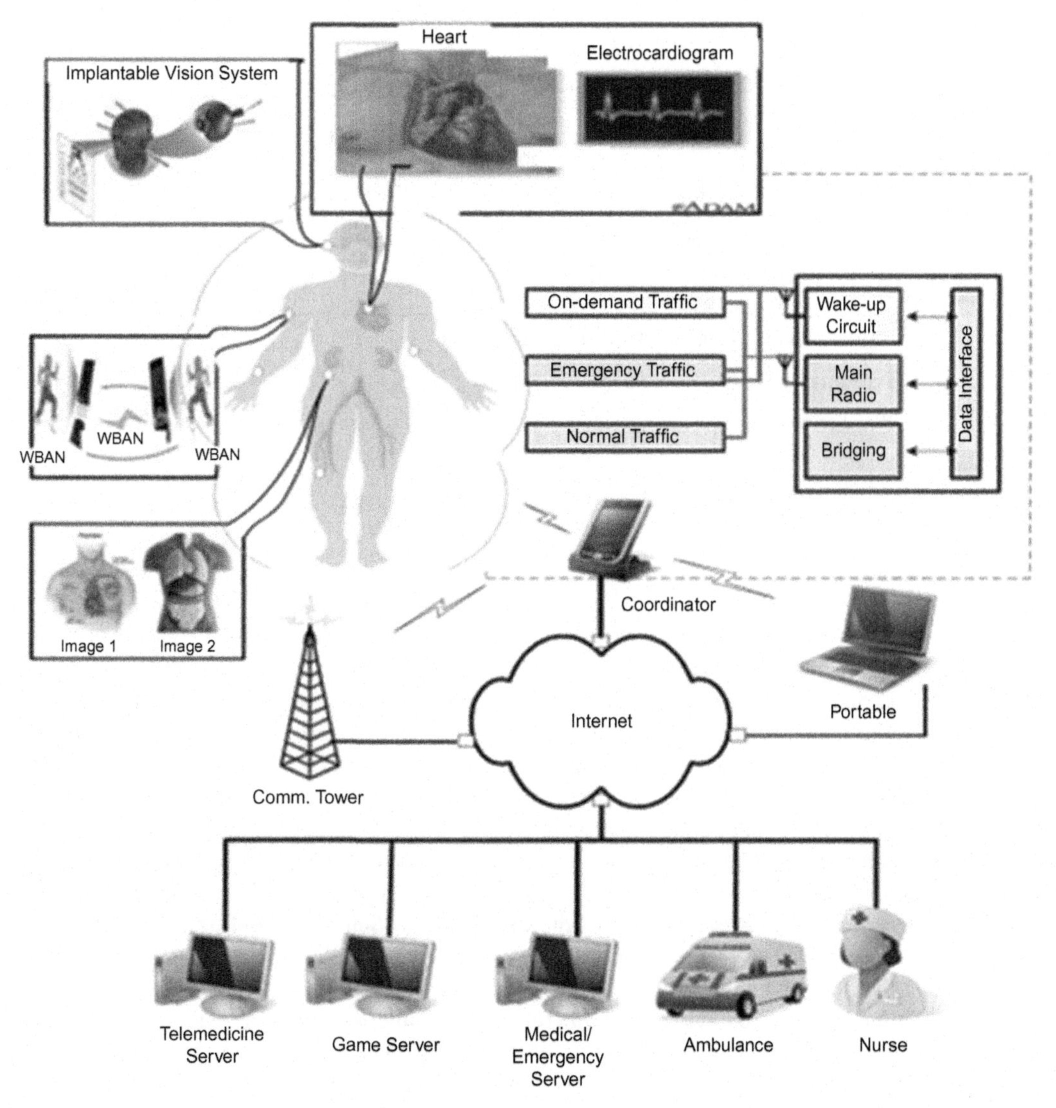


**FIG. 1**

The general structure of WBAN.

- An area sensor (e.g., GPS) to track client's area
- Accelerometer-based movement sensors to gauge sort and level of client's exercises

## 2.2 WBAN AND WSN NETWORK TOPOLOGY

According to a report by IEEE, WBANs operate in two format (1) one-hop and/or (2) two-hop star topology, and the node is often located in the center similar to a connector (Shah and Yarvis, 2006).

**Table 1 Wearable WBAN Examples**

| Wearable WBAN | Purpose |
|---|---|
| Sleep apnea | For detecting health disorders |
| Hearing aids | Hearing assisting device |
| Smart shoes and watches | For healthy life purpose |
| Insulin pump | Use for pumping insulin |

**Table 2 Implantable WBAN Examples**

| Implantable WBAN | Purpose |
|---|---|
| Blood pressure sensors | To measure and report critical level of blood pressure to doctor |
| Advanced insulin pumps | To measure and report critical level of level and save diabetic data for doctors observations |

Sukor et al. (2008) noted that there are two types of data transmission, the one-hop star topology, one that is initiated from the coordinator to the device and the other type is usually a transmission from the device to the coordinator. Furthermore, the communication methods that exist in the star topology are beacon mode and nonbeacon mode. The "beacon mode occurs when the node in the center of the star topology controls the communication. It transmits periodic beacons to define the beginning and the end of a super frame to enable network association control and device synchronization" (Movassaghi et al., 2014). Whereas the "nonbeacon mode occurs when a node in the network is capable of sending data to the coordinator and can use Carrier Sense Multiple Access with Collision Avoidance (CSMA/CA) when required. The nodes need to power up and poll the coordinator to receive data" (Movassaghi et al., 2014).

WBANs are considering as extraordinary type of a Wireless Sensor System or a Wireless Sensor and Actuator Network (WSAN) with its own requirements. Nonetheless, traditional sensor systems cannot handle the special challenges associated with human body monitoring. The human body comprises of a muddled inward environment that reacts to and interfaces with its external surroundings (Javaid et al., 2013). The human body and its environment does not just have a smaller scale, as it requires a different kind of checking and monitoring activities, that comes along with different challenges than those confronted by WSNs. Sensing and monitoring of health related data is very relevant and prevalent in most health institutions as such expanding the interest for dependability. See Tables 3 and 4.

## 2.3 EXISTING WBAN APPLICATIONS

WBAN applications are mostly heterogeneous in nature. WBAN standards are design to permit connections "between different devices because each sensor has different characteristics (e.g., a temperature sensor is necessarily different from a heartbeat sensor). The type, frequency, and amount of data are also different. A protocol able to deal with these heterogeneous systems may be important"

**Table 3 Comparison of WBANS and WSNS**

| Criteria Wireless | Wireless Sensor Network | Body Area Networks |
|---|---|---|
| Dimension of Network | Few to several thousand nodes over an area from meters to kilometers | Dense distribution limited by body size |
| Topology | Random, Fixed/Static | One-hop or two-hop star topology |
| Size of Node | Small size preferred (no major limitation in most cases) | Miniaturization required |
| Accuracy of Node | Accuracy outweighs large number of nodes and allows for result validation | Each of the nodes have to be accurate and robust |
| Replacement of Node | Easily performed (some nodes are disposable) | Difficultly in replacement of implanted nodes |
| Bio-compatibility | Not a concern in most applications | Essential for implants and some external sensors |
| Battery and Energy Supply | Accessible, Capable of changing more frequently and easily | Difficultly in replacement and accessibility of implanted settings |
| Lifetime of Node | Several years/months/weeks (application-dependent) | Months or years (depending on application) |
| Power consumption | Power demand is high and but easy to provide. | Power demand is low but difficulty to provide |
| Alternative Source of Energy | Solar and Wind energy are good option | Motion and Thermal energy are good option |
| Data Rate | Data rate is homogenous | Date rate is heterogenous |
| Data Loss Impact | Data loss over wireless transfer is compensated by the large number of nodes | Data loss is considered more significant (may need additional measures to ensure real time data interrogation capabilities and QoS) |
| Security Level | Lower (application-dependent) | Higher security level to protect patient information |
| Traffic | Application specific, Modest data rate, Cyclic/sporadic | Application specific, Modest data rate, Cyclic/sporadic |
| Wireless Technology | WLAN, GPRS, ZigBee, Bluetooth, and RF | 802.15.6, ZigBee, Bluetooth, UWB |
| Context Awareness | Insignificant with static sensors in a well-defined environment | Very significant due to sensitive context exchange of body physiology |
| Overall Design Goals | Self-operability, Cost optimization, Energy Efficiency | Energy Efficiency, Eliminate electromagnetic exposure |

(Manirabona et al., 2017). This is because information rates will shift firmly, extending from straightforward information at a couple kb/ps to video floods of a few Mb/ps, alternatively, data can be transfer in blasts, which implies that it is sent at higher rate amid the blasts.

Tables 5 and 6 present the volume of information for varieties of applications and method meant for sampling rate, the reach and the sought exactness of the estimations (Jaff, 2009; Kim et al., 2016). In general, it can be observed that the volume of information for the application are not high. In rare occasions, WBAN with fewer gadgets (i.e., twelve motion sensors, ECG, EMG, and glucose checking)

**Table 4 Medical and Nonmedical Applications for WBAN**

| Challenges | Wireless Sensor Network | Wireless Body Area Network |
|---|---|---|
| Scale | Monitored environment (Meters/Kilometers) | Human body (centimeters/meters) |
| Node number | Many redundant nodes for wide area coverage | Fewer, limited in space |
| Result accuracy | Through node redundancy | Through node accuracy and robustness |
| Node tasks | Node performs a dedicated task | Node performs multiple tasks |
| Node size | Small is preferred, but not important | Small is essential |
| Network topology | Very likely to be fixed or static | More variable due to body movement |
| Data rates | Most often homogenous | Most often heterogonous |
| Node replacement | Performed easily, nodes even disposable | Replacement of implanted nodes difficult |
| Node lifetime | Several years/months | Several years/months, smaller battery capacity |
| Power supply | Accessible and likely to be replace more easily and frequently | Inaccessible and difficult to replace in an implantable setting |
| Power demand | Likely to be large, energy supply easier | Likely to be lower, energy supply more difficult |
| Energy scavenging source | Most likely solar and wind power | Most likely motion (vibration) and thermal (body heat) |
| Biocompatibility | Not a consideration in most application | A must for implants and some external sensors |
| Security level impact of data loss | Lower Likely to be compensated by redundant nodes | Higher, to protect patient information More significant, may require additional measures to ensure QOS and real time delivery |
| Wireless technology | Bluetooth, ZigBee, GBRS, WBEN... | Lower power technology require |

when aggregated, the volume of information can reach it peak and almost achieve a couple Mbps, which is a higher than the overall piece rate of most existing low power radios (Lont, 2013). The unwavering quality of the information transmission is given as far as the fundamental piece blunder rate (BER) which is utilized as a measure for the quantity of lost parcels. For a medicinal gadget, the dependability relies on upon the information rate (Khudri and Sutanom, 2005).

"Low information rate gadgets can adapt to a high BER, while gadgets with a higher information rate require a lower BER. The required BER is too subject to the criticalness of the information" (Singh, 2013).

<table>
<caption>Table 5 Existing WBAN Applications</caption>
<tr><td rowspan="4">WBAN Applications</td><td rowspan="3">Medical</td><td>Wearable WBAN</td><td>Assessing Soldier Fatigue and Battle Readiness<br>Aiding Professional and Armature Sport Training<br>Sleep Staging<br>Asthma<br>Wearable Health Monitoring</td></tr>
<tr><td>Implant WBAN</td><td>Cardiovascular Diseases<br>Cancer Detection</td></tr>
<tr><td>Remote Control of Medical Devices</td><td>Ambient Assisted Living (AAL)<br>Patient Monitoring<br>Tele-medicine Systems</td></tr>
<tr><td colspan="2">Nonmedical</td><td>Real Time Streaming<br>Entertainment Applications<br>Emergency (nonmedical)</td></tr>
</table>

**Table 6 Examples of Medical WBAN Applications and Characteristics**

| Application | Data Rate | Bandwidth | Accuracy |
|---|---|---|---|
| ECG (12 leads) | 288 kbps | 100–1000 Hz | 12 bits |
| ECG (6 leads) | 71 kbps | 100–500 Hz | 12 bits |
| EMG | 320 kbps | 0–10,000 Hz | 16 bits |
| EEG (12 leads) | 43.2 kbps | 0–150 Hz | 12 bits |
| Blood Saturation | 16 bps | 0–1 Hz | 8 bits |
| Glucose monitoring | 1600 bps | 0–50 Hz | 16 bits |
| Temperature | 120 bps | 0–1 Hz | 8 bits |
| Motion Sensor | 35 kbps | 0–500 Hz | 12 bits |
| Cochlear implant | 100 kbps | – | – |
| Artificial retina | 50–700 kbps | – | – |
| Audio | 1 Mbps | – | – |
| Voice | 50–100 kbps | – | – |

## 2.4 TYPICAL WBAN SENSOR SPECIFICATIONS

WBANs have incredible potential for modern applications including remote restorative analysis, intelligent gaming, and military applications. However, this study is only interested in the medical side of WBAN. Table 7 presents a portion of the in-body and on body applications list (Ibraheem, 2014; Kwak et al., 2010). In-body applications designates checking scheduled programs and changes for pacemakers and implantable cardiovascular defibrillators, control of bladder capacity, and reclamation of appendage development (Khan et al., 2008). Whereas, on-body applications entails checking heart rate, circulatory strain, temperature, and breathing. Consequently, on-body nontherapeutic applications

**Table 7 Typical WBAN Sensor Specifications**

| Application Type | Sensor Node | Date Rate | Duty Cycle (per Device) % per time | Power Consumption | QoS(Sensitive to Latency) | Privacy |
|---|---|---|---|---|---|---|
| In-body Applications | Glucose Sensor | Few Kbps | <1% | Extremely Low | Yes | High |
| | Pacemaker | Few Kbps | <1% | Low | Yes | High |
| | Endoscope Capsule | >2 Mbps | <50% | Low | Yes | Medium |
| On-body Medical Applications | ECG | 3 Kbps | <10% | Low | Yes | High |
| | SpO2 | 32 bps | <1% | Low | Yes | High |
| | Blood Pressure | <10 bps | <1% | High | Yes | High |
| On-body Nonmedical Applications | Music for Headsets | 1.4 Mbps | High | Relatively High | Yes | Low |
| | Forgotten Things Monitor | 256 Kbps | Medium | Low | No | Low |
| | Social networking | <200 Kbps | <1% | Low | No | High |

entails checking sensitive information pieces that are easily overlooked, this include setting up an informal data and organizational activities, such as surveying officer and fight status (Pote, 2012).

## 3 MESSAGE INTEGRITY IN WIRELESS COMMUNICATION

Message integrity depicts the validity, trustworthiness, and authenticity of the information in a message transmitted from a source to a given destination. Message uprightness assurance systems like authentication mechanism in which the message verification codes guarantees that the receiver can easily observe or know if received information was transmitted by authorize party (Liu et al., 2015). Message validation codes influence cryptographic primitives such as cryptographic hash capacities. While message authenticity covers the sensible information substance of messages, different controls amid the transmission of the message cannot be distinguished. For instance, transfer and delay assaults can be mounted without changing the information content (Abouei et al., 2011).

Integrity mechanism guarantees that information is delivered at a sensor (the destination) is not undermined or tempered. There is a need for reliability control of the information to make sure it is not altered in anyway (Taparugssanagorn et al., 2008). Message authenticity is essential in biosensor systems because of the sensitive of healthcare-related data, altered wrongly captured information may bring about wrong drug administration, treatment, and subsequently health damage to the patient (Saleem, 2009). In this view, security seems inevitable and has become a vital aspect of biotechnology in particular when body biosensor systems are to be embraced. Securing a system incorporates the vital issues of authentication (character confirmation of conveying gadget), confidentiality (bargain of

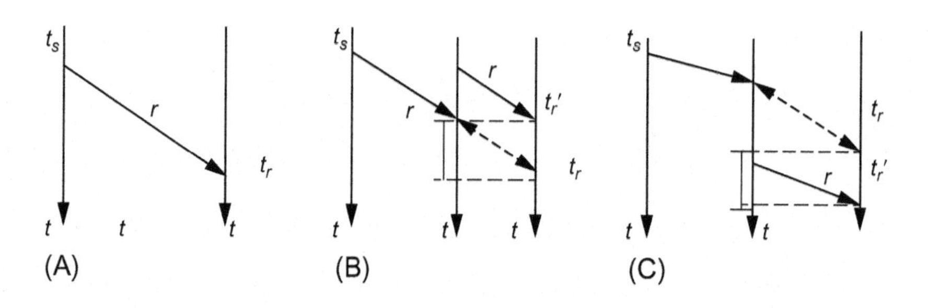


**FIG. 2**

Transmission and message integrity (Bangash et al., 2014). (A) Normal transmission. (B) Advancement attack. (C) Delay attack.

private data), integrity (the information/message must be honest to goodness and blunder free) and availability (loss of vitality and different assets).

Despite this, the system can postures more prominent significance and test to guarantee information and hub verification, information privacy, and freshness. Since we consider a system of moment bits embedded inside the body: portability and physical catch are lesser dangers (Amir and Jim, 2012). Challenging controls on the remote correspondence channel exist; such controls could target distinctive parts of physical-layer qualities of the message, for example, seen in edge of-landing, sign quality, piece blunder rate, range, and time of entry. In the meantime, assaults can likewise focus on a few viewpoints. For example, a postponed replay from an alternate area. Specifically, we focus on the part of time-of-entry, such as assaults on the transient uprightness, as depicted in Fig. 2.

According to Bangash et al. (2014), attacks or assaults on message and/or information trustworthiness can either defer or propel a message. Such that the aggressor can perform a message progression assault by transmitting a message early, that is transmits fake information. To play out a message delay assault, the aggressor needs to keep the gathering of the first message and replay. Fig. 2 shows that (A) Transmission without an assailant. (B) The assailant sends r prior to propel the message. (C) The assailant jams r and re-sends it later to defer the message gathering.

## 3.1 ENCRYPTION ALGORITHM IN WBAN

The typical WBAN incorporates different sorts of restorative sensors that remotely relate with other restorative sensors and/or associate itself with the control hub (e.g., PDAs) such as WiMAX or Wi-Fi to convey the captured data (Koopman and Driscoll, 2012). Prior studies have emphasized the importance and relevance of secure correspondence medium from the web to control hubs (Padmavathi, 2009). That is, the research highlighted the need to create a secure medium and platform by which information can be safely transmitted from/to the web control hubs. Technically, this dissertation concentrate on the securing entomb sensor correspondence over the body range. In WBAN, key circulation is constantly defenseless against man in the middle assault. The dangers can be classified: active assault and detached assault. The dynamic assailants can drop messages and replay them with old messages. The detached assailants fit for listening the communication over WBAN (Abarna, 2012).

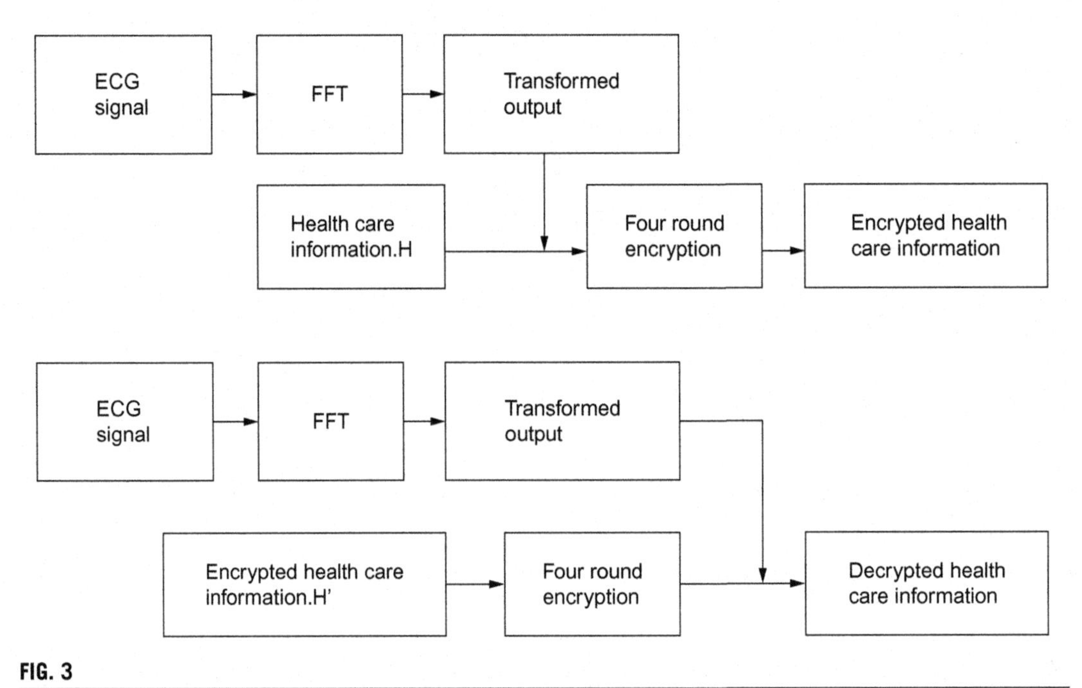


**FIG. 3**

Encryption system based on HB (Seshabhattar et al., 2011).

Accordingly, Seshabhattar et al. (2011) introduced a secure key establishment mechanism using an ultra-lightweight cipher called Hummingbird (HB). HB is a rotor-based encryption calculation intended for asset obliged gadgets. Murmuring flying creature is an ultra-lightweight cryptographic primitive for encryption and validation in extremely asset compelled situations. We propose an ECG-Humming winged animal plan for secure correspondence over WBAN. Murmuring winged animal is a blend of a piece figure and stream figure with:

- 16-bit piece size
- 256-piece key size
- 0-bit interior state

Humming-bird encrypts animal encodes 16-bit pieces of information utilizing a 256-piece key (Seshabhattar et al., 2011). The fundamental engineering of murmuring winged animal is unique and cross breed (with components of piece and stream ciphers). The encryption and unscrambling methodology (Fig. 3) can be spoken to as a constantly working rotor-based machine. Four indistinguishable interior square figures assume a part of virtual rotors.

Subsequently, Kumar et al. (2012) developed not only effective but efficient encryption protocol for medical purpose WSNs, the system ensures that patients' medical data are not exposed to unauthorized users. The system supports and works on session base such that health professionals can access patient data through mutual agreement with the patient and users can easily change their password. A dominant study proposed a two-user authentication protocol, one that uses one-way hash functions and XOR operations to reduce communication and computational overheads. A similarity and strength analysis suggested that this authentication protocol is more robust and provides better security than the existing protocols (Vaidya et al., 2009).

A deeper examination of the authentication protocols depicts that Perrig et al. (2002), peer to peer data transmission (Sensor Network Encryption Protocol) developed based on RC5 (Rivest Cipher) symmetric cryptography for confidentiality and MAC (Message Authentication Code), provided basic security attributes and playback protection. Their scheme was strong enough but largely resource-constrained. A certain number of recent security protocols in WSN was suggested to improve the data and security in the network as well as the devices itself (Amrita and Sipra, 2015; Wenbo and Peng, 2013).

## 3.2 HASH FUNCTION TECHNIQUE

Algorithms in cryptography are divided into three categories namely: private key, public key, and hash functions. As opposed to private key and public key estimations, hash expressions often request for one-way encryption of message that requires no key. Rather, a settled length hash functions is figured in light of the plaintext that makes it unthinkable for either the substance or length of the plaintext to be recouped (Laccetti and Schmid, 2007). The essential use of hash capacities in cryptography is message trustworthiness. The hash technique has a computerized unique mark of a message's substance, which guarantees that the message has not been adjusted by a gate crasher, infection, or by different means. The commonly used hash functions:

- Hashed Message Authentication Code (HMAC): Combines verification through a mutual mystery with hashing.
- Message Digest 2 (MD2): Byte-arranged, produces a 128-piece hash esteem from a discretionary length message, intended for savvy cards.
- MD4: Similar to MD2, planned particularly for quick handling in programming.
- MD5: Similar to MD4 yet slower on the grounds that the information is controlled more. Created after potential shortcomings were accounted for in MD4.
- Secure Hash Algorithm (SHA): Modeled after MD4 and proposed by NIST for the Secure Hash Standard (SHS), produces a 160-piece hash esteem.

In today's cryptography world, what appear to the client as a solitary framework really contains numerous calculations utilized as a part of conjunction to shape a half breed cryptosystem. Numerous calculations are utilized on the grounds that each is improved for a particular reason. For instance, Alice needs to make an impression on Bob. The message should be private, the message uprightness checked, and Alice's personality affirmed. Alice knows a few things, including the message, her own private key, and Bob's open key. Alice begins by sending the message through a hash capacity to acquire hash esteem. She scrambles the hash esteem with her private key utilizing an awry calculation. This structures the advanced mark (Maxwell et al., 2003).

Alice likewise makes an arbitrary session key for use by the symmetric encryption, which is utilized to encode the message. The mystery key is encoded with Bob's open key utilizing unbalanced encryption. The scrambled message and encoded session key shape an advanced envelope.

The computerized envelope and advanced mark are sent to Bob as noted by Crosby (2012). Bounce acquires the symmetric session key by unscrambling it with his private key utilizing uneven encryption. The session key is then used to decode the message. The unscrambled message goes through the hash capacity, and the worth is contrasted with the advanced mark's hash esteem that was decoded with Alice's open key (Raja, 2013).

Now, Bob knows: the substance of the private message (symmetric encryption). That the message was intended for him (since he could get the mystery key). That the message was not modified (in light of the fact that his hash esteem coordinated with Alice's hash esteem) as noted by Nabi et al. (2011). That the message was sent by Alice (since he could recuperate the hash esteem utilizing Alice's open key).

Be that as it may, why do we require these crypto calculations? Why not simply utilize Hilter Kilter encryption for everything? The answer is preparing speed: symmetric encryption is around 1000 times quicker than Hilter Kilter encryption for mass encryption. Diffie-Hellman and RSA were initially seen by their designers as an approach to scramble and unscramble data utilizing a split key, in this manner, killing the key trade issue of topsy-turvy encryption. In the mid-1980s, Lotus Notes Creator Ray Ozzie and PGP engineer Phil Zimmermann autonomously noticed that topsy-turvy encryption was much slower than symmetric encryption and utilizing uneven encryption for extensive volumes of information would be infeasible (Nabi et al., 2011; Braem and Blondia, 2011). They planned their product to utilize symmetric encryption for encryption of information and uneven encryption for key trade. Different calculations were included, for example, hash values for uprightness and marked hash values for verifying the sender (Otto et al., 2006).

We can utilize a hash whenever we need to demonstrate message uprightness. Hash values have been imperative in episode reaction for quite a while. They can be utilized to put a "carefully designed seal" on computerized proof as it is gathered. Case in point, numerous occurrence responders lean toward Polaroid cameras since advanced photographs can be effortlessly changed (Devi and Nithya, 2014). Be that as it may, computerized cameras are a great deal more helpful, so best practice is to make a hash of the advanced photograph at the earliest opportunity to diminish the time window one could assert the photograph was changed. A few cameras, for example, Nikon D200 and past can "verify" the pictures they shoot; this, obviously, is finished with a hash (Meena et al., 2014).

## 3.3 ELLIPTIC CURVE CRYPTOGRAPHY TECHNIQUE (ECC)

Koblitz (1987) developed an algorithm that later became the backbone of Elliptic Curve Cryptography (ECC). According to Lenstra and Verheul (2001), ECC is a cryptographic technique, which is asymmetric in nature that can give a complementary security provided by RSA system utilizing less lengthy keys. Recently, Marzouqi et al. (2015) noted that "the fundamental operation of the ECC is point scalar multiplication, where a point on the curve is multiplied by a scalar." "A point scalar multiplication is performed by calculating a series of point additions and point doublings, inferring from geometrical properties, points are added or doubled through series of additions, subtractions, multiplications, and divisions of their respective coordinates."

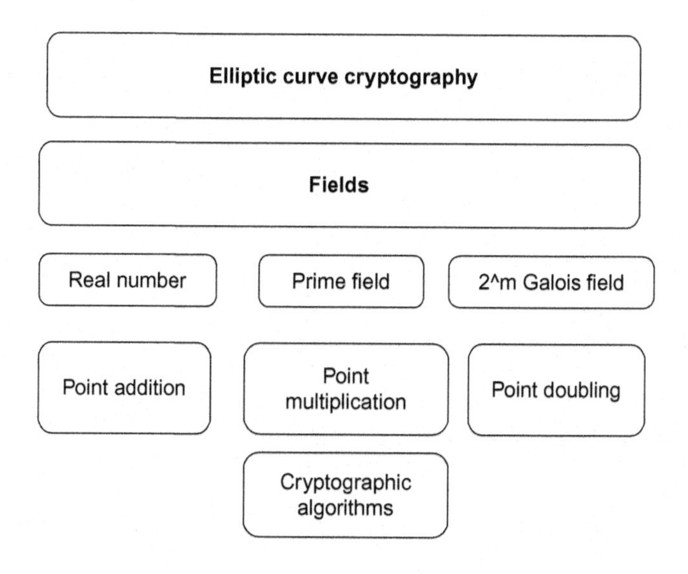


**FIG. 4**

ECC hierarchical model.

Rebeiro et al. (2012) added that the point coordinates are components of finite fields closed under prime or irreducible polynomial; hence, modular operations are necessary. A number of scholars have proposed various kinds of ECC processors in the literature based on dual field operations (Lai and Huang, 2011), binary extension fields (Wang and Li, 2011), and prime fields (Mane et al., 2011). Fig. 4 demonstrates a progressive model of ECC. ECC is isolated into three sorts of fields which are field over genuine numbers, field over prime numbers, and a paired Galois field. The fundamental operations in ECC are Point Multiplication, Point Expansion and Point Doubling (Rashwand and Misic, 2012). These operations can be performed over a wide range of fields, be that as it may this usage bargains just with the prime field, which is more qualified for programming execution purposes (Devita et al., 2014).

ECC is a multilayer system with increased hardware implementation complexity. A wide range of parameters and design choices affect the overall implementation of ECC systems (Marzouqi et al., 2015). Additionally, ECC requires smaller keys contrasted with non-ECC cryptography (taking into account plain Galois fields) to give proportional security. Elliptic bends are relevant for encryption, computerized marks, pseudo-arbitrary generators and different errands. They are additionally utilized as a part of a few whole number factorization calculations that have applications in cryptography, for example, Lenstra elliptic bend factorization (Movassaghi et al., 2014).

The fundamental difference between RSA and ECC is that ECC offers the same level of security for smaller key sizes. Elliptic Curve Cryptography is exceptionally scientific in nature. While routine open key cryptosystems (RSA, Diffie—Hellman and DSA) work specifically on extensive whole numbers, an Elliptic Curve Cryptography works over focuses on an elliptic bend. Vanstone (2003) stated that by default a subexponential time. Moreover, the algorithm is not a necessity for ECDLP, which suggest that smaller number of parameters can be used in ECC than with RSA or DSA. The size of the key depicts

**Table 8 Computational Performance Advantage of ECC Versus RSA**

| Security (Bits) | Symmetric Encryption Algorithm | Minimum Size (Bits) of Public Keys | | |
|---|---|---|---|---|
| | | DSA/DH | RSA | ECC |
| 80 | | 1024 | 1024 | 160 |
| 112 | 3DES | 2048 | 2048 | 224 |
| 128 | AES-128 | 3072 | 3072 | 256 |
| 192 | AES-192 | 7680 | 7680 | 384 |
| 256 | AES-256 | 15,360 | 15,360 | 512 |

the strength of the algorithm as in hash technique; as such the benefit reaped from smaller parameters includes speed and smaller keys or certificates. According to Vanstone (2003), the advantages that ECC has over other cryptographic techniques include the following:

- Processing power
- Storage space
- Bandwidth
- Power consumption

Therefore ECC is suitable for applications and environments with lots of constrained, e.g., smart cards, cellular phones, PDAs, digital post marks and other constrained environments especially in health services and care. As noted earlier, the key size in ECC is relatively small. In order not to compromise the security an application of an NIST's FIPS 140-2 standard should be followed. For instance, a symmetric cipher such as AES has to match the strength by public key algorithms such as RSA and ECC. For example, A 128-bit AES key demands an RSA key size of 3072-bits for equivalent security but only a 256-bit ECC key. Table 8 suggests that in ECC, key size scale linearly and not applicable to RSA, this is a result of key size increase. This is especially relevant to implementations of AES where at 256-bits you need an RSA key size of 15,360 bits compared to 512 bits for ECC.

Malan et al. (2004) at Harvard University conducted a study on the performance of ECC. The researchers evaluated ECC with 163-bit keys, per NIST's recommendation. They also compare SKIPJACK and the MICA2, and Diffie-Hellman and the MICA2, and ECC and the MICA2. At the end of their experiment, they concluded that ECC offers better performance and smaller keys (163 vs. 1024 bits). The summary of their research are given later. The costs of generating a 163-bit public or shared key are lower.

- Cost in tim: 34.161 s
- Costs in space: 1140 B of SRAM and 34,342 B of ROM
- Cost in energy: 0.816 Joules ($2.512 \times 10^8$ cycles)

ECC TinySec's shared secrets do allow for efficient, secure communications among nodes, hence there is a need to add additional crypto mechanism like Hash technique (Malan et al., 2004). ECC deployment may require so much RAM that it would be impossible to fit both the sensor network application and the ECC implementation on the same node (Gura et al., 2004). However, Liu and Ning (2008) proposed TinyECC with lower RAM consumption and power demand to meet wireless sensor applications demand. In addition, there is a need for security apparatus.

Xue et al. (2013) developed a special security scheme based on the concept of temporal credential for WSNs, the aim was to create a low cost cryptographic technique, such as hash and XOR. The authors suggested that their scheme was good enough. However, Jiang et al. (2015) found a loophole in Xue et al.'s work, such that unauthorized usage and attacks like insider attack, guessing and tracking attack of information can be done. Jiang et al. provided remedy to this flaw by proposing a much better scheme without involving public key cryptography. The authors also argued and claimed that their work is valid and safe. Nevertheless, Wang and Wang (2014) found another loophole in Xue et al. (2013) work that is the scheme is unable to provide the feature of user untraceability. In addition, He et al. (2015) also identified a loophole in Xue et al.'s (2013) and Jiang et al.'s (2015) work, depicting that the protocol is susceptible to offline guessing attack, and sensor node impersonation attack.

The authors suggested that resource constraints is a major drawback with sensors, in addition to this, authentication is another pitfall. In their review they highlighted Yeh et al. (2011) contribution with respect to using ECC for WSNs. Two years later, Shi and Gong (2013) proposed an improved authentication scheme using ECC for WSNs. Choi et al. (2014) further proposed an enhanced authentication scheme to enhance Shi et al.'s scheme. However, these three schemes cannot provide user anonymity or un-traceability. In the same year, Nam et al. (2014) proposed an ECC based authentication scheme which achieves user anonymity and perfect forward secrecy.

# 4 WIRELESS STANDARDS FOR WBANS

Scholars like (i.e., Filipe et al., 2015; Wong et al., 2013) suggested that the Wireless Standards in WBAN communication subsumes IEEE802.15.6, and IEEE 802.15. However, Ragesh and Baskaran (2012) listed three most commonly used standards, namely:

- **IEEE802.15.6:** It has been reported that a group of experts known the IEEE 802.15 task group 6 (BAN) working hard to create a communication standard aim at powering low powered devices.
- **IEEE 802.15.4**: Ragesh and Baskaran (2012) added that scientist exert so much effort on MAC protocol, but the issue here is that the performance of IEEE802.15.4 is adequate to power WBANs devices and applications.
- **Bluetooth:** Bluetooth is a widely used technology in most service and manufacturing sector, including the medical industry. Nevertheless, Bluetooth technologies were developed to work in and with high data rate networks and high powered devices. Again this failed to fully support WBAN requirements.

These standards are designed for low-power gadgets whose function are remote well-being care checking, shopper hardware, and intelligent gaming. Recently various biotechnological associations have agreed to enforce this standard on all medical gadgets, so all devices used in healthcare should be accredited for Quality of Service (QoS), to a great degree of low power, and information rates up to 10 Mbps (Begum et al., 2014). Apart from the standards enforced on WBANs, the technology is application-dependent, categorized into event detector, and periodic event recorder. In the event detection of function, the nodes capture and send information once an event occurs (e.g., cardiac arrest or stroke). For a periodic event, the nodes capture and transmit information within stated intervals usually specified periods (e.g., body temperature and sugar level). Given this, it is obvious that the system and functionality differs significantly, as such node resources, data gathering, transmitting ability, and capacity varies (Marinkovic and E., 2012).

Chen et al. (2011) categorized BAN communication architecture into three dimensions namely: (1) intra-BAN communications, (2) inter-BAN communications, and (3) beyond-BAN communications. The scope of their division and categorization is that a WBAN (Tier 1) is meant for the sensors and local gateway in patient, While WBAN Tier 2 functions as a connector between the patients and the sensors, and last but not the least the Tier 3 depicts the network framework, logically the connection medium, such as the Internet. Table 9 presents existing projects on wireless body area networks.

## 4.1 IEEE 802.15.6—WBAN

IEEE 802.15 has several approved standards within the group and other standards are in different phases of the standardization procedure. In Fig. 5, the different 802.15 standards are depicted with its subgroups.

Some experts consider IEEE 802.15.6 as a MAC convention and exploration, a thorough debate and analysis suggest that the IEEE 802.15.6 spotlights are not adequate for WBANs (Tjensvold, 2007). Consequently, Wu and Long (2015) added that this is because the execution of this convention in a multi responsive environment is exceptionally poor. IEEE 802.11 WLAN is not a viable option for WBANs, as it is not designed for low-power devices, although it can be used with laptops running on batteries. But this is only a temporary solution until an electric socket is found again. Given this, there is no specific standard for WBANs, as all the existing have limitations. The utilization of remote innovation, particularly to convey well-being care, likewise carries with it a large group of worries about security (Timmons and Scanlon, 2004). The security instrument of the framework is in charge for giving the accompanying security administrations on indicated biomedical information when asked for to do as such by the applications (Callaway, 2003).

- Information encryption—The information is scrambled with the goal that it is most certainly not uncovered while in travel. Information encryption administration gives secrecy against listening in assaults.
- Information integrity—Data integrity administration comprises of information uprightness and information starting point confirmation. Legitimate information uprightness instruments at the BN and the BNC guarantee that the got information is not changed by an enemy (Argyriou et al., 2015).
- Freshness protection—Data freshness guarantees that the information casings are all together and are not reused.
- Confirmation—This is an effective technique against mimic assaults.

A considerable measure of exploration concerning the physical layer has been researched. At the start of WBAN examination various creators proposed Ultra-Wide Band (UWB) as a physical layer for WBANs. Roy and Bhaumik (2015) asserted that UWB has the upside of low vitality utilization, great cooperation with existing remote systems and an extent sufficiently huge to bolster the whole body. UWB has failed to achieve its full potentials, as a result of institutionalization issues and troubles conveying high speeds. Instead of the wide groups proposed by UWB, different specialists propose the small, Industrial; scientific what's more, Medical (ISM) groups of the IEEE 802.15.4 and

**Table 9 Existing Projects on Wireless Body Area Networks**

| Project | Target Application | Intra-BAN Comm. | Inter-BAN Comm. | Beyond BAN Comm. | Sensors |
|---|---|---|---|---|---|
| ***MobiHealth*** | Ambulatory Patient Monitoring | Manually | ZigBee/Bluetooth | GPRS/UMTS | ECG, Heart rate, Blood Pressure |
| ***AID-N*** | Mass Casualty Incident | Wired | Mesh/ZigBee | WiFi/Internet/Cellular Networks | Blood, Pulse, ECG, Temperature |
| ***CodeBlue*** | Medical Care | Wired | ZigBee/Mesh | N/A | Motion, EKG, Pulse Oximeter |
| ***CareNet*** | Remote Healthcare | N/A | ZigBee | Internet/Multihop 802.11 | Gyroscope, Tri-axial accelerometer |
| ***UbiMon*** | Healthcare | ZigBee | WiFi/GPRS | WiFi/GPRS | 3Leads ECG, 2Leads ECG strip, SpO2 |
| ***SMART*** | Health Monitoring in Waiting Room | Wired | 802.11.b | N/A | $SpO_2$ sensor, ECG |
| ***WHMS*** | Healthcare | Wired | WiFi | N/A | EKG, ECG |
| ***MITHri*** | Healthcare | Wired | WiFi | N/A | EKG, ECG |
| ***Health Service*** | Mobile Healthcare | Wired | UMTS/GPRS | UMTS/GPRS/Internet | ECG, EMG, SpO2, Pulse rate, Respiration, Skin temperature, Activity, Plethysmogram |
| ***Life GUARD*** | Ambulatory physiologic monitoring for space and terrestrial applications | Wired | Bluetooth/Internet | Bluetooth/Internet | ECG, Respiration Electrodes, Pulse Oximeter, Blood Temperature, Built-in Accelerometer |
| ***LifeMinder*** | Real time daily self-care | Bluetooth | Bluetooth | Internet | Galvanic Skin Reflex (GSR) Electrodes, Accelerometer, Pulse Meter, Thermometer |
| ***Tele-medicare*** | Home-based Care and Medical Treatment | Bluetooth | Internet | Internet | Blood pressure, Temperature, ECG, Oximeter |
| ***WiMoCA*** | Sport/Gesture Detection Bluetooth | Star Topology and Time table-based | MAC protocol | WiFi/Internet/Cellular Networks/Bluetooth | Tri-axial Accelerometer |
| ***ASNET*** | Remote Health Monitoring | Wired or Wireless Interface (WiFi) | WiFi/Ethernet | Internet/GSM | Blood Pressure, Temperature |

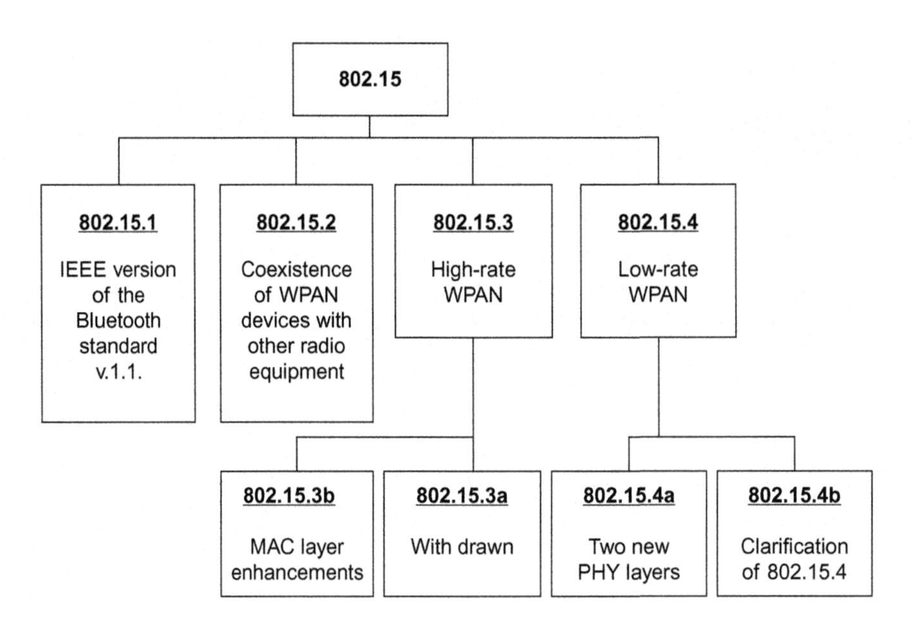


**FIG. 5**

The 802.15 standards.

IEEE802.15.6. Currently most of the existing work related to WBAN models are taking into account by ISM groups, with the aim of improving it (Wu and Long, 2015).

## 4.2 MEDICAL IMPLANT COMMUNICATION SERVICE (MICS)

MICS are short-range and wireless link use to communicate with implanted devices used in the medical field; such devices are low-power devices use for monitoring purposes. Specialists can monitor from outside a patient what is going on inside the patient body using planted medical devices like MICS. More specifically, MICS allows two way communications from the examiner to the body and from the body to the examiner. The frequency range for MICS usually is within 402–405 MHz, this range is adequate as it can provide reasonable amount of propagation in the human body. Medical implanted devices subsume cardiac pacemakers, implantable cardioverter/defibrillator (ICD), and neuro-stimulators.

To give solid correspondence low-power utilization for conveying insert information, the IEEE 802.15.6 standard considers the MICS band (Ahmed et al., 2015). The MICS band is separated into 10 channels and correspondence is performed in light of the listen-before-talk (LBT) convention, and backings divert exchanging in the MICS band. Because of this channel synthesis, a WBAN can at the same time transmit with their neighbor WBANs in the meantime when they do not utilize the same channel of the WBAN. Furthermore, interchanges in the MICS band don't endure obstruction from various correspondence innovations due to the certainty that the MICS band is an authorized band for just the medicinal insert correspondence administration. Accordingly, in-body correspondence in the MICS band is steadier than any other recurrence groups, for example, Industrial, Scientific, and Restorative (ISM) band and Ultra-Wide Band (UWB) (Ahmed et al., 2015).

Due to the previously stated favorable circumstances of the MICS band, IEEE 802.15.6 characterizes correspondence arrangement on the MICS band, alluded to as MICS band correspondence. To give dependable medicinal insert correspondence administration with low correspondence multifaceted nature and low-control utilization, MICS band correspondence utilizes surveying based channel access component with a solitary channel, which implies that the MICS band correspondence does not give an approach to choose another channel when current channel is congested because of impedance from neighboring WBANs. In any case, WBANs are thickly conveyed in a populated range, for example, a healing facility alternately a social insurance focus (Sarra et al., 2014). Also, impedance powerfully changes because of system level versatility of WBANs. In this circumstance, WBAN fundamentally experiences execution debasement regardless of previously stated favorable position of the MICS band as noted by Yang et al. (2016).

## 4.3 PATH LOSS (PL)

Path loss (or path attenuation) delineates a decline in power density of any given electromagnetic wave as it propagates through space. There are various cause of path loss ranging from natural expansion of the radio wave, diffraction path-loss occurs due to obstruction, to absorption path loss that occurs due to presence of a medium that is not transparent to electromagnetic waves. It is important to note that even when path loss occurs, the transmitted signal may travel along other paths to the intended destination, such process is called multipath. Since, these waves or transmitted information travels along other paths, the wave may regroup at the destination point resulting in received signals that vary significantly.

Recently in their study, Ding et al. (2016) argued that path loss leads to higher energy consumption, the authors proposed that this should be reduced. The idea here is because WBANs and WSNs consume more energy during data transmission. Therefore the concurrent occurrence of path-loss creates clusters and routing paths have to be rebuilt. Therein this regrouping activity of data causes large amount of similar data to be transmitted leading to higher percentage of energy consumption, technically wasted wasting. An interesting aspect of radio waves in WBAN is that propagations from devices situated inside the human body tends to be unique and multiplexing nature in contrast to devices outside the human body. A major difference results from the fact that the human body is composite, in that it has various organs and tissues working together.

For instance, our basic activities include breathing, heartbeat, walking, and running; these activities have a profound impact on the wireless propagation (Nie et al., 2012). From a pragmatic point of view, the propagation wave will not pass through the patient body, instead it will divert and split into fragments around patient body. Custodio et al. (2012) noted that "the path loss is very high especially when the receiving antenna is placed on the side opposite the transmit antenna." If we are to consider the limited memory in WBANs, we should note that there is need for efficient admission and concession, data conveyance, enable data correction and also ensures the detection of errors. Moreover, for any WBAN application to meet QoS requirements without opting out performance and/or increasing complexity, then path loss, loss-sensitive and delay sensitive.

Then, the real-life situation for WBAN applications should be taken into consideration as a WBAN deployed in a patient body can be applied to service devices with varying number of specifications, such as data rates, frequency, reliability, and power usage. Therein, WBANs must enforce stabilize message transmission between various wireless technologies be it for the purpose of scalability, or to support data transmission between two parties, adapt and work with plug and play systems, support stable and consistent connection and also enforce effective migration between networks. Therein, the technology choice should have the capability to handle the mixture of these requirements. Filipe et al. (2015) noted that previous research works and experiments focused more on layer protocol stack and have turned away from and ignored the transport layer which is frequently in contact with application layer of WBAN technologies.

This dissertation, attempts to show that this layer is indeed important, as there are lots of drawbacks ranging from congestion control, security, allocation of bandwidth, packet-loss recovery, and energy efficiency. WBAN is extraordinarily impacted by the measure of the way path loss happens because of various weaknesses. Path loss for WBAN are by and large set inside or on the body surface, path loss between these gadgets would influence the correspondence and can corrupt the execution observing in UHC (Savci et al., 2013). We examine in insight about WBAN correspondence and path loss that happens in it and how it influences the execution of UHC (Cho et al., 2009).

An incessant decline in the power density of a given electromagnetic wave can result to path loss (Yuce et al., 2007). Generally speaking, path loss occurs as a result of free space impairments of propagating signal more likely from activities ranging from attenuation, reflection, absorption, and refraction. Foerster et al. (2001) discuss that another important factor that may lead to cause path loss is the distance range between the sending and receiving antennas, in addition, location, position, height, structures (rural and urban areas, such as magnetic waves, skyscrapers, and tall building) and even the propagation channel like weather conditions, like moist or dry air. Path loss in WBAN differs from the classical wireless network, as it is contingent upon two main factors: distance and frequency. Distance is important because of the nature and design of WBAN does not allow for long range. On the other hand, frequency is also an important facet as mankind body tissues can be affected easily by the prevalence and the amount of work done by the sensor device.

Path loss model in *dB* among sending and accepting radio wires as a component of the separation *d* is registered by (Cavallari et al., 2014) as:

$$\mathrm{PL(d)} = \mathrm{PL(do)} + 10n\log_{10}\left(\frac{d}{do}\right) + \sigma_s, \tag{1}$$

where *PL(do)* "is the path loss at a reference distance *d*, *n* is the path loss exponent, and $\sigma s$ is the standard deviation, in WBAN, path loss is of great importance." Furthermore, "WBAN UHC carries out its function properly only when the path loss between the sender and the receiver end is at its lowest level" (Wac et al., 2009). In WBAN, several factors can cause path loss, among which subsumes reflection, refraction, and absorption from the patient's body. All of the above factors that cause path loss to occur are said to interfere and distort the signal at receiving especially when there is a significant distance between the transmitter and the receiver.

In a nutshell, "data can easily be distorted as a result of path loss which in turn creates problems for healthcare workers who attempt to retrieve data from a remote location" (Bienaime, 2005). Path loss in UHC might reduce the efficiency of human body events observation both from the patient and the healthcare team perspective (Khan et al., 2012). This unit aim is to diminish or reduce the number of path loss occurrence in various lifecycles of any given WBAN. More specifically, a rise in the efficiency of UHC monitoring BAN is the aim of the whole process. Path loss reliance on separation and in addition recurrence is given in Eqs. (2), (3):

$$\mathrm{PL} = 20\log_{10}\left(\frac{4\pi d}{\lambda}\right), \tag{2}$$

where $L$ is the path loss in decibels, $\lambda$ means wavelength and $d$ determines separation between transmitter and collector (Nnamani and Alumona, 2015). As, we realize that: $\lambda = cf$; in any case, Eq. (2) can be revamped as under: 2.5 Antenna Design in WBANs.

$$\mathrm{PL} = 20\log_{10}\left(\frac{4\pi df}{c}\right). \tag{3}$$

## 4.4 PERFORMANCE PARAMETERS

The parameters use for execution in a remote patient checking framework usually utilize database systems to record all information and activities concerning a specific patients. Database system is used for storing, and as such there is a need to remit data from a remote location (technically) the patient location to the central healing database (Rawat et al., 2014). The information base is stacked by the WBAN expert hub by tolerating information from different sensor nodes. A typical example of the working protocol WBAN and database system is depicted in Fig. 6. Therapeutic data is captured,

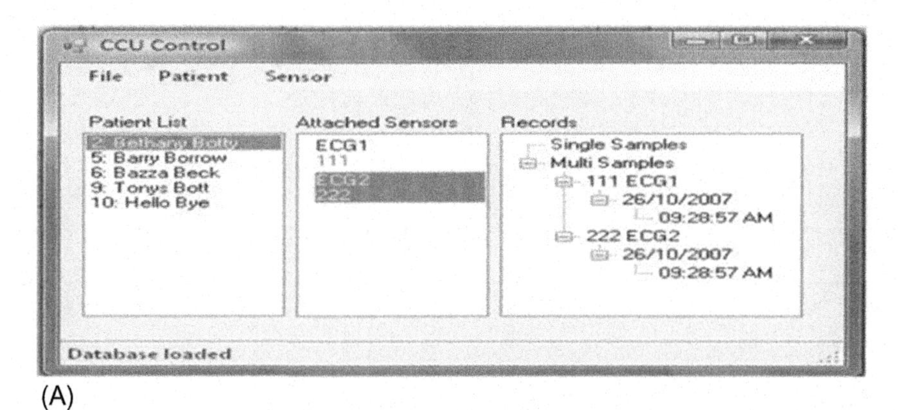


(A)

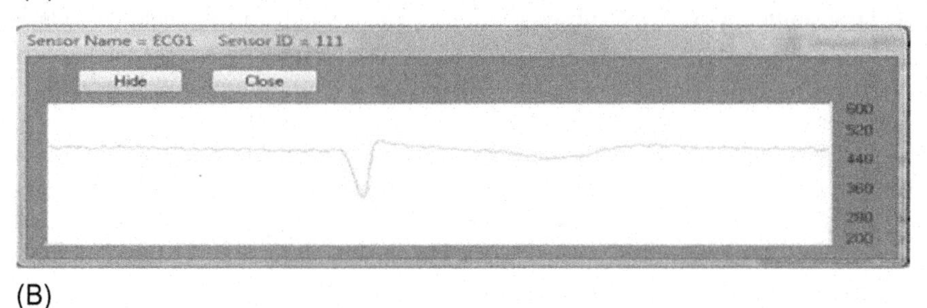


(B)

**FIG. 6**

(A) Database upload window. (B) Retrieved medicals record from database.

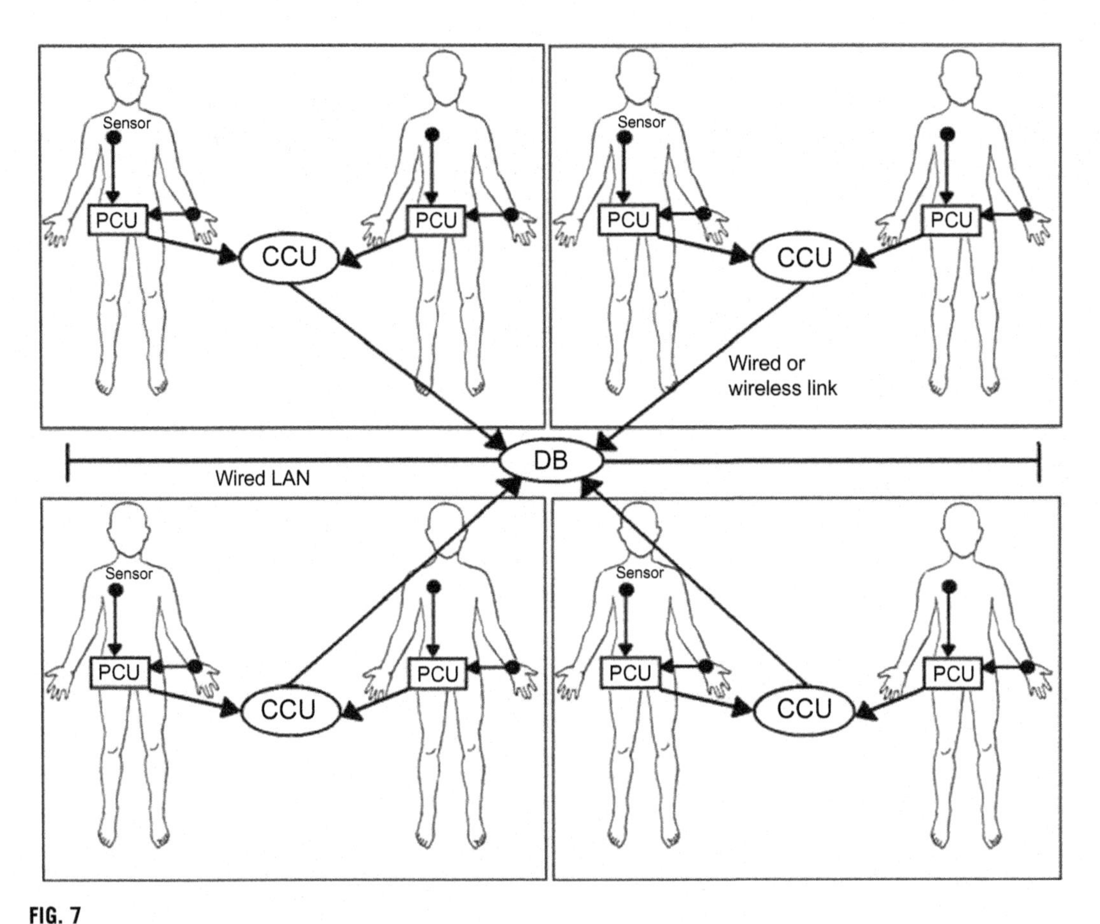


**FIG. 7**

A WBAN-based multihop patient monitoring system.

processed, and transmitted by WBAN, subsequently the data, if not distorted when it travels, is received and stored in the healing facility record PC (a typical server). Nevertheless, the system has the ability to capture and store multiple data, experts can chose the patient of interest and sensor to recover and monitor progress in real-time. See Fig. 7.

The WBAN expert sensor occasionally transfers information from different sensors. Subtle elements outline and execution consequences of the WBAN can be found in the reference (Ghamari et al., 2016). We re-enacted a doctor's facility system based on the Ethernet standard. The system associates the database server alongside different servers and applications inside the doctor's facility system (Ghamari et al., 2016). These parameters have an exchange between each other so it is difficult to meet all necessities at the same time. The impact of changing the transmitted yield force is watched further through reproduction results. Throughput is corresponding to the quantity of effectively got bundles at system organizer per sensors (Curry and Smith, 2016)

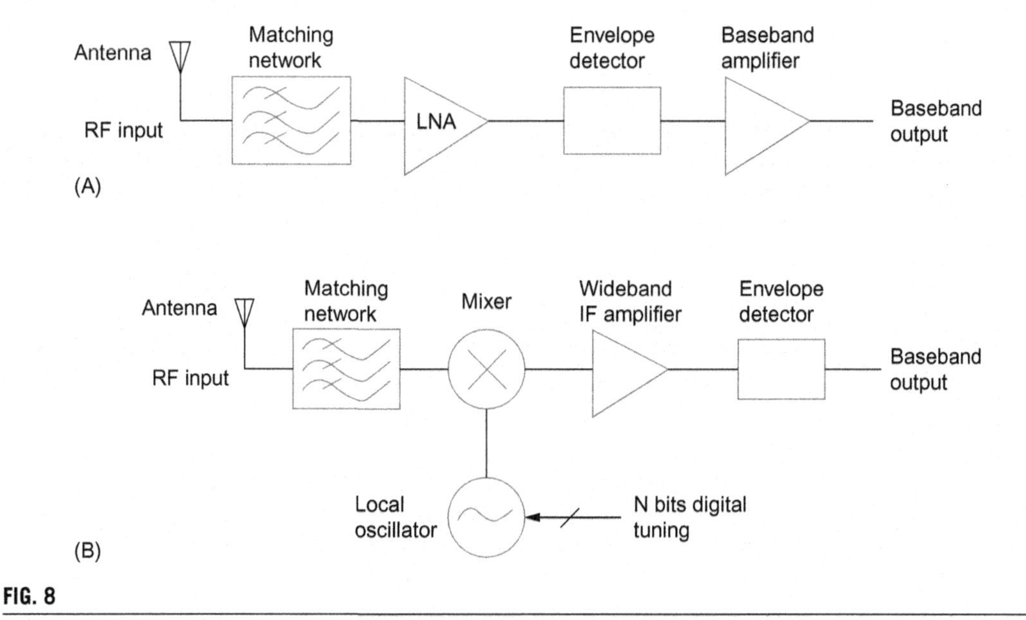


**FIG. 8**

Two envelope detector-based receiver architectures.

## 4.5 RECEIVER DESIGNS

To solve the issue of the high necessities of a nearby oscillator, vitality location recipients can be utilized. In this sort of structure, the coordinating system is just taking after by enhancer and envelope identifier. The envelope finder does not have strict necessity of the information signal recurrence. Hence, the force hungry nearby oscillator can be removed. For the vitality identification recipient, there are two structures that are the most normally utilized as represented in Fig. 8. The primary downside of this vitality recipient is that it does not have sifting capacity at the baseband. So the selectivity must be accomplished specifically at RF band. The primary force utilization is expense by the high recurrence enhancer, which needs to give high pick-up due to the affectability prerequisite of the envelope indicator. In any case, when contrasted with PLL and customary nearby oscillator, the force utilization is diminished significantly.

# 5 CONCLUSION

Recently, interest in the WBAN for patient monitoring has gained significant attention from the medical industry and the research world. Given this, the majority of modern research mostly focus on how to enhance patient monitoring systems, in order to provide and ensure service quality provisioning in terms of communication between WBAN nodes. In these views, efficient analytical examinations of recent WBAN technologies were conducted in this dissertation. Consequently, to illustrate the performance of the recently drafted IEEE 802.15.6 standard, this dissertation focused on two different scenarios to simulate and test how integrity can be assured, how energy consumption can be reduced to its minimal level, and how path loss can be mitigated in relation to this standard.

The security environment is a critical factor in ensuring the effectiveness of all the operations in a hospital department. The path loss problem is considered dynamic according to hospital preference and the demands of their nurses. In today's hospital environment, it is more demanding and challenging to manage secure level in transfer. There is a definite difference between the previous and current state-of-the-art approaches in WBAN research. In the future, the important scientific goal of WBAN protocol will be to fully address the needs and the requirements of the real world. In WBANs, fatal errors may arise when there is incompatibility between the device characteristics and the routing algorithm, which is when the device does not supports the routing algorithm of the WBAN. For instance, if a lengthy message is to be transmitted utilizing tiny sensor device, technically the lifetime of the network lifetime decreases automatically. Thus the proposed algorithm considers not only the design goals of the efficiency sensed message but also the device characteristics.

Due to the requirements of secure WBAN, the current protocol approach alone is not enough to finalize production of the transfer patient data. The MAC protocol and IEEE 802.15.6 condition must also be involved in order to achieve the best path transfer for data. The mathematical approach is useful in order to calculate (value) the pattern of the schedule. Then the values of the best path loss itself represents the best fitness. Thus a routing algorithm for a WBAN should support not only a variety of devices, but also satisfy these design requirements. Regarding Message Integrity, even though received messages are authenticated and encrypted, an adversary can intentionally tamper with the message. In such cases, the receiving node should be able to detect such data corruptions and reject the data or information sent. Nevertheless, harsh weather condition have the propensity to disrupt or even corrupt data in a wireless channel. A common way to deal with message integrity is to give identification with a description for every sensor node; the proposed algorithm follows and applies the following modules namely: "Routing Table," "Error Detector," and "Path Selector."

## REFERENCES


Abarna, K.T.M., 2012. Light-weight security architecture for IEEE 802.15.4 body area networks. Int. J. Comput Appl. 47 (22), 1–8.

Abouei, J., Brown, J.D., Plataniotis, K.N., Member, S., 2011. Energy efficiency and reliability in wireless biomedical implant systems. IEEE Trans. Inform. Technol. Biomed. 1–12.

Ahmed, S.S., Hussain, I., Ahmed, N., Hussain, I., 2015. Driver level implementation of TDMA MAC in long distance Wi-Fi. In: International Conference on Computational Intelligence & Networks, pp. 2375–5822.

Amir, Z., Jim, W., 2012. PRC/EPRC: Data Integrity and Security Controller for Partial Reconfiguration. Xilinx Int., Application Note, 887, pp. 1–17.

Amrita, G., Sipra, D., 2015. A lightweight security scheme for query processing in clustered wireless sensor networks. Comput. Electr. Eng. 2015 (41), 240–255. doi:10.1016/j.compeleceng.2014.03.014.

Argyriou, A., Breva, A., Aoun, M., 2015. Optimizing data forwarding from body area networks in the presence of body shadowing with dual wireless technology nodes. IEEE Trans. Mobile Comput. 14 (3).

Avi, K., 2016. Lecture 15: Hashing for Message Authentication. Lecture Notes on Computer and Network Security. Retrieved from https://engineering.purdue.edu/kak/compsec/NewLectures/Lecture15.pdf. (Accessed December 2016).

Bangash, J.I., Abdullah, A.H., Anisi, M.H., Khan, A.W., 2014. A survey of routing protocols in wireless body sensor networks. Sensors (Basel, Switzerland) 14 (1), 1322–1357. doi:10.3390/s140101322.

Begum, F., Sarma, M.P., Sarma, M.P., 2014. Preamble aided energy detection based synchronization in non-coherent UWB receivers. In: IEEE International Conference on Signal Processing and Integrated Networks, Noida, India, February.

Bellovin, S., Rescorla, E., 2006. Deploying a new hash algorithm. In: Proceedings of NDSS 06.

Bh, P., Chandravathi, D., Roja, P.P., 2010. Encoding and decoding of a message in the implementation of elliptic curve cryptography using Koblitz's method. Int. J. Comput. Sci. Eng. 02 (05), 1904–1907.

Bienaime, J., 2005. 3G/UMTS: an evolutionary path towards mobile broadband & personal Internet. Retrieved from https://studylib.net/doc/13187117/3g-umts-an-evolutionary-path-towards-mobile-broadband-and. (Accessed December 2016).

Braem, B., Blondia, C., 2011. Supporting mobility in wireless body area networks: an analysis. In: 2011 18th IEEE Symposium on Communications and Vehicular Technology in the Benelux (SCVT), pp. 1–6.

Callaway, E., 2003. Low Power Consumption Features of the IEEE 802.15.4/ZigBee LR-WPAN. Florida Communication Research Lab Motorola Labs.

Cavallari, R., Flavia, M., Rosini, R., Buratti, C., Verdone, R., 2014. A survey on wireless body area networks: technologies and design challenges. IEEE Commun. Surv. Tut. 16 (3), Third Quarter.

Chen, M., Gonzalez, S., Vasilakos, A., Cao, H., Leung, V.C.M., 2011. Body area networks: a survey. Mobile Netw. Appl. 16 (2), 171–193. doi:10.1007/s11036-010-0260-82-s2.0-79956094375.

Cho, N., Bae, J., Kim, S., Yoo, H.J., 2009. A 10.8 mW body-channel communication/MICSdual-band transceiver for a unified body-sensor-network controller. In: IEEE International Solid-State Circuit's Conference—Digest of Technical Papers (ISSCC 2009), San Francisco, CA, 8–12 February, pp. 424–425.

Choi, Y., Lee, D., Kim, J., 2014. Security enhanced user authentication protocol for wireless sensor networks using elliptic curves cryptography. Sensors 14 (6), 10081–10106.

Crosby, V.G., 2012. Wireless body area networks for healthcare: a survey. Int. J. Ad Hoc Sens. Ubiquitous Comput. 3 (3), 1–26. doi:10.5121/ijasuc.2012.3301.

Curry, R.M., Smith, J.C., 2016. A survey of optimization algorithms for wireless sensor network lifetime maximization. Comput. Ind. Eng. 101, 145–166. doi:10.1016/j.cie.2016.08.028.

Custodio, V., Herrera, F.J., López, G., Moreno, J.I., 2012. A review on architectures and communications technologies for wearable health-monitoring systems. Sensors 12 (10), 13907–13946. doi:10.3390/s1210139072-s2.0-84868218707.

Devi, L., Nithya, R., 2014. Wireless body area sensor system for monitoring physical activities using GUI. Int. J. Comput. Sci. Mobile Comput. 3 (1), 569–577.

Devita, G., Wong, A., Dawkins, M., Glaros, K., Kiani, U., Lauria, F., Madaka, V., Omeni, O., Schiff, J., Vasudevan, A., Whitaker, L., Yu, S., Burdett, A., 2014. A 5 mW multi-standard Bluetooth LE/IEEE 802.15.6 SoC for WBAN applications. In: ESSCIRC, pp. 283–286.

Ding, Y., Chen, R., Hao, K., 2016. A rule-driven multi-path routing algorithm with dynamic immune clustering for event-driven wireless sensor networks. Neurocomputing 203 (26), 139–149. doi:10.1016/j.neucom.2016.03.052.

Esat, D.E., 2004. Analysis and Design of Cryptographic Hash Functions, MAC Algorithms, Ph.D. Thesis, Katholieke Universiteit Leuven, pp. 1–259. ISBN 90-5682-527-5.

Filipe, L., Fdez-Riverola, F., Costa, N., Pereira, A., 2015. Wireless body area networks for healthcare applications: protocol stack review. Int. J. Distrib. Sens. Netw. 11 (10), 213705.

Foerster, J., Green, E., Somayazulu, S., Leeper, J., 2001. Ultra-Wideband Technology for Short or Medium Range Wireless Communications. Retrieved from http://developer.intel.com/technology/itj. (Accessed December 2016).

Ghamari, M., Janko, B., Sherratt, R., Harwin, W., Piechockic, R., Soltanpur, C., 2016. A survey on wireless body area networks for eHealthcare systems in residential environments. Sensors 16 (6), 831. doi:10.3390/s16060831.

Gura, N., Patel, A., Wander, A., 2004. Comparing elliptic curve cryptography and RSA on 8-bit CPUs. In: Proceedings of the 2004 Workshop on Cryptographic Hardware and Embedded Systems (CHES 2004), pp. 119–132.

He, D., Kumar, N., Chilamkurti, N., 2015. A secure temporal-credential-based mutual authentication and key agreement scheme with pseudo identity for wireless sensor networks. Inform. Sci. 321, 236–277.

Ibraheem, A.A.Y., 2014. Implanted antennas and intra-body propagation channel for wireless body area network. Doctor of Philosophy in Electrical Engineering.

Jaff, B.T.H., 2009. A Wireless Body Area Network System for Monitoring Physical Activities and Health-Status via the Internet. Master's Thesis. Department of Information Technology, Uppsala University.

Javaid, N., Khan, N.A., Shakir, M., Khan, M.A., Bouk, S.H., Khan, Z.A., 2013. Ubiquitous healthcare in wireless body area networks—a survey. J. Basic Appl. Sci. Res. 3 (4), 747–759.

Jiang, Q., Ma, J., Li, G., Li, X., 2015. Improvement of robust smartcard-based password authentication scheme. Int. J. Commun. Syst. 28 (2), 383–393.

Khan, J.Y., Yuce, M.R., Karami, F., 2008. Performance evaluation of a wireless body area sensor network for remote patient monitoring. In: Proceedings of the 30th Annual International Conference of the IEEE Engineering in Medicine and Biology Society, Vancouver, BC, pp. 1266–1269.

Khan, N.A., Javaid, N., Khan, Z.A., Jaffar, M., Rafiq, U., Bibi, A., 2012. Ubiquitous healthcare in wireless body area networks. In: IEEE 11th International Conference on Trust, Security and Privacy in Computing and Communications (TrustCom), pp. 1960–1967.

Khudri, W., Sutanom, S., 2005. Implementation of El-Gamal elliptic curve cryptography. In: International Conference on Instrumentation, Communication and Information Technology (ICICI), pp 1–6.

Kim, T., Kim, Y., 2014. Human effect exposed to UWB signal for WBAN application. J. Electromagnet. Waves Appl. 8 (12), 1430–1444.

Kim, B., Cho, J., Kim, D.Y., Lee, B., 2016. ACESS: adaptive channel estimation and selection scheme for coexistence mitigation in WBANs. In: 10th International Conference on Ubiquitous Information Management and Communication (IMCOM 2016). ACM, New York, NY, USA, p. 96.

Koblitz, N., 1987. Elliptic curve cryptosystems. Math. Comput. 48 (177), 203–209.

Koopman, P., Driscoll, K., 2012. Tutorial: Checksum and CRC Data Integrity.

Kumar, P., Lee, S.S., Lee, H.J., 2012. E-SAP: efficient-strong authentication protocol for healthcare applications using wireless medical sensor network. Sensors (Basel) 12 (2), 1625–1647. doi:10.3390/s120201625.

Kwak, K.S., Ullah, S., Ullah, N., 2010. An overview of IEEE 802.15. 6 standard. In: Proceedings of IEEE ISABEL, pp. 1–6.

Laccetti, G., Schmid, G., 2007. Brute force attacks on hash functions. J. Discrete Math. Sci. Cryptography 10 (3), 439–460.

Lai, J.Y., Huang, C.T., 2011. Energy-adaptive dual-field processor for high performance elliptic curve cryptographic applications. IEEE Trans. VLSI Syst. 19 (8), 1512–1517.

Lenstra, A.K., Verheul, E.R., 2001. Selecting cryptographic key sizes. J. Cryptol. 14 (4), 255–293.

Liu, A., Ning, P., 2008. TinyECC: a configurable library for elliptic curve cryptography in wireless sensor networks. In: Proceedings of the 7th International Conference on Information Processing in Sensor Networks (IPSN 2008), SPOTS Track, pp. 245–256.

Liu, D., Ning, P., Zhu, S., Jajodia, S., 2015. Practical broadcast authentication in sensor networks. In: Proceedings of the 2nd Annual International Conference on Mobile and Ubiquitous Systems: Networking and Services (MobiQuitous, 2015).

Lont, M., 2013. Wake-Up Receiver Based Ultra-Low-Power WBAN. Technische Universiteit Eindhoven, Eindhoven. doi:10.6100/IR762409.

Malan, D., Welsh, M., Smith, M., 2004. A public-key infrastructure for key distribution in TINYOS based on elliptic curve cryptography. In: Proceedings of IEEE Conference on Sensor and Ad Hoc Communications and Networks (SECON), pp. 71–80.

Mane, S., Judge, L., Schaumont, P., 2011. An integrated prime-field ECDLP hardware accelerator with high-performance modular arithmetic units. In: Proceedings of the 2011 International Conference on Reconfigurable Computing and FPGAs, IEEE Computer Society, New York, pp. 198–203.

Manirabona, A., Fourati, L.C., Boudjit, S., 2017. Investigation on healthcare monitoring systems: innovative services and applications. Int. J. E-Health Med. Commun. 8 (1), 1–18. doi:10.4018/IJEHMC.2017010101.

Marinkovic, S., Popovici, E., 2012. Ultra-low power signal oriented approach for wireless health monitoring. Sensors 12 (6), 7917–7937. doi:10.3390/s1206079172-s2.0-84863207168.

Marzouqi, H., Al-Qutayri, M., Salah, K., 2015. Review of elliptic curve cryptography processor designs. Microprocess. Microsyst. 39 (2), 97–112. doi:10.1016/j.micpro.2015.02.003.

Maxwell, B., Thompson, D.R., Amerson, G., Johnson, L., 2003. Analysis of CRC methods and potential data integrity exploits. In: Proceedings of the International Conference on Emerging Technologies, Minneapolis, Minnesota, 25–26 August.

Meena, R., Ravishankar, S., Gayathri, J., 2014. Monitoring physical activities using WBAN. Int. J. Comput. Sci. Inform. Technol. 5 (4), 5880–5886.

Movassaghi, S., Abolhasan, M., Lipman, J., Smith, D., Jamalipour, A., 2014. Wireless body area networks: a survey. IEEE Commun. Surv. Tut. 16 (3), 1658–1686.

Nabi, M., Geilen, M., Basten, T., 2011. MoBAN: A Configurable Mobility Model for Wireless Body Area Networks. SIMUTools, Barcelona, Spain.

Nam, J., Kim, M., Paik, J., 2014. A provably-secure ECC-based authentication scheme for wireless sensor networks. Sensors 14 (11), 21023–21044.

Nie, Z.D., Ma, J.J., Li, Z.C., Chen, H., Wang, L., 2012. Dynamic propagation channel characterization and modeling for human body communication. Sensors 12 (12), 17569–17587. doi:10.3390/s1212175692-s2.0-84871673244.

Nnamani, K.N., Alumona, T.L., 2015. Path loss prediction of wireless mobile communication for urban areas of Imo state, south-east region of Nigeria at 910 MHz. Sensor Netw. Data Commun. 4 (1).

Otto, C.A., Jovanov, E., Milenkovic, E.A., 2006. WBAN-based system for health monitoring at home. In: IEEE/EMBS International Summer School, Medical Devices and Biosensors, pp. 20–23.

Padmavathi, G., 2009. A survey of attacks, security mechanisms and challenges in wireless sensor networks. Int. J. Comput. Sci. Inform. Security 4 (1), 1–9.

Perrig, A., Szewczyk, R., Wen, V., Culler, D., Tygar, J., 2002. SPINS: security protocols for sensor networks. Wirel. Netw. 8, 521–534.

Pote, S.K., 2012. Elliptic Curve Cryptographic Algorithm. 978–981. doi:10.3850/978-981-07-1403-1.

Preneel, B., 2003. Analysis and Design of Cryptographic Hash Functions. Graduate Theses and Dissertations. KatholiekeUniversiteit Leuven.

Ragesh, G.K., Baskaran, K., 2012. An overview of applications, standards and challenges in futuristic wireless body area networks. Int. J. Comput. Sci. Issues 9 (1), 180–186.

Raja, S.K.S., 2013. Level based fault monitoring and security for long range transmission in WBAN. Int. J. Comput. Appl. 64 (1), 1–9.

Rashwand, S., Misic, J., 2012. Bridging Between IEEE 802.15.6 and IEEE 802.11e for Wireless Healthcare Networks. University of Manitoba, Department of Computer Science. Ryerson University, Department of Computer Science, pp. 303–337.

Rawat, P., Singh, K.D., Chaouchi, H., Bonnin, J.M., 2014. Wireless sensor networks: a survey on recent developments and potential synergies. J. Supercomput. 68 (1), 1–48.

Rebeiro, C., Roy, S.S., Mukhopadhyay, D., 2012. Pushing the limits of high-speed GF (2m) elliptic curve scalar multiplication on FPGAs. In: Cryptographic Hardware and Embedded Systems, CHES, vol. 7428, pp. 494–511.

Roy, U.K., Bhaumik, P., 2015. Enhanced ZigBee tree addressing for flexible network topologies. In: Applications and Innovations in Mobile Computing (AIMoC).

Ryckaert, J., De Doncker, P., Meys, R., de Le Hoye, A., Donnay, S., 2004. Channel model for wireless communication around human body. Electron. Lett. 40, 543544. doi:10.1049/el:20040386.
Saleem, 2009. On the security issues in wireless body area networks. Int. J. Digital Content Technol. Appl. 3 (3). doi:10.4156/jdcta.vol3.issue3.22.
Sana, S.U., Khan, P., Ullah, N., Saleem, S., Higgins, H., Kwak, K.S., 2009. A review of wireless body area networks for medical applications. Int. J. Commun. Netw. Syst. Sci. 1 (7), 797–803.
Sarra, E., Moungla, H., Benayoune, S., Mehaoua, A., 2014. Coexistence improvement of wearable body area network (WBAN) in medical environment. In: 2014 IEEE International Conference on Communications (ICC), pp. 5694–5699.
Savci, H.S., Arvas, S., Dogan, N.S., Arvas, E., Xie, Z., 2013. Low power analog baseband circuits in 0.18 $\mu$m CMOS for MICS and body area network receivers. In: Southeast con, 2013 Proceedings of IEEE.
Seshabhattar, S., Yenigalla, P., Krier, P., Engels, D., 2011. Hummingbird key establishment protocol for low-power ZigBee. In: 2011 IEEE Consumer Communications and Networking Conference (CCNC), Las Vegas, NV, pp. 447–451. doi:10.1109/CCNC.2011.5766509.
Shah, R., Yarvis, M., 2006. Characteristics of on-body 802.15.4 networks. In: 2nd IEEE Workshop on Wireless Mesh Networks (Wi-Mesh), pp. 138–139.
Shi, W., Gong, P., 2013. A new user authentication protocol for wireless sensor networks using elliptic curves cryptography. Int. J. Distrib. Sens. Netw. 9 (4), 730831.
Singh, V., 2013. Performance analysis of MAC protocols for WBAN on varying transmitted output power of nodes. Int. J. Comput Appl. 67 (7), 32–34.
Sukor, M., Ariffin, S., Fisal, N., Yusof, S.S., Abdallah, A., 2008. Performance study of wireless body area network in medical environment. In: Asia International Conference on Modelling and Simulation, pp. 202–206.
Taparugssanagorn, A., Rabbachin, A., Matti, H., 2008. A Review of Channel Modelling for Wireless Body Area Network in Wireless Medical Communications. Centre for Wireless Communications.
Timmons, N.F., Scanlon, W.G., 2004. Analysis of the performance of IEEE 802.15.4 for medical sensor body area networking. In: Proceedings of First Annual IEEE Communications Society Conference on Sensor and Ad Hoc Communications and Networks (IEEE SECON 2004), pp. 16–24.
Tjensvold, J.M., 2007. Comparison of the IEEE 802.11, 802.15.1, 802.15.4 and 802.15.6 Wireless standards. https://janmagnet.files.wordpress.com/2008/07/comparison-ieee-802-standards.pdf.
Ugent, B.L., Braem, B., UGent, I.M., Blondia, C., Ugent, P.D., 2011. A survey on wireless body area networks. Wirel. Netw. 17 (1), 1–18.
Vaidya, B., Rodrigues, J.J.P.C., Park, J.H., 2009. User authentication schemes with pseudonymity for ubiquitous sensor network in NGN. Int. J. Commun. Syst. 23, 1201–1222.
Vanstone, S.A., 2003. Next generation security for wireless: elliptic curve cryptography. Comput. Security 22 (5), 412–415. doi:10.1016/S0167-4048(03)00507-8.
Wac, K., van Beijnum, B., Bults, R., Widya, I., Jones, V., Konstantas, D., Vollenbroek, M., H., Hermens, H., 2009. Mobile patient monitoring: the MobiHealth system. In: Proceedings of the 31st Annual International Conference of the IEEE Engineering in Medicine and Biology Society: Engineering the Future of Biomedicine. EMBS Engineering in Medicine and Biology Society, New Jersey, pp. 1238–1241.
Wang, Y., Li, R., 2011. A unified architecture for supporting operations of AES and ECC. In: 2011 Fourth International Symposium on Parallel Architectures, Algorithms and Programming (PAAP), pp. 185–189.
Wang, D., Wang, P., 2014. On the anonymity of two-factor authentication schemes for wireless sensor networks: attacks, principle and solutions. Comput. Netw. 73, 41–57.
Wong, A.C.M., Dawkins, M., Devita, G., Kasparidis, N., Katsiamis, A., King, O., 2013. A 1 V 5 mA multimode IEEE 802.15.6/Bluetooth low-energy WBAN transceiver for biotelemetry applications. IEEE J. Solid-State Circuits 48 (1), 186–198.
Wu, Y., Long, Y., 2015. A planar antenna for GPS/WLAN/UWB applications. In: Wireless Symposium (IWS), 2015 IEEE International. doi:10.1109/IEEE-IWS.2015.7164573.

Xue, K., Ma, C., Hong, P., Ding, R., 2013. A temporal-credential-based mutual authentication and key agreement scheme for wireless sensor networks. J. Netw. Comput. Appl. 36 (1), 316–323.

Yang, P., Yan, Y., Li, X., Zhang, Y., Tao, Y., You, L., 2016. Taming cross technology interference for Wi-Fi and ZigBee coexistence networks. IEEE Trans. Mobile Comput. 1009–1021. doi:10.1109/TMC.2015.2442252.

Yeh, H.L., Chen, T.H., Liu, P.C., Kim, T.H., Wei, H.W., 2011. A secured authentication protocol for wireless sensor networks using elliptic curves cryptography. Sensors 11 (5), 4767–4779.

Yuce, M.R., Nag, P.C., Lee, C.K., Khan, J.Y., Wentai-Liu, L., 2007. A wireless medical monitoring over a heterogeneous sensor network. In: Proceedings of the International Conference of the IEEE Engineering in Medicine and Biology Society, 22–26 August, pp. 5894–5898.

CHAPTER

# DESIGNING RESILIENT AND SECURE LARGE-SCALE CRISIS INFORMATION SYSTEMS

14

**Marcello Cinque*, Domenico Cotroneo*, Christian Esposito*, Mario Fiorentino*, Stefano Russo***

*National Inter-university Consortium for Information Technology (CINI), Naples, Italy**

## 1 INTRODUCTION

Recently, we have witnessed the increasing occurrence of disasters characterized as having affected large geographic regions, such as those involving large portions of a country (e.g., the March 11 earthquake and tsunami in Japan) or portions of several different countries (e.g., the 2004 Indian Ocean tsunami), and having huge economic and social impacts and costs. Disasters of this magnitude have had a short-term destabilizing effect on the macro-economy (e.g., the Great Flood caused by the Mississippi River in 1993 caused a total loss estimated to be between $10.5 and $20.1 billion or the September 11th attacks in New York resulted in approximately $40 billion in insurance losses). The scale, impact, and severity of these disasters are so devastating that a single first responder organization is unable to properly react and recover from them, as traditionally occurs in the management of a crisis. Instead, these disasters require a collaboration among different organizations at the national and international level, which must act together, and may be coordinated by a given organization such as OCHA (Office for the Coordination of Humanitarian Affairs), paving the way for a Collaborative Crisis Management (CCM). Since decision-making must be flexible and responsive (Organization for Economic Co-operation and Development (OECD), 2004), there is a great demand for CCM intensive information sharing among the organizations involved in the crisis management. The accelerating rates at which disasters have unfolded in recent years, and the unprecedented levels of economic losses associated with them indicate we have entered a new era of catastrophes (NatCatSERVICE, 2012), where severe natural disasters are increasing in their frequency and consequences as depicted in Fig. 1.

A series of technological, behavioral and organizational obstacles make it difficult to meet such a demand. Our work will focus on the technological aspects that prevent the full cooperation and orchestration of a number of heterogeneous first responder organizations. Specifically, a given first responder organization is characterized by a proper solution for collecting data on the damaged area by the volunteers and personnel deployed on the field, processing such amounts of data in order to assess the overall picture of the damages and potential risks and to exchange collected and obtained data among all the involved people participating in the chain of command within the organization and the volunteers and personnel deployed on the field. However, there is not a single product that dominates the market of crisis management that may be designed and implemented internally and/or acquired from an external provider rather, each organization adopts a particular solution for crisis data management. Therefore the first difficulty is working with different technological products, each

Security and Resilience in Intelligent Data-Centric Systems and Communication Networks. https://doi.org/10.1016/B978-0-12-811373-8.00014-8

Copyright © 2018 Elsevier Inc. All rights reserved.

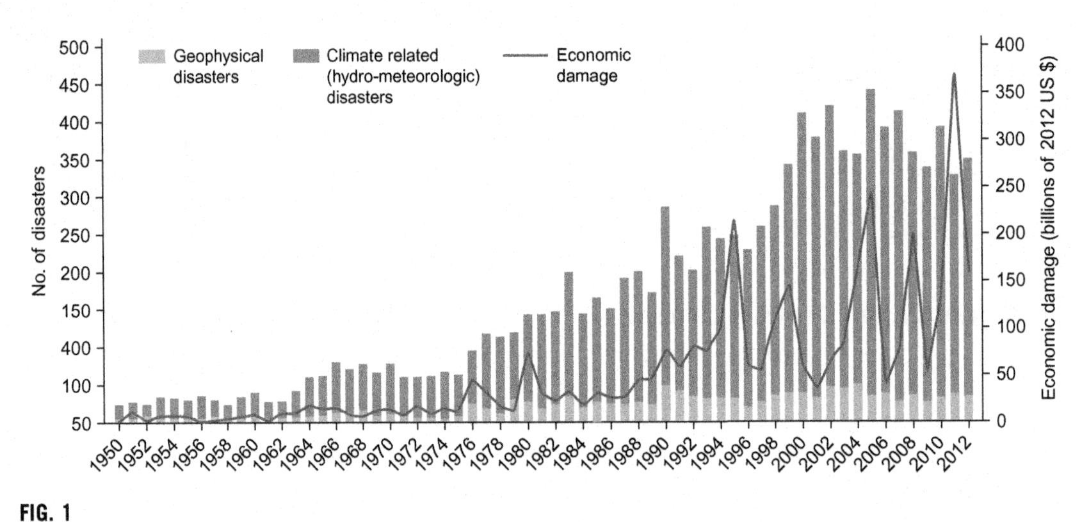


**FIG. 1**

Increasing frequency and cost of natural disaster.

*Source: http://hubpages.com/education/Worlds-worst-natural-disasters#.*

designed upon a different middleware, programming language and/or target computing resources. Apart from technological heterogeneity, other obstacles may come from syntactical and semantical heterogeneity, such as the use of a different format and/or semantics for expressing given common concepts and data. Therefore we can conclude that there is a lack of proper technological solutions that enables cross-border networks of crisis management organizations to set-up flexible support systems for a joint operation in which information is continuously updated and shared between organizations, leveraging on the information systems existing within the involved organizations, and in which progress is monitored and resource sharing is facilitated. DESTRIERO and SECTOR are two EU-funded projects that aim at providing an answer to the increasing demand for data sharing and cooperation in large-scale disaster management by offering a systematic, holistic, intergovernmental, and multidisciplinary approach to the collaboration of heterogeneous first responders when managing large-scale disasters. Specifically, a middleware platform for crisis information systems (Cinque et al., 2016, 2015) has been prototyped in the context of DESTRIERO as a mean to provide data sharing and cooperation capabilities to integrate heterogeneous information sources and to support damage and needs assessment as well as recovery planning. In SECTOR the focus is on the interoperability of Crisis Management Systems (CMSes) used across Europe, by means of a Common Information Space platform, is currently under development.

Apart from the required means to resolve the obstacles of the heterogeneity at the technological level, a series of nonfunctional requirements are to be satisfied in order to use such integration platforms during a concrete crisis situation. Among these requirements, in this context of crisis management, reliability becomes a key technical concern, given the critical role of information sharing in crisis management, and the potential adverse condition that the product may be deployed, which may negatively affect the correct dissemination of information. In fact, in the presence of catastrophic events, which carry heavy societal and financial impacts, the stability and survivability of the networks, such as the Internet and/or cellular networks, may be compromised, causing service disruptions and/or phenomena of several packets losses and congestions (Palmieri et al., 2013a). The importance that reliability exhibits in the communications for crisis management has been also confirmed by stakeholders and end-users of the platform envisioned within DESTRIERO and SECTOR, by means of interviews and technical workshops. However, there is a lack of common understanding of the key requirements to be guaranteed in order to assume a crisis information sharing to be reliable, and the relative means to satisfy such requirements. In addition, another inalienable requirement is also security, such as the capability of providing data of interest only to authorized entities and to protect the communication infrastructure against attacks aimed at affecting its availability, correctness, and privacy. Despite the fact that security is a composite concept, characterized by several attributes and formalizations, in this work we have limited our attention to one of the undeniable property, which is sometime the foundation for providing other ones: confidentiality.

In this chapter, we identify the most suitable reliability solutions for a crisis information system, as the one developed in DESTRIERO, and integrated it within a concrete platform for crisis information sharing. Furthermore, we also discuss how achieving the previously mentioned security attribute (a key requirement to integrate different CMSes, emerged within SECTOR), and propose a preliminary solution to be added to the platform for crisis information sharing.

The chapter is structured as follows. In Section 2, we start by reporting the collection of reliability and confidentiality requirements from different end-users and technology experts in the domain of crisis management. Then, in Section 3, we analyze the available literature on the topic and investigate the main platforms for crisis information exchange so as to identify the requirements that are satisfied and the ones that are neglected. Section 4 presents the high level design of the Crisis Information Sharing Platform (CISP) used as the base of our solutions. Knowing what is lacking in the current literature and practice, we propose in Section 5 a set of means to tolerate the faults that may occur during the life of a crisis information system with reaching an invalid state and compromising the mission of the system, and to provide confidentiality guarantees within the crisis information sharing through an interoperability framework among first responders. Section 6 presents empirical results proving the validity of our approach. We conclude with final remarks in Section 7.

## 2 RELIABILITY AND CONFIDENTIALITY REQUIREMENTS

A widely accepted definition of reliability conceives it as the "continuity of correct service," despite the occurrence of possible faults that may compromise the correct behavior of the system (Avizienis et al., 2004). In some papers dealing with crisis information systems, it is also possible to find references to data reliability, as the ability of an information system in handling accurate, trustworthy, and honest pieces of information about a crisis event (Ley et al., 2012; Kamel Boulos et al., 2011). Such a kind of reliability requires means to verify accuracy and reputation in data acquisition and processing. In this work, the first reliability definition is assumed.

A term that is frequently coupled with reliability in articles and vendor documents is resiliency. A system is defined resilient when it is able to adapt under stress or faults in order to avoid failure and to continue to offer some level of performance (possibly not the original level of performance—degraded mode) (Haimes, 2011). Therefore a reliable system is essentially one that functions as the designer intended it to, when it is expected to, and wherever the customer is connected; while, a resilient system is able to withstand certain types of failure and yet remain functional from the customer perspective. In the literature of crisis management, there are some references to societal resilience (Boin and McConnell, 2007) where citizen, first-responders and operational commanders are trained to act efficiently and independently in crisis events based on a set of core values, ethics, and priorities to guide them in their decisions and actions after a disaster. The intentions of this work is not to address such a broad concept, but to build the bases to create resilient data sharing networks able to have governance frameworks that enhance societal resilience.

In order to determine the requirements that make a crisis information system resilient, and therefore, reliable, we have to determine a fault model for this kind of systems, such as identifying what kinds of faults may occur and compromise the overall system. A crisis information system can be sketched as in Fig. 2, and is composed of several nodes, which, without loss of generality, run a single process that may acquire, store, process, and/or visualize data, and such nodes are interconnected by communication channels, forming the Data Sharing Network. Both processes and channels may expose faulty behaviors, such as they diverge from what is considered correct or specified behaviors, which can be modeled as follows (Cinque et al., 2012). Interruptions may occur when processes and links suddenly stop working. Such faults can be further classified in Node crashes, where nodes stop to produce data or react to incoming data due to hardware/software faults, like aging phenomena, software bugs and/or hardware malfunctioning; and Link crashes, where links experience complete loss of connectivity, such as packets are lost for a certain time interval.

Even when links are working properly, the network of interconnected routers and nodes may lead to several communication faults, such as (1) Data Loss, where a loss happens when single or multiple packets are lost (e.g., corrupted, discarded by congested routers, and so on), and do not reach

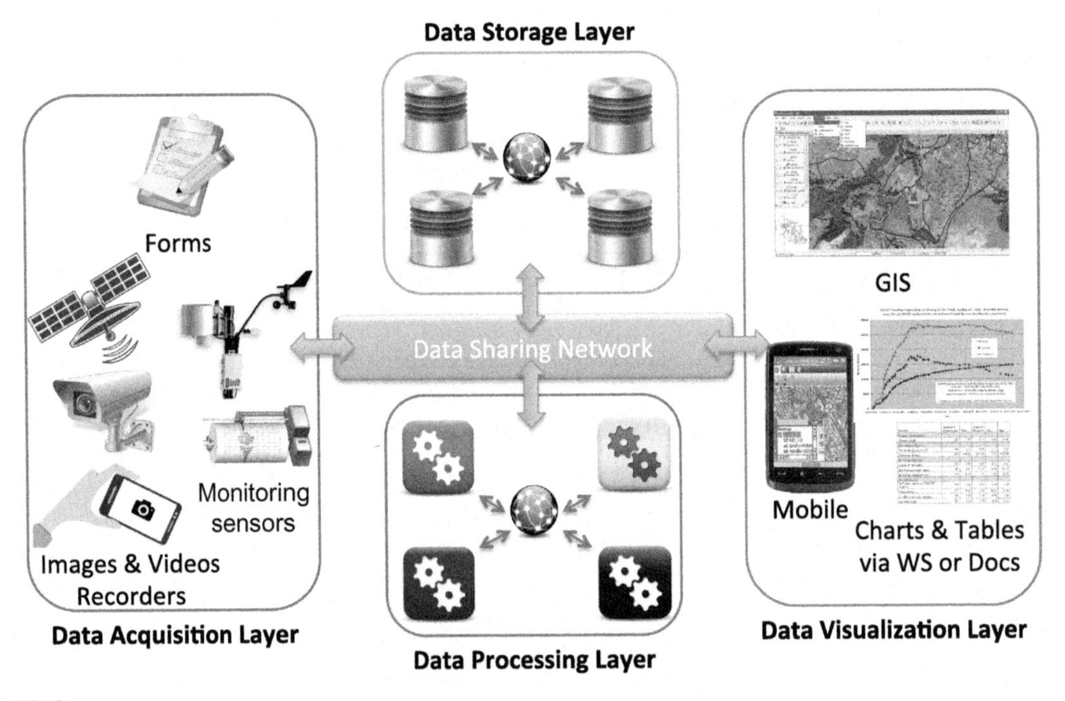


**FIG. 2**

Schematic architecture of a crisis information system.

their intended destinations; (2) Unexpected Delays, where the time needed to exchange a message along a channel is greater than the maximum expected delivery time (e.g., due to router congestion or overload); (3) Message Corruption, when packets with corrupted contents (e.g., due to physical interference, especially over wireless channels) are detected by means of error codes and discarded, and/or a fraction of them could elude the detection and be delivered to upper layers; (4) Unordered reception, for several reasons messages could arrive in a different order than the sending one; and (5) Partitioning, when the network may be segmented in several isolated partitions, due to malfunctioning network devices or broken network connections. Nodes located within the same partition are able to exchange messages, messages exchanged among nodes on different partitions are always lost. Given such a fault model, the required fault-tolerance means needed by a crisis information system can be summarized as follows (Cinque et al., 2012):

- To handle losses and incorrect network behavior, proper recovery means are needed so as to assure that messages are delivered to the intended destinations at the right time and despite of network conditions.
- To deal with interruptions, the design of the system, at any of its levels, has to assume a proper degree of redundancy, so that in case of the unavailability of a node or link, another is ready to take its place. Redundancy in the design of the crisis information system is felt important in Hernantes et al. (2013) within the context of an expert panel on resiliency in crisis management.
- A proper logging system is required to monitor that the crisis information system is behaving as expected, and to raise alarms in case of deviations in order to apply proper mitigation actions able to bring the system in a safe state and in a correct behavior. This represents a continuous maintenance of the crisis information system, deemed as a prerequisite for reliability in Hernantes et al. (2013).
- Crisis management always requires that decisions must be taken at real-time so as to timely and promptly adopt the needed countermeasures to face consequences of a disaster and to reduce the number of causalities and injuries. This also means that fault-tolerance has to be timely and does not have to affect the performance of the system.

Table 1 summarizes previous concepts as a list of system reliability requirements.

**Table 1 User Requirements for Reliable Crisis Information Systems**

| Id | Name | Description | Addressed Faults |
|---|---|---|---|
| REL_REQ_001 | Network Resiliency | The possible malfunctions and/or routing failures have not compromised the lossless, ordered and efficient dissemination of data among the nodes. | Communication faults |
| REL_REQ_002 | System Redundancy | The platform has to be designed in a redundant manner, so that in case of component or link crashes, spare components can take over the lost ones and perform their job. | Interruptions |
| REL_REQ_003 | System Logging | The platform has to log its activities, self-diagnose possible deviations from the correct behavior and mitigate them. | ALL |
| REL_REQ_004 | Timely Fault-tolerance | The platform has to tolerate faults without paying an excessive cost in terms of performance degradations. | ALL |

Without loss of generality, we assume that the communications among different systems are conveyed by large scale networks such as the Internet. However, it is reasonable that during a crisis event, the network may be compromised and affected by malfunctioning and unavailability (Palmieri et al., 2013b). To this aim, it is possible that an ad hoc wireless network may be put in place within the damaged area so as to allow the interconnection for the personal deployed on the field and substituting and augmenting the existing, and potentially damaged, Internet. This makes the communications vulnerable to possible eavesdropping and a malicious adversary may be able to intercept the exchanged signals and access to the message content. Crisis information are important data that should be accessible only to authorized entities in order to avoid the possibility of having attacks compromising the effective management of the crisis.

- The adversary may know where the first responders are deployed and avoid such areas so as to steal left goods within the damaged areas.
- The adversary may access to classified information about the damaged area, not intended to be available o the public access, and distribute it so as to cause panic.
- The adversary may know where the people are being evacuated and plan a terrorist attack so as to maximize the causalities.

To avoid the occurrences of these, and others, situations, communications among first responders must be guaranteed to be confidential. For example, only authorized entities, which have a certain set of attributes or claimed a certain identity, can access to a given message, or even a part of it by retrieving sensitive information related to the crisis and recovery plans.

## 3 STATE OF THE ART FOR RELIABILITY AND CONFIDENTIALITY APPROACHES

In the current literature and practice on crisis information systems, there is scarce information on fault tolerance. The reason is that the research community has focused on different topics, such as the integration of heterogeneous data sources, improving the visualization of data by integrating GIS and sensor data, or identifying scalable architectures that embody the adoption of the cloud computing paradigm. Reliability concerns are left to the middleware technology used for the integration of the different parts of a crisis information system, since most of these middleware solutions come with options for reliability. As a practical example, Kolozali et al. (2014) describes a solution to realize a network of sensors to realize large interconnected ecosystems in a Smart City to monitor and influence the performance of the operational efficiency of the city services and infrastructure. The solution is built on top of Advanced Message Queuing Protocol (AMQP), which is equipped with a proper protocol to partially support network resiliency. In the intentions of the authors this is sufficient to achieve reliability. However, while the reliable data distribution protocol of AMQP is able to recover message losses, the other communication faults are left uncovered. Moreover, the solution in Kolozali et al. (2014) has the requirement of real-time data processing, which is not achieved if using the reliable data distribution protocol in AMQP. Apart from REL_REQ_001, the architecture proposed in Kolozali et al. (2014) does not contain any details on how to deal with the remaining requirements.

As a practical example, the Common Information Platform for Natural Hazards—GIN[1] (developed by the Swiss government to collect, visualize, and aggregate natural hazard information within the Swiss Confederation) is built upon several prediction and analysis models fed by information from different observation networks on the national, as well as on the regional level. It was not possible to gain a comprehensive picture on the overall architecture of GIN; however, the related web pages[2] are rich of details on the analysis tools and visualization means that the platform exhibits, neglecting the resiliency methods that GIN uses to provide reliability. Another example is ResilienceDirect,[3] which is a collaboration and information-sharing platform for the emergency response community in the UK. The platform guarantees the reduction of the communications disruption during emergencies so as to achieve resiliency.[4] The adopted solution is based on the guidelines of the Centre for the Protection of National Infrastructure (CPNI) on enhancing the resilience of telecommunications networks and services.[5] However, the solution deals only with the resiliency of the adopted telecommunication infrastructure by the use of end to end separation on all components, so as to achieve decoupling, and of components with inherent resilience, for example using a Synchronous Digital Hierarchy (SDH) ring rather than point-to-point SDH or Plesiochronous Digital Hierarchy (PDH), so as to remove possible single point of failures which may cause incorrect network behaviors. This solution is able to reduce communications outages after the occurrence of a disaster, but is not able to cope with message losses and/or any possible failures at the other components of crisis information system depicted in Fig. 2.

These practical examples are representative of how resiliency and reliability are considered in the context of crisis information systems by building on reliable and resilient networking and neglecting the other system reliability requirements. In fact, having resilient communication channels working during crisis for local blue light services (e.g., fire department and ambulance) is necessary to have response preparedness during a crisis of emergency services. However, it is not sufficient because other factors are also needed to be considered. In fact, it is crucial to also achieve a high degree of reliability for the various kinds of data required to make timely decisions during crisis management (i.e., reliability of the data storage layer), for the various data processing engines able to transform and aggregate the required data (i.e., reliability of the data processing layer), and/or for the different visualization operations (i.e., reliability of the data visualization layer).

Encryption is among the best known security means, and consists in encoding a given information so as to make it understandable without a valid decoding key. Specifically, a certain destination allowed to access such information has to properly decode the encrypted data so as to obtain the original one. From the literature, we can distinguish between two main cryptographic schemes (Stallings, 2010), based on the type of keys used in the encryption and decryption operations, where a key is the parameter passed as input of a complex cryptographic algorithm or cipher so as to perform one of the mentioned operations and to calculate its output (Stallings, 2013). These schemes can be symmetric or asymmetric (Stallings, 2010), respectively if a single shared key is used for both the encryption and decryption or two distinct keys are adopted. Most of the symmetric and asymmetric schemes

[1]https://www.gin.admin.ch/gin/index.html.

[2]http://www.wsl.ch/fe/warnung/warn_informationssysteme/informations-systeme/gin/index_EN or http://www.gin-info.admin.ch/gin_short_en.htm.

[3]https://www.ordnancesurvey.co.uk/business-and-government/case-studies/resilience-direct.html.

[4]https://www.gov.uk/resilient-communications.

[5]https://www.gov.uk/telecoms-resilience.

are very complex and cannot be easily described, so we refer interested readers to Stallings (2010); Schneier (1996) for more details. Having two keys implies a simpler management of the used keys and the confidentiality of the exchanged information, because in symmetric schemes the publisher has to distribute its key to all the interested subscribers or to use a trusted centralized server. This is a severe limitation because the communication can be insecure if the key exchange phase is compromised. In asymmetric schemes, the subscriber locally generates the couple of keys, and makes publicly available only the encryption key to all potential publishers. On the contrary, in asymmetric schemes the privacy of notifications is assured thanks to the fact that the decryption key remains secret. However, there is no evidence that asymmetric schemes provide a higher degree of security, because the achievable security depends only on the key length and on the complexity of breaking the encryption algorithm. However, the asynchronous schemes exhibit a higher computational overhead than symmetric ones (Petullo et al., 2013; Al Hasib and Haque, 2008).

## 4 A PLATFORM FOR CRISIS INFORMATION SHARING

The platform developed within the DESTRIERO project, here indicated as Crisis Information Sharing Platform (CISP), has been designed and developed to fit crisis management recovery and reconstruction needs to facilitate cross-border information exchange and to support decision makers in the selection and prioritization of the activities to be conducted in the field where the disaster has occurred. Thanks to a clear, documented, and standardized specification of the exposed interfaces, the CISP platform is easily accessible from legacy systems (external systems), which can join the CISP network for boosting information sharing (Fig. 4). In this case, the main scope of the CISP is not to substitute such systems by realizing new ones, but to let the heterogeneous and distributed legacy systems to interact and cooperate among each other by invoking the proper services offered by the CISP platform. Organizations involved in the crisis management can choose if they want to access to platform functionalities directly from their own systems or from a dedicated CISP human machine interface. To this aim, the platform has to enable interoperability among different existing systems at the integrated stakeholder, constructing a common model for the their collaboration and the communications among them, tolerating possible heterogeneity at the technologies, syntax and semantics used by the single integrated systems, and providing capabilities to improve their performance in crisis situations and recovery processes. The solution adopted is to realize an adapter that allows a legacy system to access the CISP network, which we have called as CISP Node. Such a node is installed within the stakeholder premises and represents the front-end of the platform by offering a set of services for communications, interoperability, data management and collaboration. Such a design choice has been drawn by the FINMECCANICA experience in Mission Critical Systems context, as Air Traffic Management, where its new generation of ATM Information Bus, the Swim-Box Technical Infrastructure (Di Crescenzo et al., 2010), has been realized based on open standards for information sharing with complete support to Security and Reliability (Carrozza et al., 2010). DESTRIERO platform extends Swim-Box approach providing Crisis Collaboration Management domain specific services.

The CISP node deals with the heterogeneity among the integrated nodes by proper transformations from/to a common model for the communications along the CISP network. To this aim, within the

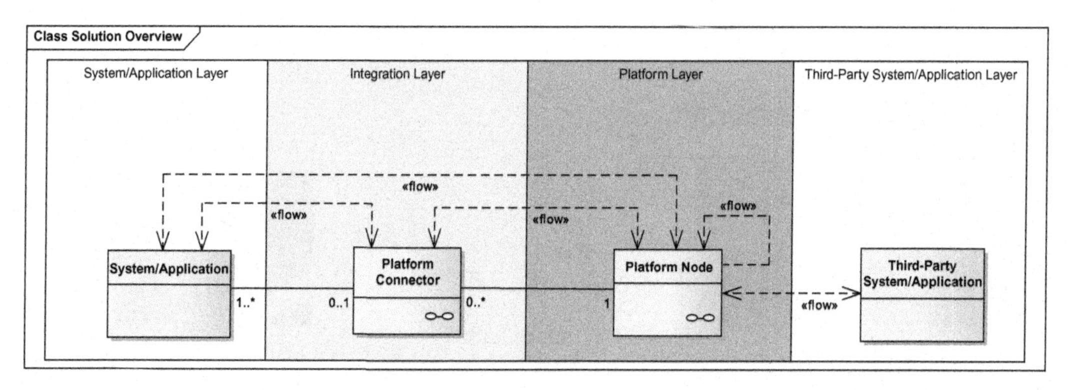


**FIG. 3**

CISP platform.

CISP node, we can find two different logical components, as illustrated in Fig. 3: a first one contains the integration logic for the conversion back and forward the given model of a system to the common one so as to provide interoperability, while a second one implements all the logic for the platform services of interconnection, collaboration data management. Specifically, the internal design of the platform layer of the CISP node is shown in Fig. 4, where we can identify a layered hierarchy with three tiers:

- The bottom of the design is represented by a series of basic services for messaging (both with a request-reply and a publish/subscribe manner), data and knowledge management and data transformation, for comparing data with a specific schema and transforming data from one representation to another. Such services are provided by the enabling technology used to develop our solution, such as JBoss, Web Services, and OASIS Web Service Notification.
- CISP Core Layer: It is composed of all those functionalities that are accessible directly by the system adapters, DESTRIERO-compliant applications and/or the higher layer. These functionalities have been realized on top of the basic services, which are generic, and have the responsibility to make the basic services more domain-specific. This layer has to allow Data Dissemination, Data Persistency and Durability, Data Validation and Transformation.
- CISP Methodology Layer: It is composed of all those functionalities providing a support for coordination/collaboration, information integration and events management. This layer offers API to support both CRUD (create, read, update, and delete) basic functions and domain specific Web Services for interactions between involved connected systems. Such API has been designed based on the Emergency Data Exchange Language (EDXL) standard to enhance semantic and syntactic interoperability between all the involved Crisis Management Systems.
- Transversal Capabilities Layer: It is composed of all functionalities with impact on both the previous two layers and are therefore, vertical to them. This layer offers Reliability Management, and it is enhanced with Supervision and Security Management capabilities, coming from SECTOR needs.

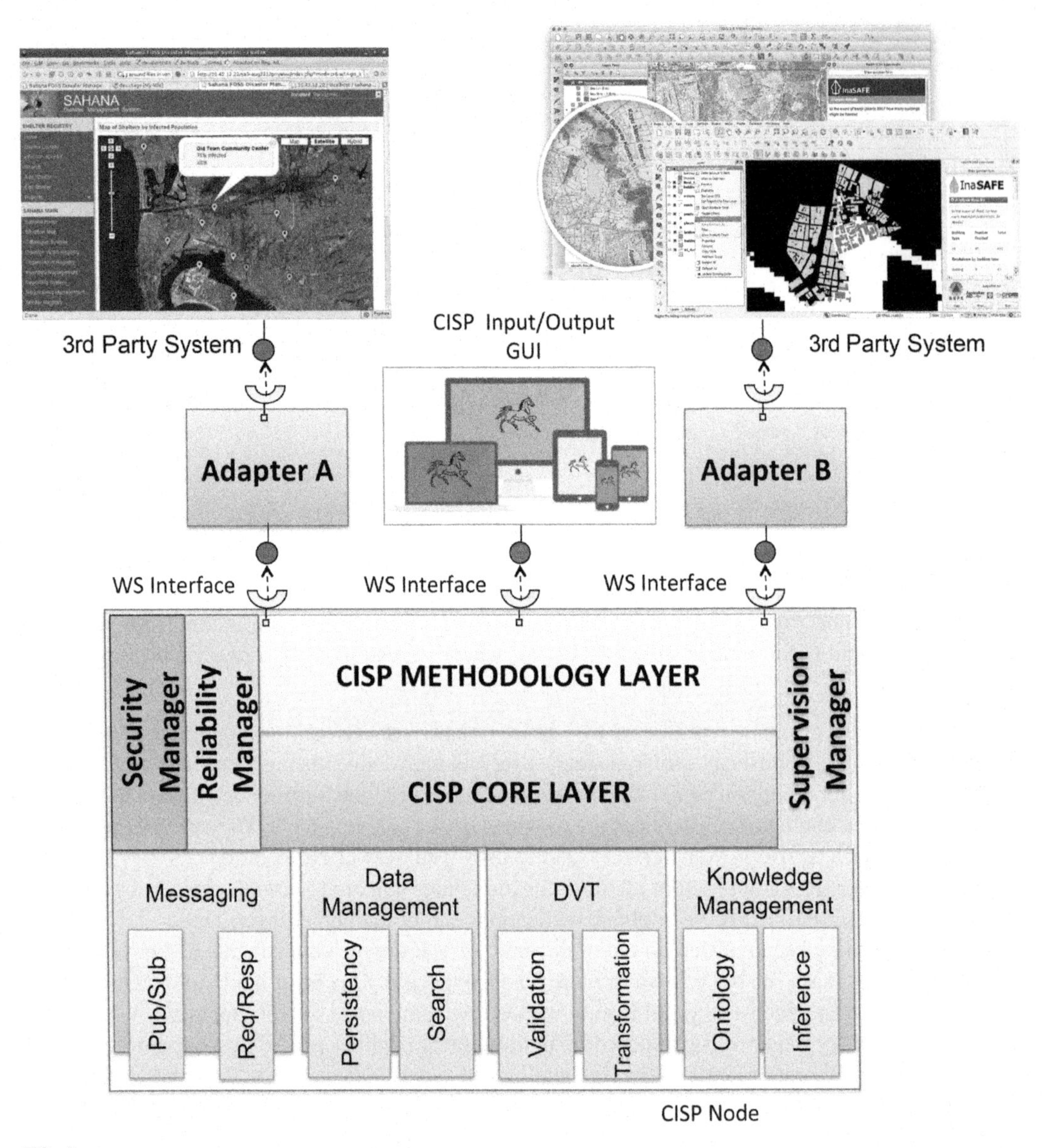


**FIG. 4**

Scheme of the CISP Node.

Crisis Management Systems might be enabled to offer specific collaboration management services through the platform, directly implementing the API to interact with the CISP Node of reference, or realizing an interoperability layer by means of an adapter, which is able to transform data from legacy format to adopted one and vice-versa, providing also services mapping between the involved entities.

# 5 PROPOSED SOLUTION

The CISP platform has been designed based on a set of widely-known standards. A number of standards have been investigated to be used as basic services for both reliable information sharing and data synchronization/persistency, upon which developing the core layer of the platform. Among the analyzed standards, OMG Data Distribution Service (DDS) (OMG, 2012) has been adopted in order to support mechanisms for information distribution and distributed data and metadata persistency. Both Messaging and Registry core services have been designed and developed on top of the OMG DDS in order to hide to domain specific components all "steps" for a reliable platform. In particular:

- Messaging Component distributes information to all the Messaging core Components peers interested in such information, according to the Quality of Services (QoS) specified for each information to be exchanged. In particular, key QoS are (but are not limited to) durability and reliability.
- Registry Component provides reliable communication and periodic discovery process aiming at synchronize specific metainformation among all interested CISP Nodes. Furthermore, the service provides also metadata verification and coherency checking. In order to avoid single points of failure and to give a proper level of redundancy (REL REQ 002), metainformation are shared between all the CISP nodes.

The metainformation sharing has required a further investigation to make the nodes able to maintain the consistency in an environment subject to different faults. Periodic discovery process, together with realignment activities, solve most of the common problems without significantly impacting performance in accordance with the requirements for reliable Crisis Information Systems. However, the use of DDS is not able to cover all the reliability requirements of interest for this work. Specifically, OMG DDS is only able to provided link-by-link of the communications over the overlay formed by the CISP nodes, due to the use of TCP-like retransmission means. We are more interested in end-to-end assurances for reliable communications. Moreover, we need to properly configure the data redundancy by placing the data replicas where needed and suitable. To this aim, we have built on top of the basic pub/sub communication and data persistency means provided by OMG DDS a proper layer for reliability, as illustrated in Fig. 5.

The CISP platform has been built on top of a service-oriented architecture. When security, and in particular confidentiality, should be enforced within the context of Web services, it is possible to make use of the WS-Security specification (Nordbotten, 2003; Naedele, 2009), which is typically natively supported in the programming environment and containers of Web services (a flag of the used communication protocol should be enabled in order to have a secure transport). WS-Security is a composite standard made by combining other different specifications and methods, and specifies two different levels of mechanisms to enforce the provided security level. The first is implemented at the message level by defining a SOAP header that carries out extensions to security. The second is realized at the service level to perform higher-level security mechanisms, such as access control or authentication.

Without loss of generality, considering the design of the CISP platform, we can schematically see a large-scale crisis information system as the integration of different systems by means of large-scale networks as in Fig. 6. In each system, we can find a series of nodes belonging to the first responder's organization (indicated in orange with Nx in the figure) and responsible for different operations.

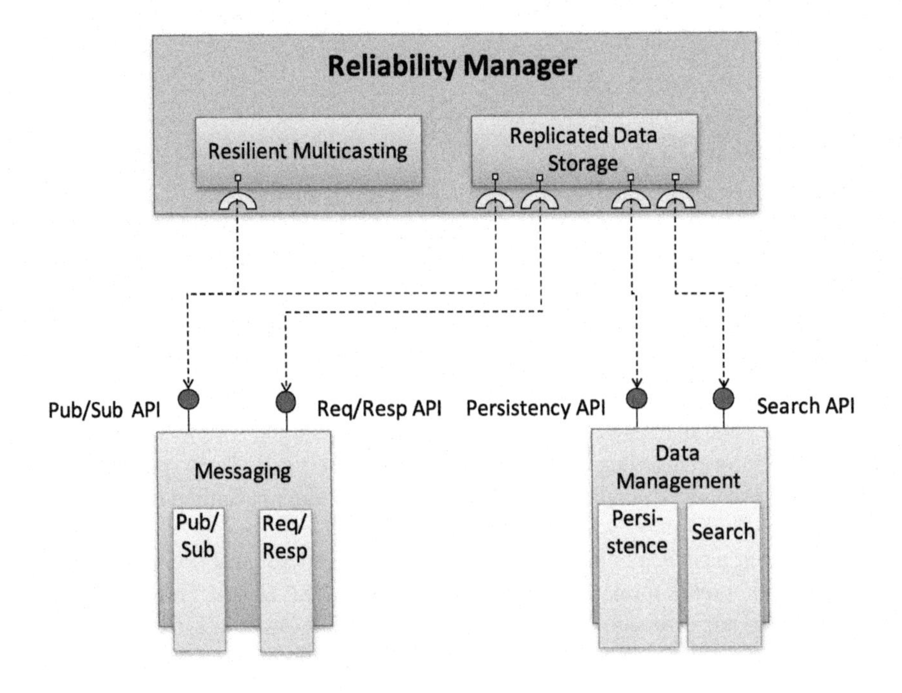


**FIG. 5**

Scheme of the integration of the reliability enforcement solution within the CISP Node.

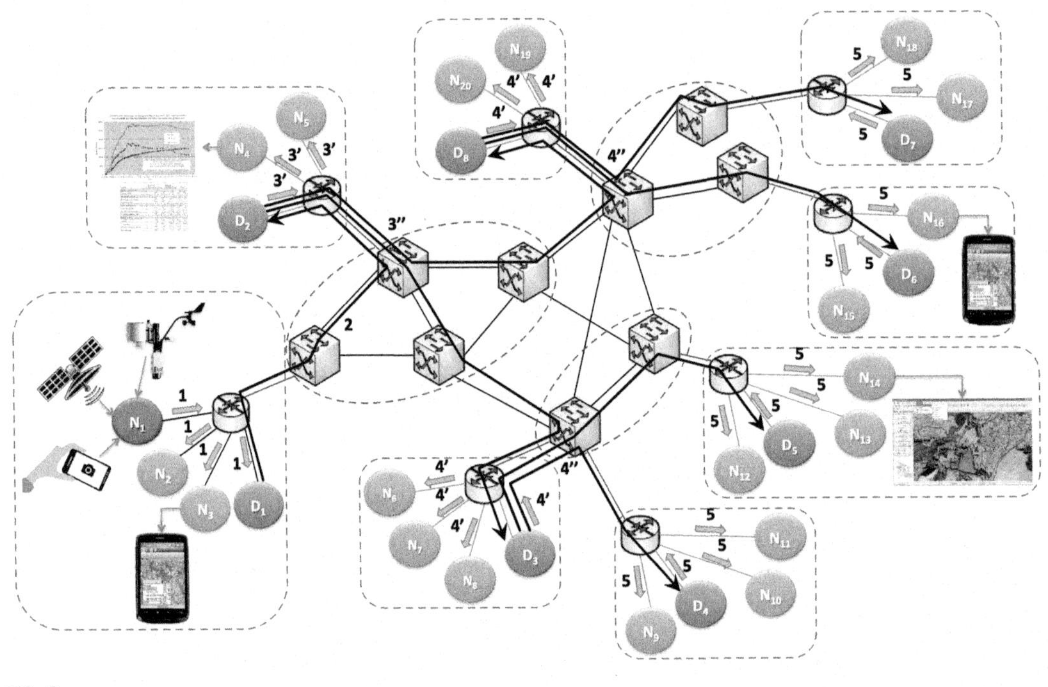

**FIG. 6**

Collaboration in a large-scale crisis information system.

For a concrete example, node N1 collects all the sensing data from smartphones, sensors and satellite about the damaged area. Nodes N3, N4, N14, and N16 have duties of properly presenting some sensing data, sometimes contextualized in maps, to the organization's operators by means of GIS, charts and tables, or apps. The other nodes, in yellow in the figure, can execute storing or processing operations on the collected sensing data and/or on historical data about past disasters. Within each of the organization's infrastructure we can find a CISP node (indicated in red with Dx in the figure), which hosts the functionalities of the CISP platform made available to the services and operators of the hosting organization. Each time the information is delivered to a CISP node, it may be distributed locally to the other nodes (indicated with green in the figure), considering the needs of such nodes and providing an illusion that the information has been internally produced and not received from another organization. In order to satisfy the system reliability requirements mentioned in Table 1, we have to guarantee that such an information sharing mechanism is successful despite of possible fault occurrences. For this aim, the information of interest should not be hold only within a single organization, but distributed among the organizations so that in case of a failure it is still available within the overall infrastructure and its access latency is optimized since it can be retrieved from a more convenient location than the one where it has been produced. Therefore in the reliability layer of the DESTRERIO node we have a replicated data storage that decide the best place to put a data replica so as to maximize its availability and minimize its retrieval time and memory consumption.

Moreover, we have to equip the CISP platform of a proper resilient multicast communication protocol, so that the information of interest is guaranteed to reach the intended destinations. To this aim, we also have a component for resilient multicasting so as to reliably share crisis information. The following subsections will provide detailed descriptions of the first two solutions, leaving as a future work the design of a logging scheme.

## 5.1 REPLICATION SCHEME IN CRISIS INFORMATION SYSTEMS

A well-known solution to tolerate the crash problem is to replicate the entity for which it is crucial to tolerate its eventual crash (Wiesmann et al., 2000). On the one hand, one option is to use a passive replication scheme, in which the failed entity is replaced by one of its backups. The case that both the entity and all of its replicas fail at the same time is extremely rare, so this solution improves the availability of the entity and the relative reliability of the overall system. However, timeliness is compromised since the entity would be down for a certain time window leading to the impossibility of satisfying some requests from its end-users. On the other hand, another option is to use an active scheme, in which the system exposes a virtual entity made up of a set of distinct replicated objects with equal responsibilities. In this case, a failed entity is instantaneously replaced without causing the unavailability of certain system functionalities. So timeliness is achieved; however, there is still the possibility of a failure of all the coordinators due to common mode errors.

In the literature, there are hybrid schemes, such as semi-active replication (Defago et al., 1998), where the entity is actively p-redundant, such as there are p replicated objects active at the same time; moreover, there are k backups for each active object. The system designer is free to choose the robustness of the system by varying $\langle p, k \rangle$. Such a replication scheme allows the system to tolerate entity crashes without having certain functionalities unavailable for a certain period of time. Thanks to backups, such a solution is not vulnerable to common mode errors as active replication. Hybrid schemes allow taking the best of the combined replication schemes without being affected by the same issues.

In our work, the application of a redundancy degree within a typical large-scale crisis information system is realized by adopting the above introduced hybrid model with active and passive replicas of the key components in the infrastructure. Specifically, we have considered as critical the components in each first responders' organizations that composes the data storage layer in Fig. 2, since the loss of such amount of data can be catastrophic for the overall system. Therefore the data storage within the CISP node deployed at the premises of an organization is schematically structured as a set of nodes, each identified by a proper string and interconnected as in Fig. 7. A set of nodes for data storage is running and receiving all the incoming requests arrived to the organization. The write requests, which are the ones that alter the data stored in these nodes, are executed by all the nodes so as to keep consistent their state among each other, such as all retain the same pieces of data at the same version and content. At the conclusion of each write request, the node logs an entry in a proper file that keeps the history of past activities for the node. Such an entry contains the identifier of the changed datum, its previous value, and its current value and who requested the change. Such a mechanism is required so as to undo any possible change at any time and to know who is the responsible for each change. If a write request fails, then a special message is exchanged among the nodes to cancel possible changes that such a request have caused in the other nodes and restore consistency among the nodes. The read requests, which aim to return a copy of certain stored data to the requesting organization, are executed only by one of the active data storage nodes, such as the first available one. This allows evading the incoming read requests in parallel and reducing the overall retrieval latency. Moreover, due to nodes capabilities to synchronize themselves, read requests can be executed in parallel among each other and in series with a write request, which is preemptive with respect to the reads.

A set of nodes for data storage is running and assumed to be the backup of one of the active nodes in the previous set. Such nodes receive only the writing requests, and change their internal state (i.e., the stored data), accordingly. It is important that the backups are consistent with the active nodes, so

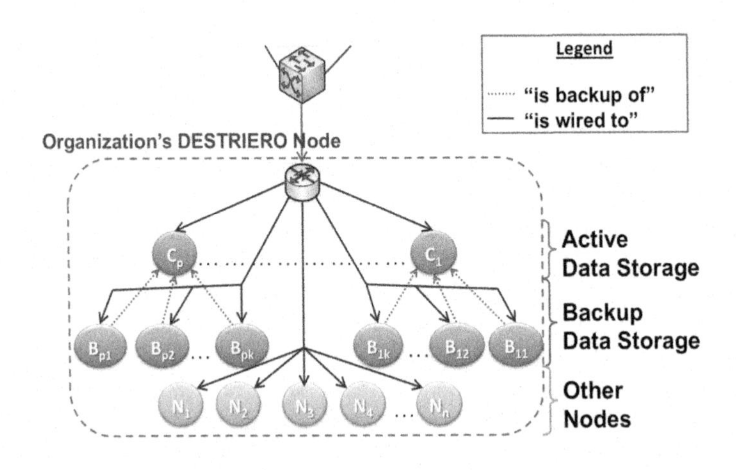


**FIG. 7**

Replicated data storage within the CISP Node.

any time one node changes its internal state a special message is distributed within the organization among active and backup nodes. Such a message contains the content of the entry in the history of past activities related to the write request. A node that receives such a message checks if it holds the same entry in its own history of past activities. In the positive case, the consistency is assured; while in the negative case, an inconsistency is detected and the change is requested to be undone by all the nodes and the requesting organization is informed of the unsuccessful execution of the operation. Whenever an active node is compromised, it is substituted by one of the available backups. For example, the node with the greater identifier is the new active node.

Each organization keeps the data for which it is the producer in its local data storage. Such a strategy implies that in the case of unreachability of an organization, due to routing misbehaviors or crashes of the data storage, all the data held by the unresponsive organization is not available. The solution is to have that each organization keeps a copy of all the data of the other organization. However, this implies a huge waste of storage resources and implies a considerable workload for keeping the consistency. A better solution is to place data at given organizations so as to achieve optimal data availability and retrieval latency, while keeping a lower storage resource consumption and consistency agreement workload.

We can assume two kinds of overlays, the one within each organization that interconnects the nodes storing data, as illustrated in Fig. 7 with active and backup nodes, and an additional level with a single active node per organization being interconnected with its peers in the other organizations, as illustrated in Fig. 8. Such a node is called super-peer and has the duty of determining which data is needed to be kept by its organization of reference. The other active nodes behave as backup in case the super-peer may result unavailable. When a super-peer is detected as unavailable, by means of keep-alive messages or timeout-base mechanism, the replica with the higher identifier is automatically elected as the super-peer of the organization. When data is replicated, write and read requests are not only directed toward the organization that has produced the data of interest, but all the organizations that hold in their data storage system such a data.

To help find such information, the data available in the platform is classified in different types by placing a special string indicating such a classification, and the super-peers, and their replicas, holds a map that associate each class of data to the address of the different nodes holding a replica of such data. The remaining issue left to be solved is how to determine where to place replicas of each data class. Such a decision is taken without the intervention of any centralized managers, but the super-peers collaborate among each other to find a solution to this problem in a distributed manner. Specifically, each super-peer keeps a set of statistics for each class of data requested by any entity in its organization. We have envisioned two classes of statistics: data availability and retrieval time. In the first case, the super-peers counts the number of successful write/read request over the total number of requests, in the second case, computes the average time to retrieve instances of the given class of data. Periodically, the super-peer computes a satisfaction degree per each own $i$th organization knowing the number of available replicas and their placement within the overall CISP platform:

$$\delta_i(n, \overline{P}) = \omega_A \left( \rho_A - \frac{\kappa_{Succ}}{\kappa_{Tot}} \right) + \omega_\Lambda - \kappa_\Lambda \tag{1}$$

where $\omega_A$ and $\omega_\Lambda$ are weights chosen by organization administrator in the 0, 1 interval and stating the importance of availability over latency, $\kappa_{Succ}$ is the total number of successful requests, $\kappa_{Tot}$ is the number of requests, $\kappa_\Lambda$ is the computed average retrieval latency, $\rho_A$ and $\rho_\Lambda$ are the required level of

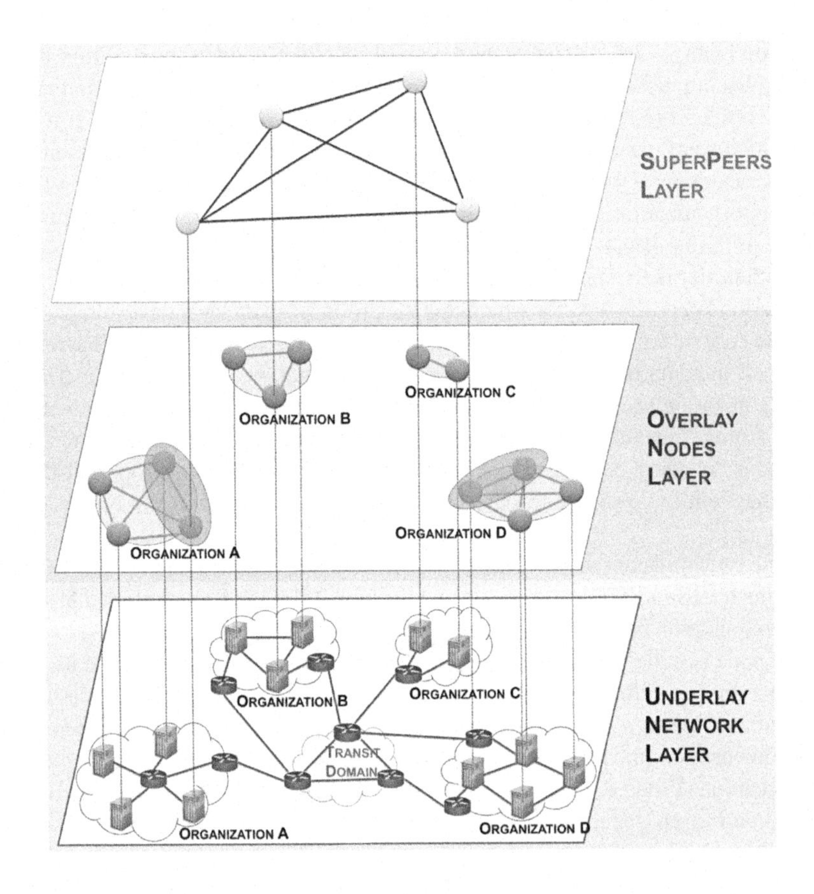


**FIG. 8**

Two-level overlay of the nodes storing data in each organization interconnected by the CISP platform.

data availability and retrieval latency chosen per each class of data by the organization administrator. If the satisfaction degree is positive, it means that the current setting of the platform satisfies the requirements of the users of the given organization and no further actions are needed. On the contrary, a negative value states that the current setting is not satisfactory and changes are required. The costs for an organization for holding a replica for the given class of data can be formulated as follows:

$$cost_i = \omega_M \left( \frac{\theta_{data}}{\theta_{Tot}} \cdot 100 \right) + \omega_\Lambda \left( \frac{\iota_{data}}{\iota_{Tot}} \cdot 100 \right) \tag{2}$$

where the first contribution measure the fraction of the storage resources used to hold the replica, while the second one indicates the fraction of requests received for the hold replica over the total number of served requests. The driving idea is to maximize the satisfaction degree at each organization and minimizing the relative costs by determining an optimal number and placement of replicas for the class of data of interest:

$$\begin{array}{c} \max_{i \in 0,N} \delta_i(n, \overline{P}) \\ \min_{i \in 0,N} cost_i \cdot P_i \end{array} \tag{3}$$

subject to:

$$\begin{array}{l} n = \sum_{i=0}^{N} P_i \leq \max \leq N \\ P_i = \begin{cases} 0 & \text{if no replica at the } i\text{th organization} \\ 1 & \text{if replica at the } i\text{th organization} \end{cases} \end{array} \tag{4}$$

where max indicates the maximum number of replicas to be placed in the infrastructure made of $N$ organizations integrated by the CISP platform. This is an example of multiobjective optimization problem (Marler and Arora, 2004), since the two objective functions to satisfy are opposing. For example, a solution for the first maximization is not a solution for the second minimization, and a trade-off is needed. Such a trade-off is called Pareto solution, such as a configuration able to keep in balance the two objectives and a different solution is not able to improve such a situation. Classical solutions (Jones et al., 2002) are not viable in our case, since the CISP platform is large and global knowledge in such systems is not achievable. Therefore we have considered a distributed approach to the resolution of this problem by means of a noncooperative game (Cardinal and Hoefer, 2010).

More formally, let us consider the super-peer overlay as an undirected graph $(X, E)$, consisting in a set of nodes, namely $X$, and a set of edges connecting two nodes, namely $E$. Within the set of the available nodes, we define a subset of all the nodes, namely $Y \subset X$, containing the nodes where a player can be located. We consider a set of players $P := \{c1, c2, \ldots, cp\}$ of finite size $p \leq 2$. Formally, the strategy set for each player $c \in P$ is defined as $S^c = Y$, such that a strategy of a player is the selection of a node $s^c \in Y$. Combining the strategy sets of all the players, namely $S = S^{(c_1)} \times S^{(c_2)} \times \cdots \times S^{(c_p)}$, a strategy profile $s \in S$ implies a certain payoff to each player $c$, namely $\Phi^{c(s)}$, which are aggregated in the so-called profile of payoffs denoted as $\Phi^{S}$. The payoff is the gain achievable by a player to host a given replica considering the store the replica and to manage the incoming requests to read/write it, as mentioned previously.

The scope of the game is to determine the best strategy profile that implies the maximum payoff for all the players. Despite the several possible formulations that came out within the literature, we have described such a game in terms of a noncooperative game, where players are selfish (i.e., there is no direct communication between the players) and each one only cares to maximize its own profit or to minimize its own costs without considering the state of the other players (with the eventuality of damaging them, even if it is not intentional). Then, the normal form for the noncooperative game for our replica placement problem is given by $\Gamma = (P, S, \pi)$, with the objective of maximizing the payoff for all the players, for which we are interested in finding Nash equilibria (i.e., given a certain strategy $s \in S$, it is not profitable for a player to select a different replica placement pattern than the one in the current strategy profile since adding or removing a replica will not change or even reduce the achievable payoff) so a player has no incentive to change strategy. The demonstration of the existence of such equilibria is a known NP-hard problem and is resolved by means of theorems. For a concrete example, the authors in Vetta (2002) demonstrate the existence of at least a Nash equilibrium for games as ours and the conditions to induce such equilibrium are presented.

We can model our replica placement problem as a noncooperative game with N players, whose strategy is represented by a binary decision to hold a replica or not. Rather than formalizing the payoff of each player, we consider its costs, according to the previous formulation of $cost_i$, properly assigned to each player. Specifically, let us indicate with Si the binary value representing the strategy chosen by the $i$th player (which is 1 if the $i$th player decides to hold a replica; otherwise, it is 0). The cost paid by the $i$th player to follow its strategy can be formalized as follows:

$$C_i(S_i) = cost_i \cdot S_i + \delta_i \cdot (1 - S_i) \tag{5}$$

where $cost_i$ and $\delta_i$ are expressed in the previous equations. The game can start with a random strategy profile and evolve over the time where each player changes its strategy so as to minimize its costs formulated in the previous equation. Such an evolution will bring to a stable solution represented by the Nash Equilibrium, where no player has an incentive to change its strategy. Based on the definition of a Nash Equilibrium, it is possible to see that a strategy profile s represents a Nash Equilibrium if and only if the two following conditions are guaranteed:

$$\exists i \in Y s.t.\ \delta_i \leq cost_i \tag{6}$$

and

$$\nexists i \in Y s.t.\ cost_i - \delta_i > 0 \tag{7}$$

The first condition indicates that the satisfaction generated by a replica placed at the $i$th node is never greater than the cost of placing it; so, none of neighboring nodes has an incentive to act as a codec. While, the second condition states that, when the $i$th node holds a replica, it is not convenient to stop holding it since the paid cost is already minimized. The condition in the first equation defines the control behavior of the super-peers to decide holding a replica or not.

## 5.2 RESILIENT MULTICASTING IN CRISIS INFORMATION SYSTEMS

Communications among CISP nodes in different organizations are conveyed by the Internet, so they are affected by link crashes and bursty loss patterns. In addition, due to the introduction of the replication scheme in the platform, the data exchange pattern adopted within the system is a multicast one. The general literature of reliability means to tolerate losses in a multicast communication infrastructure can be roughly classified in two big classes (Esposito et al., 2013): the one based on temporal redundancy, and the one based on spatial redundancy. In the first case, eventual losses caused by failures are detected somehow (e.g., by means of timeouts or incorrect order of arriving of the packets associated with a sequence identifier), and lost packets are recovered by means of retransmissions. On the other hand, additional information is sent along a notification so that eventual losses can be resolved without requiring retransmissions.

Since in crisis information systems performance matters, we have focused our attention on proactive methods based on spatial redundancy, among which the main ones are Forward Error Correction (FEC) (Rizzo and Vicisano, 1998), and Path Redundancy (Birrer and Bustamante, 2007), both affected by the following drawbacks, which have limited their applicability in real middleware products: Path Redundancy is effective only if path diversity is guaranteed, i.e., the multiple paths do not share any routing component, which is difficult to obtain over the Internet (Han et al., 2006). In FEC approaches, coding is typically performed at the source, and the applied redundancy degree is tailored on the worst-case loss rate experienced by one of the destinations. This may result not optimal and even dangerous for other destinations with lower loss rate, but also capacity, and can cause congestion phenomena. If such a rule is not followed and redundancy degree is lower, then a subset of the destinations will experience losses and reliability is compromised.

We have decided to apply a hybrid approach by properly combining FEC and Path Redundancy into a proper communication protocol, which is aware of the underlay topology when building the multiple diverse trees to be used to disseminate information as to achieve path diversity. To this aim, we have applied a protocol we have designed in Esposito et al. (2009) to build multiple diverse trees by selecting the paths from any new node to its parents that expose the lowest measure of diversity, and keeping the paths to the children of a given a parent to maintain a measure of diversity closer to the value they had before the inclusion of the new node as a child. For measuring the path diversity, we have used the Sørensen-Dice coefficient (Dice, 1945), which is a statistic used for comparing the similarity of two samples.

We have shown that this protocol is able to achieve a multiple tree with a degree of diversity that is close to the intrinsic diversity of the topology at the network level, implying a robust delivery of multicast messages despite of link crashes, since it allows circumventing the crashed link and making all the nodes reachable even if link crashes occur. However, such a solution is not able to build multiple trees that are completely diverse, if such a diversity is not present at the network level, as in practical Internet configurations (Esposito et al., 2009).

Disseminating over multiple trees is not so effective when the dissemination infrastructure is affected by losses. In fact, it is not negligible that a given packet of the stream may be dropped along all the paths from the producer to a destination. To lower such a probability, the message publisher can adopt several strategies that imply a different redundancy degree to the packet flow exchanged between source and destination. The most effective strategy is to send replicas of the message through all the trees, but this presents a troublesome side-effect: it generates a strong traffic load that can strengthen the loss patterns experienced by the network. A more network-friendly solution is to generate redundant packets by using FEC technique, and to forward a portion of the encoded packets per each tree. Even if a FEC-enhanced multiple-tree approach can theoretically reduce the probability that a node experiences the loss of a given packet, in practice it is not achieved since opportune tuning of the redundancy degree applied by the message producer is not possible. In fact, an optimal tuning of FEC requires gathering global knowledge of the loss patterns within the system, and this is impractical for Internet-scale systems.

Since they present a dual behavior (i.e., one technique is vulnerable toward failures that are effectively tolerated by the other one and vice-versa), a suitable solution to have a dissemination strategy that proactively tolerates both link crashes and message losses is combining them. Specifically, in each tree FEC is performed so that the packets dispatched by the root along the tree has a high probability to be delivered to all the nodes despite of losses under the assumption that no link may crash. On the other hand, the producer of a message applies a coding technique to generate r redundant packets from the k information packets (so that the packets to deliver to each node is $n = k + r$). Then, it equally disseminates the n packets though the $t$ multiple trees (i.e., each tree conveys a number of packets equal to $n/t$). If there may happen only a single link crash that compromises the message delivery along a single tree, each node will receive only $n - n't$ packets, so the system will tolerate the crash if the number of received messages is greater or equal to the capacity, namely C of the adopted coding technique:

$$n - \frac{n}{t} \leq C \rightarrow \left(1 - \frac{1}{t}\right) \cdot n \geq C \rightarrow \left(1 - \frac{1}{t}\right) \cdot (k + r) \geq C \tag{8}$$

Considering this equation, we can formulate a condition on the applied redundancy degree $r$, based the size in packets of the message to be sent, namely $k$, so that a single link crash can be tolerated:

$$r \geq \left(1 - \frac{1}{t}\right) \cdot (C - k) \tag{9}$$

This result can be generalized for the number, namely $ft$, of faulty trees (i.e., they do not deliver their packets to a certain node due to a link crash), as follows:

$$r \geq \left(1 - \frac{1}{ft}\right) \cdot (C - k) \quad \text{iff} \quad ft < t - 1 \tag{10}$$

This equation shows that the tuning of the FEC coding used at the information producer does not depend on the loss patterns affecting the network, but on the tolerance degree that the system has to exhibit, namely $t$. This means that the tuning does not require a global knowledge of the system, so it can be realized in an Internet-scale system. Such a consideration on link crashes can be applied also to message losses, but in this case not only the message producer can apply FEC, but all the nodes in the multi forest, so as to protect each of its links from losses.

## 5.3 CONFIDENTIAL COMMUNICATIONS IN THE CISP PLATFORM

The naive solution provided by the WS-Security is not optimal in the context of the CISP platform, since it establishes long-term pairwise secure channels among end-points within the notification overlay by using Secure Sockets Layer (SSL) or Transport Layer Security (TLS). A message before reaching its intended destination has to go through a series of exchanges among the nodes composing the overall platforms and conveyed the exchanges among the federated organizations by the CISP solution. The reasons are the following ones:

- at each link of the overlay shown in the figure it is needed to pay the cost of the encryption and decryption operations and to keep the connection among the two endpoints;
- the link-by-link approach achieves end-to-end confidentiality guarantees only and only if the traversed nodes are assumed trusted, with may not be the case in a large scale platform managed by different organizations.

To have empirical evidences of the first element of the previous list, we have implemented a link-by-link and end-to-end encryption within the context of a publish/subscribe service that resemble the multilink paths used within the CISP platform, and measured the performance worsening of the communication latency with respect to the case without any encryption. Link-by-link encryption is heavier since a node of the communication platform should perform as many encryptions as the number of traversing nodes by using their keys, previously exchanged by means of a key agreement protocol. This implies a high performance penalty, as illustrated in Fig. 9.

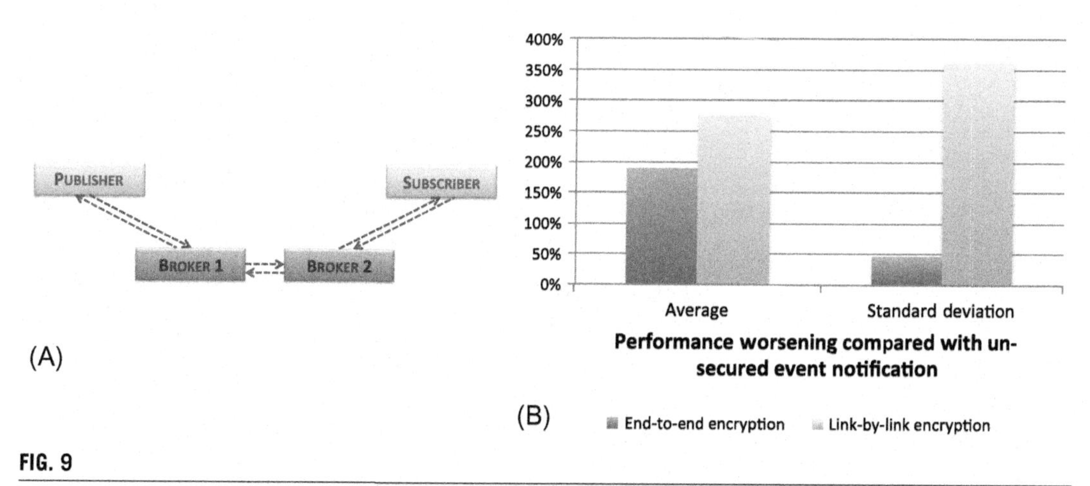


**FIG. 9**

Performance comparison of link-by-link and end-to-end encryption in event notification.

Based on these considerations, we can assume two different approaches for implementing confidentiality in the communications within the CISP platform. The first one is depicted in Fig. 10A, where secure protocols, as stated by the WS-Security specification, are applying to protect all the interactions to bring a certain piece of information from the producer within an organization to its intended destination within the same organization or at a different one. Despite the issues raised previously, such a solution has the pros to be fully supported by all Web service containers and development frameworks, and easy to apply to our implementation of the CISP platform. The other possible solution is to implement end-to-end encryption, as illustrated in Fig. 10B, where the data producer encrypts the message before passing it to the platform, while the destination decrypts the message right after its reception from the platform. In this case, we have more implementation issues, since all the approach must be realized by scratch, and a proper support of the platform to key management should also be designed and implemented.

## 6 EMPIRICAL EVALUATION

To quantify the goodness of our proposed solution, we have conducted a preliminary assessment by using a simulation approach. The scope of this section is to present the results of such assessment that (1) studies the impact of the proposed resilient multicasting on the quality of data sharing, and (2) investigates the effect of the proposed replication scheme on the data availability within the large-scale crisis information system. To achieve this aim, we implemented our proposed solutions by using the OMNET++[6] simulator. In our first simulations, the exchanged messages have a size of 23 KB, the publication rate is one message per second and the total number of nodes is 40. We have assumed that the coding and decoding time are respectively equal to 5 ms and 10 ms. We have published 1000 events

[6]www.omnetpp.org.

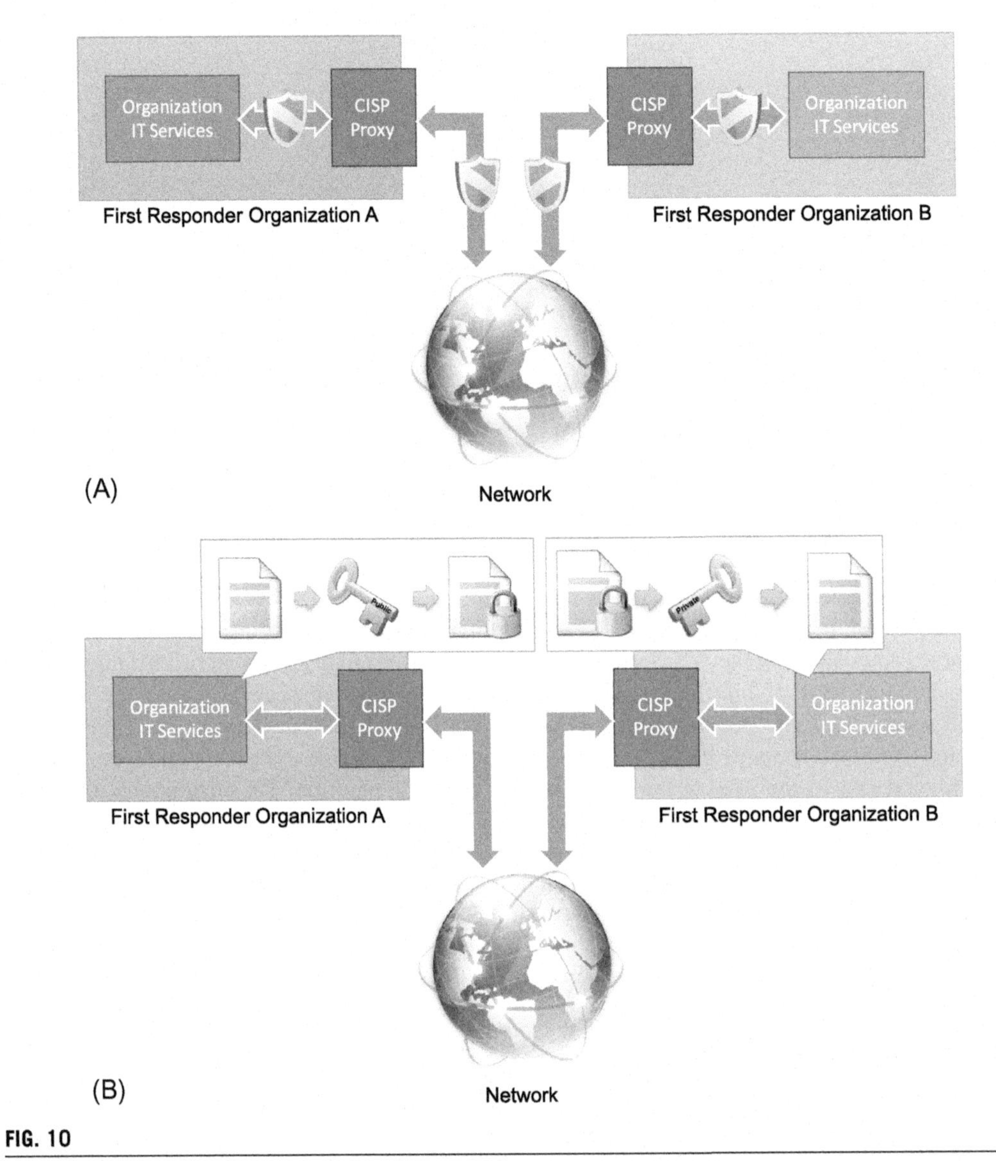


**FIG. 10**

Two possible solutions for confidential communications within the CISP platform.

per each experiment, executed each experiment three times and reported the average. In the second simulations, we assume that each organization, made of 4 nodes, periodically generate a datum of 23 KB, while others periodically make requests for a piece of data previously produced by one of the remaining organizations. We assume that nodes can crashes with a probability of 0.05. In order to keep our simulation closer to the expected behavior of our solutions when applied within the context of the

CISP platform, we have fed some of the parameters of our model with values obtained by testing the performance exhibited by the implementation of the platform. Specifically, we have obtained the link delays at the overlay level from a series of experiments aimed at measuring the latency of exchanging a message between the CISP nodes by using a request/reply communication model. We have deployed the platform implementation by means of JBoss in a series of virtual machines interconnected by means of a local LAN and running experiments among 2, 3, and 4 nodes, with one producing notifications and the other ones interested in receiving them. We applied two different communication patterns: Request/Response over SOAP CXF WebService with messages having a size of 2375 bytes, and Publish/Subscribe over OpenSplice DDS with messages of 1902 bytes, where multicast has been enabled. We can notice from Fig. 11B and D that increasing the number of nodes, the standard deviation, which measure the variability of the latency around the mean value, augments, where the publish/subscribe service exhibiting the lower values since the request/reply pays the overhead of the SOAP protocol in having XML format of the messages, known to be heavy for its internal redundancy (Esposito et al., 2010). We have exchanged 1000, however we have discarded the latency collected for the first 50 messages, since they diverge with the rest of the results due to the overhead to setup the platform and bring it to a stable state.

From Fig. 11A we may assume that increasing the number of destinations it may lower the latency, however this is untrue due to the high variability. Within the context of the publish/subscribe communications, we clearly see an increase of the latency, as illustrated in Fig. 11D, which is around

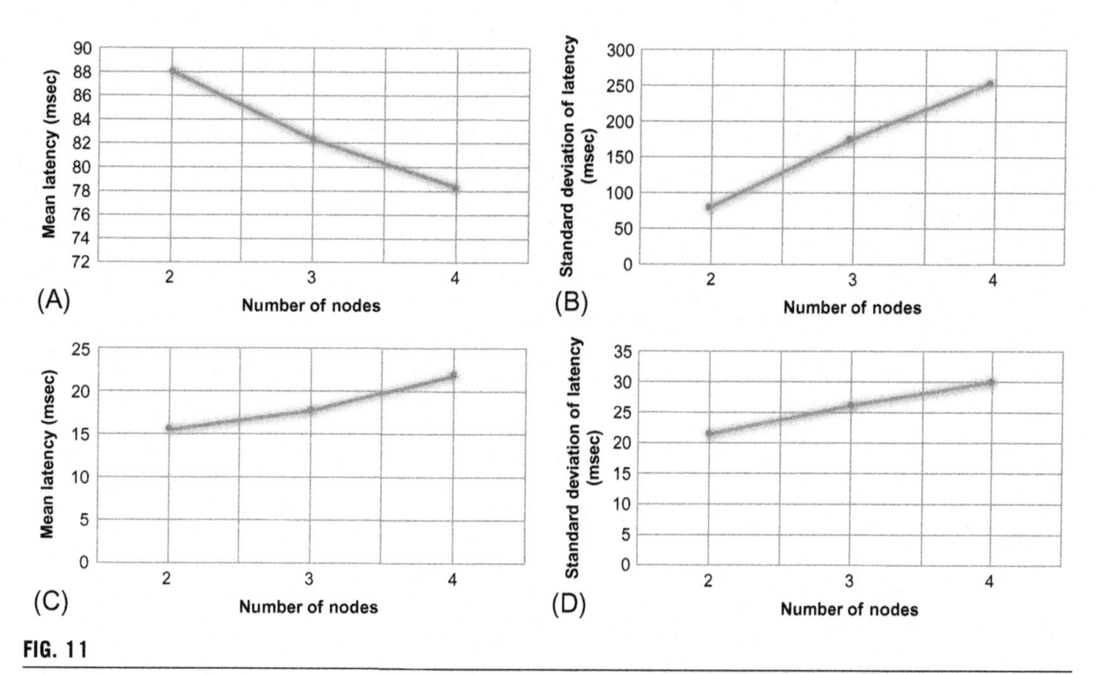


**FIG. 11**

Performance assessment of the CISP platform: the upper part shows the latency of the request/reply communications, while the lower one of the publish/subscribe communications.

15% when passing to 2 nodes, and around 40% with three nodes as destinations of the notifications. Also in the case of the request/reply there is an increase, but it is marginal and its observability in the statistics is compromised by the high variability of the measures. Based on these measures, we have given a value of the latency among the nodes of the overlay. Since we have modified Scribe to build balanced trees with no more than two children per each node, we assumed that latency is around 50 ms, which is the mean of the values obtained in the experiments with the two different kind of communications.

The metrics evaluated in our study are the following. First, the *Success rate* is the ratio between the number of the received events and the number of the published ones, and this is referred to as the reliability of the data dissemination protocol. If the success rate is 1 (i.e., complete reliability), then all the published events have been correctly received by all the interested destinations. Second, the *Performance* is expressed as the mean latency, which is a measure of how fast the given dissemination algorithm is able to deliver notifications, and the standard deviation of the latency, which indicates the possible performance fluctuations due to the applied fault-tolerance mechanisms, highlighting the timing penalties that can compromise the timeliness requirement. This is a measure of the traffic load that the dissemination strategy imposes on the network, and should be kept as low as possible, in order to avoid any congestion.

Our experiments show that without any resilient means, the data delivery protocol is not able to reach all the interested destinations, as depicted in Fig. 12C. The multiple-tree solution alone is able to increase the success rate (i.e., the ratio between the number of the received events and the number of the published ones) of the data sharing. However, it is not able to reach full success, which is achieved

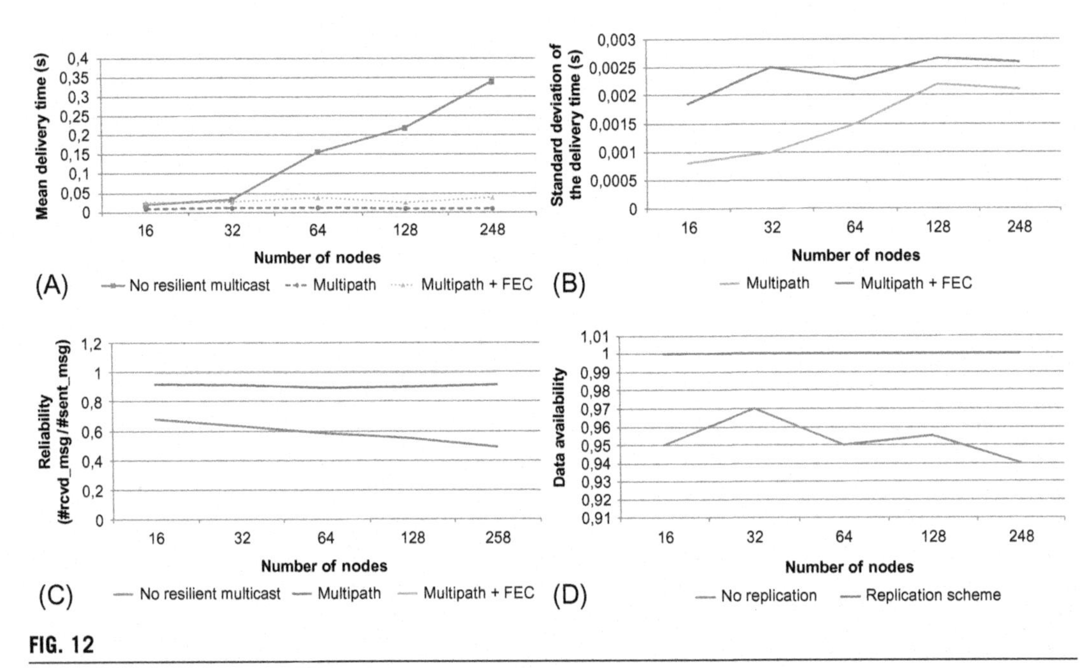


**FIG. 12**

Experimental results.

when FEC is also used. This is due to the fact that it is not possible to have complete disjoint paths and coding is able to recover losses occurred along joint links. These solutions are not only able to augment the achievable success rate, but also to reduce the delivery delay due to their intrinsic parallelism in the data dissemination, as proved in Fig. 12A. Both solutions exhibit stable performances demonstrated by very low standard deviation in the delivery delay, as illustrated in Fig. 12B.

Our experiments also investigate the data availability in our infrastructure, by determining the ratio between data request are successfully replied and the total number of made requests. Such an availability is quite low without any replication means in the infrastructure, as shown in Fig. 12D, while our replication approach is able to tolerate node crashes without compromising the data availability. For space limit, we are not able to show also the performance of our solution. Our replication approach is able to achieve a speed up in the data retrieval operation equal to 35% than the case without any replication.

## 7 FINAL REMARKS

This chapter described the key requirements to be fulfilled in order to have a reliable and confidential exchange of crisis information, and some solutions, proposed within the context of EU-funded projects named DESTRIERO and SECTOR, to meet such requirements and to be included within a Crisis Information System for the recovery from a large-scale disaster, so as to achieve a suitable level of reliability and confidentiality. Specifically, the requirements of network resiliency and systems redundancy has been the focus of the first part of the presented work, and a replication scheme to deal with the possible interruptions of nodes and network partitions, and a resilient multicasting to cope with network failures and crashes have been detailed. These two solutions have been assessed by means of simulation-based experiments, whose parameters have been obtained by on-field performance measurements taken from the preliminary implementation of the platform under development in the two projects. In the second part of the presented work, We have described the possible solutions to guarantee he demanding issue of confidential communications for the data exchanges required by a collaborative solution to crisis management.

In order to completely cover all the identified requirements for reliable crisis information, we have planned as possible future work the design of a logging strategy in order to detect possible failures and determine their causes. Moreover, we are also working on implementing the envisioned solutions for confidential communications within SECTOR's Common Information Space platform, and to empirically evaluate the efficiency and effectiveness of them with a measurement campaign on a realistic deployment of the platform.

## ACKNOWLEDGMENTS

This work has been partially supported by the European Commission in the framework of the Collaborative Project "A DEcision Support Tool for Reconstruction and recovery and for the IntEroperability of international Relief units in case Of complex crises situations, including CBRN contamination risks" (DESTRIERO, http://www.destriero-fp7.eu/—Grant agreement no: 312721), and the Collaborative Project "Secure European Common Information Space For The Interoperability Of First Responders And Police Authorities" (SECTOR, http://www.fp7-sector.eu/—Grant agreement no: 607821).

## REFERENCES

Al Hasib, A., Haque, A.A.M.M., 2008. A comparative study of the performance and security issues of AES and RSA cryptography. In: Proceedings of the Third International Conference on Convergence and Hybrid Information Technology, vol. 2, pp. 505–510.

Avizienis, A., Laprie, J.-C., Randell, B., Landwehr, C., 2004. Basic concepts and taxonomy of dependable and secure computing. IEEE Trans. Dependable Secure Comput. 1 (1), 11–33.

Birrer, S., Bustamante, F., 2007. A comparison of resilient overlay multicast approaches. IEEE J. Sel. Areas Commun. 25 (9), 1695–1705.

Boin, A., McConnell, A., 2007. Preparing for critical infrastructure breakdowns: the limits of crisis management and the need for resilience. J. Conting. Crisis Manag. 15 (1), 50–59.

Cardinal, J., Hoefer, M., 2010. Non-cooperative facility location and covering games. Theor. Comput. Sci. 411 (16–18), 1855–1876.

Carrozza, G., Crescenzo, D.D., Napolitano, A., Strano, A., 2010. Data distribution technologies in wide area systems: lessons learned from SWIM-SUIT project. Netw. Protoc. Algorithm. 2 (3).

Cinque, M., Martino, C.D., Esposito, C., 2012. On data dissemination for large-scale complex critical infrastructures. Comput. Netw. 56 (4), 1215–1235.

Cinque, M., Esposito, C., Fiorentino, M., Mauthner, J., Szklarskic, L., Wilson, F., Semete, Y., Pignon, J.P., 2015. Sector: Secure common information space for the interoperability of first responders. Procedia Comput. Sci. 64, 750–757.

Cinque, M., Esposito, C., Fiorentino, M., Carrasco, F., Matarese, F., 2016. A collaboration platform for data sharing among heterogeneous relief organizations for disaster management. In: Proceedings of the ISCRAM 2015 Conference.

Di Crescenzo, A.S., Strano, A., Trausmuth, G., 2010. SWIM: a next generation ATM information bus-the SWIM-SUIT prototype. In: Proceedings of the 14th IEEE International Enterprise Distributed Object Computing Conference Workshops (EDOCW), pp. 41–46.

Defago, X., Schiper, A., Sergent, N., 1998. Semi-passive replication. In: Proceedings of the Seventeenth IEEE Symposium on Reliable Distributed Systems, pp. 43–50.

Dice, L., 1945. Measures of the amount of ecologic association between species. Ecology 26, 297–302.

Esposito, C., Cotroneo, D., Gokhale, A., 2009. Reliable publish/subscribe middleware for time-sensitive internet-scale applications. In: Proceedings of the Third ACM International Conference on Distributed Event-Based Systems (DEBS), pp. 16:1–16:12.

Esposito, C., Cotroneo, D., Russo, S., 2010. An investigation on flexible communications in publish/subscribe services. In: Software Technologies for Embedded and Ubiquitous Systems, Lecture Notes in Computer Science, vol. 6399, pp. 204–215.

Esposito, C., Cotroneo, D., Russo, S., 2013. On reliability in publish/subscribe services. Comput. Netw. 57 (5), 1318–1343.

Haimes, Y., 2011. On the definition of resilience in systems. Risk Anal. 29 (4), 498–501.

Han, J., Watson, D., Jahanian, F., 2006. An experimental study of internet path diversity. IEEE Trans. Dependable Secure Comput. 3 (4), 273–288.

Hernantes, J., Rich, E., Lauge, A., Labaka, L., Sarriegi, J., 2013. Learning before the storm: modeling multiple stakeholder activities in support of crisis management, a practical case. Technol. Forecast. Soc. Change 80, 1742–1755.

Jones, D.F., Mirrazavi, S.K., Tamiz, M., 2002. Multi-objective meta-heuristics: an overview of the current state-of-the-art. Eur. J. Oper. Res. 137 (1), 1–9.

Kamel Boulos, M.N., Resch, B., Crowley, D.N., Breslin, J.G., Sohn, G., Burtner, R., Pike, W.A., Jezierski, E., Chuang, K.Y., 2011. Crowdsourcing, citizen sensing and sensor web technologies for public and environmental health surveillance and crisis management: trends, OGC standards and application examples. Int. J. Health Geogr. 10 (67), 1–29.

Kolozali, S., Bermudez-Edo, M., Puschmann, D., Ganz, F., Barnaghi, P., 2014. A knowledge-based approach for real-time IoT data stream annotation and processing. In: Proceedings of the IEEE International Conference on Internet of Things (iThings), pp. 215–222.

Ley, B., Pipek, V., Reuter, C., Wiedenhoefer, T., 2012. Supporting improvisation work in inter-organizational crisis management. In: Proceedings of the SIGCHI Conference on Human Factors in Computing Systems

Marler, R., Arora, J., 2004. Survey of multi-objective optimization methods for engineering. Struct. Multidiscip. Optim. 26 (6), 369–395.

Naedele, M., 2009, 3rd Quarter. Standards for XML and Web services security. IEEE Comput. Mag. 11 (3), 4–21.

NatCatSERVICE, 2012. Münchener rŸckversicherungs-gesellschaft. Geo Risks Research.

Nordbotten, N.A., 2003. XML and Web services security standards. IEEE Commun. Surveys Tut. 36 (4), 96–98.

Organization for Economic Co-operation and Development (OECD), 2004. Large-scale disasters—lessons learned. Available at http://www.oecd.org/futures/globalprospects/40867519.pdf.

OMG, 2012. Data Distribution Service (DDS) for Real-Time Systems, v1.2. www.omg.org. (Accessed September 2012).

Palmieri, F., Fiore, U., Castiglione, A., Leu, F.Y., de Santis, A., 2013a. Analyzing the internet stability in presence of disasters. In: Security Engineering and Intelligence Informatics, Lecture Notes in Computer Science, vol. 8128, pp. 253–268.

Palmieri, F., Fiore, U., Castiglione, A., Leu, F.Y., Santis, A.D., 2013b. Analyzing the internet stability in presence of disasters. In: Proceedings of the CD-ARES Workshops, pp. 253–268.

Petullo, W.M., Zhang, X., Solworth, J.A., Bernstein, D.J., Lange, T., 2013. Minimalt: minimal-latency networking through better security. Available at http://eprint.iacr.org/.

Rizzo, L., Vicisano, L., 1998. RMDP: an FEC-based reliable multicast protocol for wireless environments. ACM SIGMOBILE Mobile Comput. Commun. Rev. 2 (2), 23–31.

Schneier, B., 1996. Applied Cryptography: Protocols, Algorithms, and Source Code in C, second ed. Wiley, New York, NY, USA.

Stallings, W., 2010. Network Security Essentials—Applications and Standards, fourth ed. Prentice Hall, Upper Saddle River, NJ.

Stallings, W., 2013. Cryptography and Network Security: Principles and Practice, sixth ed. Prentice Hall, Upper Saddle River, NJ.

Vetta, A., 2002. Nash equilibria in competitive societies, with applications to facility location, traffic routing and auctions. In: Proceedings of the 43rd Annual IEEE Symposium on Foundations of Computer Science, pp. 416–425.

Wiesmann, M., Pedone, F., Schiper, A., Kemme, B., Alonso, G., 2000. Understanding replication in databases and distributed systems. In: Proceedings of the 20th International Conference on Distributed Computing Systems, pp. 464–474.

# Index

Note: Page numbers followed by *f* indicate figures and *t* indicate tables.

## D

## J

## Z

Printed in the United States
By Bookmasters